PROPERTY of
PAUL R. GOODIN
WD8JOM

REFLECTIONS

Transmission Lines and Antennas

By M. Walter Maxwell, W2DU

Published by the
American Radio Relay League
225 Main Street
Newington, CT 06111

Printed in USA

$20 in the USA

First Edition
First Printing

ISBN: 0-87259-299-5

Diskette Availability

Chapter 15 in this book contains listings of BASIC programs suitable for use with an IBM PC or compatible computer. The programs aid in designing T and pi networks, in calculating transmission-line constants from measurements, and in determining complex impedances encountered in a line having attenuation loss. Use of the programs is described in Chapter 15.

The ARRL offers a 5¼-inch (360k) computer diskette for the IBM PC/XT/AT as an optional supplement to this book. The diskette contains nine BASIC programs in ASCII text format, and a file containing measured data in the format required for input by one of the BASIC programs. The diskette is copyrighted but not copy protected.

The program files have been supplied by Walt Maxwell, W2DU, the author of this book, and the diskette is made available as a convenience to the purchaser. ARRL Headquarters has verified that the programs run properly and produce the results described herein.

Foreword

Tune in a QSO in the amateur bands where the subject is transmission lines, standing waves, antenna matching, reflected power or antenna tuners, and what do you hear? A lively discussion, most likely! The chances are high that you'll also hear some information that is only half true, or perhaps just plain wrong. Myths and old wives' tales about these related subjects prevail, even though some years ago *QST* carried a series of articles to dispel such misinformation. The series, "Another Look at Reflections," by Walt Maxwell, W2DU, appeared as seven parts in *QST* issues from April 1973 through August 1976. The series was to be concluded with two more parts, rewritten from Maxwell's initial submission. But Walt was unable to complete the rewriting then because of family and health reasons. The series was never concluded in *QST*.

Walt is now retired, his health has improved, and he is active in writing again. Because of the years that have passed since Part 7 of his series appeared, it would be inappropriate to publish concluding parts in *QST* now. Rather, the entire series has been used as a basis for the first seven chapters of this book. The subsequent chapters contain concluding material, and also present a wealth of related information on matching networks, antennas, and the use of the Smith Chart.

Every amateur will find meaningful facts and thought-stimulating discussions in this volume. With a knowledge of the information in these pages, you can be one who knows the difference between myth and reality in those lively on-the-air discussions.

David Sumner, K1ZZ
Executive Vice President
Newington, Connecticut
July 1990

About the Author

M. Walter Maxwell, W2DU, is an ARRL Technical Adviser (TA) in the specialty field of antennas and transmission lines. Walt was born in Daytona Beach, Florida in 1919, and grew up in Mount Pleasant, Michigan. A life member of both the ARRL and QCWA, he became W8KHK in 1933, and has been licensed continuously ever since. He was graduated from high school and entered Central Michigan University in Mount Pleasant in 1935, earning a BS degree in mathematics and physics. He played in professional dance bands, and specialized in auditorium and outdoor sound systems until early 1940. Then Walt joined the announcing and technical staff of WMFJ, Daytona Beach, and was assigned the call W4GWZ. Walt also copied Press Wireless News Service from WCX/WJS, 38 WPM CW, while at WMFJ.

With the FCC from late 1940 to 1944, his professional antenna experience included participation in building antenna farms at monitoring stations in Hawaii and Allegan, Michigan. Then until 1946 he was a navy instructor of Aviation Electronic Technicians at Corpus Christi, Texas. While in the navy he played trumpet in the big band of Alvino Rey, W6UK. From 1946 to 1949 in his own electronic and mobile-communications business, Walt did broadcast-engineering consulting, and was chief engineer of WCEN, Mount Pleasant, having engineered and built that AM station in 1948.

As the result of an interview with RCA's Clarence D. Tuska, cofounder of the ARRL, Walt joined the RCA Laboratories (the David Sarnoff Research Center) in Princeton, New Jersey as an engineer in 1949, later becoming a charter member of its new Astro-Electronics Division in Princeton. From 1960 until retirement in 1980 he was in charge of Astro's Space Center Antenna Laboratory and Test Range. More than 30 earth-orbiting spacecraft utilize antennas that were designed solely by Walt, which include ECHO 1 and all early TIROS-ESSA-NOAA weather satellites. He assisted in the design of many other spacecraft antenna systems, including the data-link antennas on TIROS-M and TIROS-N, and on RCA's SATCOM communications satellites. He also performed design work on the Search and Rescue (SAR) system antennas flying on TIROS-N, which are used worldwide for relaying signals from emergency locator transmitters (ELT) aboard aircraft in distress. He assisted in designing the moon-to-earth TV dish antenna used on the moon on Apollo's lunar rover—the moon buggy. (See the story and photos in Chapter 1.) He set up its test-range facilities and performed all of its pattern, gain and impedance-matching measurements. He engineered ground-based antenna systems at the Kennedy Space Center, Cape Canaveral, for prelaunch checkout of the TIROS and RELAY spacecraft. In addition he had total engineering responsibility for the receivers, transmitters and antennas of the five ground stations spread across the US, used in Project SCORE, the orbiting Atlas rocket that broadcast President Eisenhower's "Christmas Message From Space" in December 1958.

Having also been licensed as W8VJR and W2FCY, Walt has held the Extra Class license since 1967, and the call W2DU since 1968. Every full-time position in his career resulted from association with Amateur Radio. Now retired in DeLand, Florida, he has served as antenna consultant for AMSAT, as a member of FCC's advisory committee for WARC-79, and as trustee for K2BSA at National Headquarters, Boy Scouts of America. He writes and edits using an XT computer, and still enjoys music, playing string bass in a small volunteer string group that entertains at nursing homes, and in a professional 13-piece 1940s Glenn Miller style big band. He also enjoys Florida boating. A three-generation family of hams, his father was W8YNG, two of his three sons are Bill, AG2B (ex-5A4TY and WA2ETP) and Rick, WB4GNR (ex-WB2HKX). His daughter Sue is KA9AJC, and son-in-law Keith is WD9JCA. His third son, John, is an environmental scientist.

Dedication

This book is dedicated to the memory of my father, William W. Maxwell, W8YNG, whose inspiration pointed me in the scientific direction. My interest in radio originated when I was three, while watching him build radios from 1922 on. With his support I built my first one-tube broadcast receiver using a UV-200 detector when I was six.

My father operated his own engineering and machining business. He was a mechanical engineer, machinist, innovative welder, astronomer, telescope builder, and radio amateur. Growing up under his guidance in his machine shop fostered my creative ability. I learned from observing his methods for solving problems that required ingenuity, and his attention to precision and detail was not lost on my impressionable mind. In short, I owe my successful and enjoyable lifetime career in the scientific world to his early teachings and continued support.

He was a member of the ARRL for many years, was very active on all the phone bands, and for a long time before becoming a Silent Key in 1958 he was the net control station for the 75-meter Michigan Emergency Net, a position which I held as W8KHK from 1938 until 1940.

Preface

"Pictures or models that illuminate a mathematical formula in terms of a readily visualized physical mechanism often serve a valuable purpose. . . In terms of such a physically attractive picture the instantaneous potential difference at any point along a terminated transmission line is the resultant of all the contributions reaching that point simultaneously from both directions after an infinity of successive reflections at the ends. The terminated line is thus seen to play the role of an infinite line folded back and forth upon itself, with discontinuities at intervals equal to the actual length and with the potential difference in these folded parts actually superimposed and combined algebraically into a single value." [Ronold W. P. King, PhD, *Transmission Line Theory* (Dover Publications, 1965)]

The above quote from this eminent Harvard professor typifies our actual amateur operation with lines that feed antennas. This book is about the care and feeding of antennas, with special emphasis on dispelling some of the prevalent misconceptions surrounding the mismatched antenna and its feed line. It explains in great detail how the antenna tuner, or Transmatch at the input terminals of the feed line provides a conjugate match at the antenna terminals, and tunes a nonresonant antenna to resonance while also providing an impedance match for the output of the transceiver. This aspect of impedance matching is of great importance to users of modern transceivers having broadband output circuits which require a 50-Ω load to deliver full power.

The book also highlights some of the reasons many amateurs seek low values of SWR for the wrong reasons. It explains why a 1:1 SWR isn't always a necessary goal, and it also describes how to determine whether it's worth the effort to reduce the SWR below a given value based on the conditions at hand.

This is not a "how to" book. Instead, its theme is "how it works." It is tutorial in that it explains certain principles of transmission-line theory which are important to the amateur's understanding of what is occurring in routine operations with tuners, feed lines, and antennas. But yet the book is couched in language easy to understand at the amateur level. In addition, a voluminous, broad-based bibliography is included as Chapter 24 for those who wish to seek further study. There is also a smattering of space-communications nostalgia as I share some of my experiences in antenna engineering during the early years of our country's efforts in space.

I wish to thank my son Rick, WB4GNR, for generating the impedance graphs in Chapters 15 and 20 with an HP LaserJet 2000, using Hewlett-Packard's Graphics Gallery Collection software on a PC computer. I also thank my son-in-law, Keith Glasnapp, WD9JCA, for his help with the BASIC computer language that made the line-impedance programs of Chapter 15 possible. And a special thanks to my long-time friend and editor, Jerry Hall, K1TD, who assisted me in all of my writings published by the ARRL. Their help has been invaluable. Acknowledgment with thanks is also given to the late Phillip H. Smith, ex-1ANB, for his kind permission to use the Smith Chart in the studies which follow.

Walter Maxwell, W2DU
DeLand, Florida
July 1990

Contents

Introduction

The chapters in this book are a mixture of old and new. The older chapters comprise a collection of articles and other material I published previously in *QST, The Radio Amateur's Handbook* and *The ARRL Antenna Book*. Hence, the reader will find some unavoidable repetition on the subject of impedance matching between some of the chapters, because those chapters were written during different time periods, and for achieving different objectives which did not include the continuity one obtains when the original writing is intended for a book.

Chapters 1 through 7 are from my series of seven articles published in *QST* in the early 1970s entitled, "Another Look At Reflections." The articles were well received, and over the years both ARRL Headquarters and I have had many requests for reprints. It is because of these requests that the articles are reprinted in this book, along with some of my writings from other ARRL publications. Chapters 1 through 7 are essentially the same as the original articles. They are identical in substance, but I have edited them rhetorically for easier reading.

The idea for writing the original series of articles was born while I was listening to and participating in many discussions concerning mismatch and reflections on transmission lines. I made up a list of topics which appeared to be the most controversial among amateurs I heard or talked with. The topics themselves are not controversial, except as the amateurs misunderstood the factual realities. From the list I wrote out 27 true statements, which I presented as a true-false test in the *QST* series. Those statements are included here in Chapter 2. From the discussions I mentioned earlier, it was apparent that many amateurs would mark them false.

Those discussions also revealed that many experiments involving antennas, and especially transmission lines, were performed for which the results were predestined to futility. The reason is because the experimenters were unaware that they lacked sufficient understanding of the principles involved. Then the appearance of erroneous material in amateur literature motivated my desire to do something constructive about the problem. The literature implied that little is known on the subject of mismatch on transmission lines. This is not true, unless the author was referring specifically to us amateurs, because the scientific and engineering literature is replete with voluminous data on the subject, going back to the late 19th century with the works of James Clerk Maxwell, Oliver Heaviside and many others.

One of the problems that limits the understandability and quality of our experiments on antennas and transmission lines is the limited equipment available for amateurs to make precise and complete measurements at RF, as compared to dc, low-frequency ac, and audio frequencies. But this doesn't excuse us from needing to know some of the fundamental principles of transmission lines—it actually makes knowledge of the fundamentals all the more necessary, so we can diagnose and evaluate logically the limited measurement data we do obtain. Our experimental efforts can become more productive if we are able to visualize an accurate physical picture of the voltages, currents, and fields, and how they interact on the transmission line.

My original plan was to compile and publish an extensive bibliography of accurate and readily available references on the subject in *QST*. But in 1971, the late John Marsh, W3ZF, convinced me that a mere list of references would receive little attention. John then proceeded to fire real enthusiasm for writing an extensive paper which would not only complement and unify the references, but would also underscore the problem and stimulate interest in studying the subject of antennas and transmission lines. I followed John's suggestion, and as the work progressed it became apparent that additional references from *professional* engineering sources would be valuable in providing access to greater depth of study for the more advanced amateur, and also to make available reliable sources for verification of the points which have been clouded by controversy among amateurs.

During the study of the numerous references in the preparation of the manuscript for the articles, an interesting and valuable aspect emerged: *The ARRL Antenna Book* was seen to unify many of the other references, plus more, because in one handy volume it contains concise, sufficient, and well-presented material on nearly every point of importance to the amateur's needs concerning antennas and transmission lines, including the fundamental principles! For the amateur who has no access to the other references, but could manage to acquire just one book, *The Antenna Book* is it! Every amateur who feeds RF into an antenna should have a copy—and read it! One would be surprised to learn how many antenna engineers have a copy in their own personal libraries. The best edition so far is the 15th (see Chapter 24, Ref 71).

With the exception of *The Antenna Book*, the details of reflections and wave action in transmission lines seems to be somewhat obscure in the amateur literature. Therefore, Chapter 3 presents a simplified treatment of the interaction of the fields, currents, and voltages on a transmission line in explaining the development of the standing wave.

A simple introduction to the Smith Chart is presented in Chapter 10, along with a novel use of the chart to assist in understanding how the impedance changes with length on a transmission line. A new vector-type presentation using the chart is given, which enhances the *physical* picture of standing-wave development, input impedance, and other activity along the line, and forms a basis for an analytical proof that the explanations presented are correct.

For those who are unfamiliar with the Smith Chart, it is easy to learn. See Part 1 of Ref 25, Chapter 28 of Ref 71, and Chapter 10 in this book. (References are listed in Chapter 24.) You'll find the chart to be a valuable tool. It has separate sets of circles for resistance and reactance. To plot a single complex impedance R + *j*X, you simply find the point where the appropriate resistance, R, and reactance, X, circles intersect. Knowledge of the Smith Chart is helpful in understanding and devising impedance-matching circuitry; it is actually fun to use for this purpose. Additional references of interest are included in Chapter 24. Smith Charts are available at college book stores and stores that carry engineering supplies. They may also be ordered from ARRL Headquarters.

Chapter 15 includes several computer program listings for solving impedance-matching problems. The programs are available on diskette, as announced on an earlier page of this book. Included are programs for solving T and pi networks, and for obtaining the frequency response of the networks. Also included are programs for obtaining precise values of the transmission-line constants—characteristic impedance, electrical length, and attenuation—obtained from measurements made on the line. There is also a program for transferring complex impedances from one end of the line to the other for accurately determining the terminal impedance of an antenna from measurements made at the input of the feed line. The efficacy of these transmission-line programs may be seen in the figures and tables they produced from data I measured—Figs 15-1 and 20-1 through 20-3, and Tables 15-1, 15-2, and 20-1 through 20-5.

Some important aspects of baluns that are not generally appreciated are presented in Chapter 21. These are taken from my March 1983 *QST* article, "Some Aspects of the Balun Problem."

Chapter 1

Too Low an SWR Can Kill You

(Adapted from *QST*, April 1973)

Sec 1.1 Introduction

Judging by what we hear on the air, nearly everyone is looking for an SWR of one-to-one. Question why, and the answer may be, "I'm not getting out on this frequency because my SWR is 2.5:1. There's too much power coming back and not enough getting into the antenna." Or, "If I feed a line having that much SWR, the reflected power flowing back into the amplifier will burn it up." Or still, "I don't want my feed line to radiate." Any of these answers shows a misunderstanding of reflection mechanics, and are symptomatic of the present state of thinking on this subject. Rational and creative thinking toward antenna and feed-line design practice has been absent for a long time. Such thinking has been replaced with an unscientific and thought-inhibiting attitude, as in the days before Copernicus persuaded the multitudes that the universe did not revolve around the earth. This situation originated with the introduction of coaxial transmission lines for amateur use around the time we got back on the air after World War II. It has gained momentum since SWR indicators appeared on the scene and since the loading capacitor of the pi-net tank replaced the swinging link as an output-coupling control. We are in this state of mind because much misleading information has been, and is still being published concerning (1) behavior of antennas which are not self-resonant, (2) feed-line performance in the presence of reflections when mismatched to the antenna, and especially (3) the meaning and interpretation of the SWR data.

Articles containing explicitly erroneous information and distorted concepts have found their way into print, became gospel, and continue to be propagated with chain-letter effectiveness. These include such gems as (1) always requiring a perfect match between the feed line and the antenna, (2) evaluating antenna performance or radiating efficiency only on the basis of feed-line SWR—the lower the better, (3) pruning a dipole to exact resonance at the operating (single) frequency and feeding with an exact multiple of a half-wavelength (½-λ) coax—no other length will do, (4) adjusting the height—perhaps just lowering the ends into an inverted-V dipole—to make the resistive component of the antenna terminal impedance equal to the line impedance, or (5) subtracting percent reflected power from 100 to determine usable percentage of transmitter output power. Nomographs have even been published for this erroneous method (*Ref 102*).[1]

As a result of these misdirected concepts, we have been conditioned to avoid any mismatch and reflection like the plague. One-to-one all the way! Sound exaggerated? Not if your receiver is tuning the same amateur bands as mine! In the current vernacular, you could say we have a severe SWR hang-up! In many instances, from the viewpoint of good engineering, this hang-up is inducing us to concentrate our impedance-matching efforts at the wrong end of the transmission line (*Ref 16*).

It is ironic that we should be in this situation, because the amateur is generally quite practical when it comes to following theoretical considerations. In this case we have been following the perfect-match theory down the narrow path because many of the aforementioned articles have misled us to believe that all reflected power is lost. They've given never an inkling that, properly controlled, reflections can be turned to our advantage in obtaining increased flexibility concerning operating frequencies which we are presently throwing away.

That so much misinformation gained a foothold is surprising in view of the correct teachings of *The ARRL Handbook* (*Ref 1*), *The ARRL Antenna Book* (*Ref 2*), the works of Grammer (*Refs 3 through 5*), Goodman (*Ref 7*), McCoy (*Refs 8 through 13 and 41*), Drumeller (*Ref 14*), Smith (*Ref 15*), and especially two articles addressed to a subject nearly identical to this one by Grammer (*Ref 6*) and Beers (*Ref 16*). One objective of this book, therefore, is to identify some of the many erroneous concepts concerning reflection principles, with sufficient clarity to make you question your own position on the subject. Once we

---

[1] All references appear in Chapter 24.

correctly understand mismatch and reflections, we can obtain improvement in operational antenna flexibility, similar to going VFO after being rock-bound with a single crystal. And when we discover how little we gain by achieving a low SWR on the average feed line, we will avoid unnecessary and time-consuming antenna modifications. Such modifications often involve hazardous climbing and precarious operations on a roof or tower, which can result in injuries or even death. Let's kill SWR misconceptions—not ham operators!

Sec 1.2 Open-Wire Versus Coax Feed Lines

The theory behind the transmission of power through a feed line with minimum loss by eliminating all reflections—terminating the line with a perfect match—is equally valid, of course, for open-wire and coaxial lines. But in the days of open-wire lines prior to our widespread use of coax, theory was tempered with practical considerations. Open-wire line was, and still is, used with high SWR to obtain tremendous antenna flexibility relative to operating over a wide range of frequencies with high efficiency. This is because all power reflected from the feed-line-to-antenna mismatch which reaches the input source is conserved, not dissipated. The power is returned to the antenna by re-reflection in the antenna tuner (Transmatch) at the line input. But, although the loss from reflections and high SWR is not zero, this additional loss is negligible because of the low attenuation of open-wire lines. If the line were lossless (zero attenuation), no loss whatever would result because of reflections. (This is discussed further in Chapter 6, in connection with Fig 6-1.)

The error in our thinking that standing waves on coaxial line must always be completely eliminated originated quite naturally, because the permissible reflection and SWR limits are much lower than in open-wire lines. When using coax for truly single-frequency operation, it makes sense to match the load and line to the degree economically feasible. But it makes no sense to match at the load in many amateur applications where we are chiefly interested in operating over a band of frequencies! Single-frequency operators we are not, except as our misguided concern over increasing SWR restricts our departure from the resonant frequency of the antenna.

Many authors are responsible for perpetuating the unscientific and erroneous viewpoint that the coax-fed antenna must be operated at its self-resonant frequency. They have continually overemphasized the necessity for the antenna being matched to the line within some arbitrary, low SWR value to preserve transmission efficiency, and by implying that efficiency equals 100 minus percent reflected power. The viewpoint is unscientific because it neglects the most important factor in the equation for determining efficiency—line attenuation. And it is also erroneous because efficiency does not relate to reflected power by simple subtraction. Setting an SWR limit alone for this purpose is meaningless, because the amount of reflected power actually lost is not dependent on SWR alone. The attenuation factor for the specific feed line must also be included. This is because the only reflected power lost is the amount dissipated in the line because of attenuation—the remainder returns to the load. Some authors have so wrongly conditioned us concerning what happens to the reflected power that many of us have overlooked the correct approach to the subject. It is clearly presented in both *The ARRL Handbook* and *The Antenna Book* that transmission efficiency is a two-variable function of both mismatch and line attenuation. With this knowledge and by using a graph of the function appearing in *The ARRL Handbook*, presented here as Fig 1-1, the amateur can determine how much efficiency he will lose for a given SWR with the attenuation factor of each specific feed line. He can then decide for himself what the realistic SWR limit should be.

Sec 1.3 Unimportance of Low SWR Values

In our efforts to obtain low feed-line SWR values of 1.1, 1.2, or even 1.5 to 1, we have gone far past the diminishing-returns point with respect to efficient power transfer, even for single-frequency operation. It is like installing no. 4 or no. 6 wire in a house-wiring run where no. 12 wire is sufficient. Reference to the basic transmission-line equations, which have always been readily available in engineering texts and handbooks (*Refs 1, 2, 17, 18, 19 and 33*) verify this analogy. In addition, such reference makes it clearly apparent that authors who simply insist on low SWR, or find 1.5 or 2 to 1 objectionably high, have failed to comprehend the true relationship between reflected and dissipated power. From the viewpoint of amateur communications, it can be shown mathematically and easily verified in practice that the difference in power transferred through any coaxial line with an SWR of 2 to 1 is imperceptible compared to having a perfectly matched 1:1 termination. This is true no matter what the length or attenuation of the line. Further, it can be shown that many typical coaxial feed lines we use in the HF

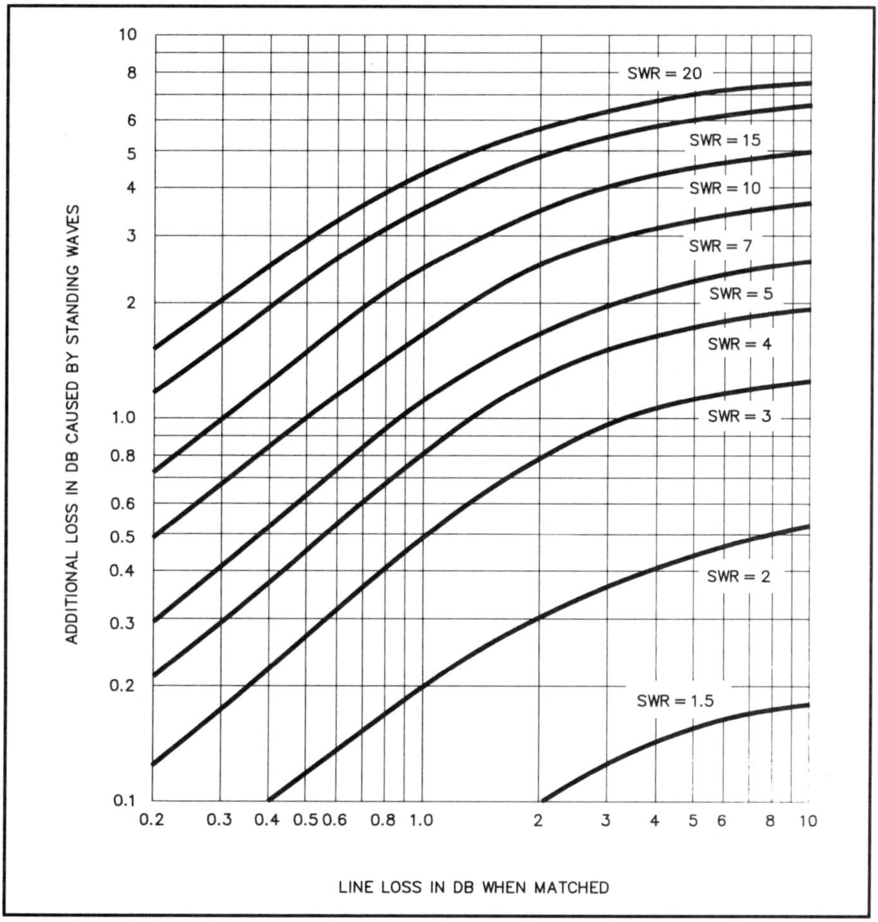

Fig 1-1—Increase in line loss because of standing waves (SWR value at the load). To determine the total loss in decibels in a line having an SWR greater than 1, first determine the loss for the particular type of line, length, and frequency, on the assumption that the line is perfectly matched. Locate this point on the horizontal axis and move up to the curve corresponding to the actual load SWR. The corresponding value on the vertical axis gives the additional loss in decibels caused by the standing wave. (Also see Fig 6-1.)

bands with an SWR of 3 or 4, and often as high as 5 to 1, have an equally imperceptible difference. When feed-line attenuation is low, allowing such higher values of SWR permits operating over reasonably wide frequency excursions from the self-resonant frequency of the antenna with the imperceptible power loss just described, in spite of the popular impression to the contrary.

The relative unimportance of low SWR when feed-line attenuation is low is demonstrated rather vividly in the following two examples of spacecraft antenna applications. First, the TIROS-ESSA-ITOS-APT weather satellites, of which the design of the entire antenna system fed by four transmitters operating simultaneously on different frequencies, was the work of the author. (See the accompanying story at the end of this chapter.) The terminal impedance of each of four crossed dipoles at the beacon-telemetry frequency (108 MHz in early models) was $150 - j100\ \Omega$, for an SWR of 4.4, and reflected power of 40%. Matching was performed at the input of four separate feed lines by a complex stripline matching network fed by two 30-milliwatt telemetry transmitters. (We can't afford much power loss here!) The combined attenuation of the feed line and matching network was 0.2 dB, and the additional loss from the SWR on the feed line was 0.24 dB (5.4%), for a total loss of 0.44 dB (only 9.6%). On the prevalent but erroneous assumption that all reflected power is lost (40%), only 18.1 milliwatts would reach the antenna. Efficiency determined on the same erroneous basis would only be 60%. But 27.1 milliwatts was measured at the antenna. Of the 2.9 milliwatts lost in total attenuation, only 1.6 milliwatts was lost because of the 4.4:1 SWR. So the real efficiency would have been 95.5% if the feed lines had been perfectly matched at the antenna, but reduces only to 90.4% by allowing the 4.4 SWR to remain on the feed line. Second, in the Navy Navigational Satellite (NAVSAT), which is used for precise position indications for ships at sea, the antenna terminal impedance at 150 MHz is $10.5 - j48\ \Omega$, for an SWR of 9.3, reflected power 65%. Also matched at the input to the feed line, the matched-line attenuation is 0.25 dB, and the additional loss from SWR is 0.79 dB, for a total system loss of 1.04 dB. This equates to approximately ⅙ of an S unit. This is an insignificant amount of loss for this situation, even in a space environment where power is at a premium.

Why did we match at the line input? Because critical interrelated electrical, mechanical and thermal design problems made it impractical to match at the antenna. Line-input matching provided a simple solution by permitting the matching elements to be moved to a noncritical location. This design freedom afforded tremendous saving in engineering effort with negligible compromise in RF efficiency, in spite of SWR levels many amateurs would consider unthinkable.

Another factor which has contributed to misunderstanding concerning power lost because of mismatched loads is the confusion between three different conditions of line usage: (1) one in which the incident, or forward *voltage* on the line is constant, independent of the load terminating the line; (2) one in which the forward *power* is constant, also independent of the load, and (3) one in which the forward power *varies* with changes in the load. (For the relative amplitudes of the SWR and line voltage in the first two cases, see *Ref 19, Fig 1.3, p 6 and Fig 3.6, p 29*).

The first condition involves laboratory and experimental work, which generally requires holding the forward *voltage* constant with variations in the load terminating the line. A constant-voltage source is usually obtained for this purpose by using a signal generator having a pad of from 10 to 20 dB of attenuation between its source and output terminals to absorb the reflected power. Absorbing the power in the pad prevents it from reaching the source in the generator where it would otherwise alter the line coupling and cause the output voltage of the source to vary. Consequently, the generator sees a perfect match for all terminating load conditions, and *all reflected power is lost.* This is a condition which is required to obtain certain laboratory test data.

The second condition involves a power source, or generator which maintains a constant forward *power* on the line independent of the load. The distinguishing feature of this condition is that the generator has an internal source resistance R_G that is equal to the characteristic impedance Z_C of the transmission line into which it delivers its power. If the line is lossless, conservation of energy demands that power reflected from a mismatched load termination must cause the generator to deliver less power to the line by exactly the amount of power reflected. Since the source resistance R_G equals Z_C, the power reflected from a mismatched load termination sees a match at the input terminals of the line, and hence there is no direct re-reflection of any reflected power. However, the phasor voltage at the input terminals of the line is the sum of the reflected and source phasor voltages at that point. The result is that, with the combination of the change in output power of the source and the amount of power dissipated in the source resistance R_G, the power entering the line remains the same as before the reflected power returned. Consequently, the forward power remains constant, the power absorbed in the load is reduced by the amount of the power reflected, and *the reflected power is lost!* (See *Ref 70, pp 204-205*.) The fact that reflected power is lost under these conditions is probably the chief reason many have been misled to believe reflected power is also absorbed in the plate resistance of the output amplifier of our transmitters, which is not true at all. The reason is total re-reflection by the reactances in the tank circuit, which is explained in great detail as we proceed.

The third condition involves the power amplifiers in our transmitters, or transceivers, in which the forward power in the transmission line varies directly in response to the power reflected from a mismatched load, such as a nonresonant antenna. The reason is that for whatever power we adjust the amplifier to deliver, any power reflected from the mismatched load is returned, either to the tank circuit of the amplifier or to an external line-matching network. There, by the action of re-reflection, the reflected power is added to the power delivered by the amplifier and returned to the mismatched load. In this third case, *no reflected power is dissipated in the plate circuit* of the amplifier, and with lossless line, *no reflected power is lost!*

As a result of these various misunderstandings, many amateurs never even wonder whether there are any benefits to be gained by not matching at the junction of the feed line and the antenna. Many even shun the use of open-wire lines (not the old-timers). They completely miss the joy of a QSY to the opposite end of the band with only a simple change in Transmatch tuning. The fear of reflections engendered by the exaggerated application of the theory to coax has crept into their thinking concerning any form of mismatched connection.

Adding still further to the confusion is the old-wives' tale that the reflected power is dissipated in the transmitter, causing tube and tank-coil heating and all kinds of other damage. This myth developed out of ignorance of the true mechanics of reflections and became the easy, but fallacious, explanation of what seems to be abnormal behavior in the transmitter when feeding a line with reflections. What really happens at the transmitter in the presence of reflected power is simply a change in coupling caused by a change in impedance at the input terminals of the feed line. This is explained in detail as we proceed from chapter to chapter. Then we may understand how to operate with absolutely no danger of damaging the amplifier while feeding into a line with high SWR. Although rigs having solid-state output amplifiers have no provision for working into any load other than 50 Ω, rigs with tubes and pi-network output coupling circuits can work into impedances that far exceed the 2:1 SWR limits the manufacturers put on their warning labels. Hence, the manufacturers are also guilty of perpetuating the low-SWR myth.

Sec 1.4 Engineering an Antenna System

Engineering is the process of making workable compromises in design goals where theories guiding different aspects of the design are in conflict, making it impossible to optimize all the goals. Good engineering is simply recognizing the correct choices in the compromises and relaxing the right goals, as in the spacecraft-antenna design described earlier. We amateurs spend many hours building and pruning antenna systems. Wouldn't it be worthwhile spending some of that time learning how to engineer the design in order to make correct trade-off decisions among related factors instead of letting old King SWR dictate the design?

FIRST, we need to improve our knowledge of reflection mechanics and transmission-line propagation to understand:

1) why reflected power by itself is an unimportant factor in determining how efficiently power is being delivered to the antenna.

2) the effect of line attenuation to discover why it is the *key* factor which tells us when and how much to be concerned with reflected power and when to ignore it.

3) why *all* power fed into the line, minus the amount lost in line attenuation, is absorbed in the load *regardless of the mismatch at the antenna terminals.*

4) why reflection loss (mismatch loss) is canceled at the input of the line by reflection gain (*Ref 19, p 38, and Ref 25 Part II, p 33*).

5) why a low SWR reading by itself is no more a guarantee that power is being radiated efficiently than a high SWR reading guarantees it is being wasted.

6) why SWR is not the culprit in transmitter-loading problems—why the real culprit is the change in line-input impedance resulting from the SWR, and why we have complete control over the impedance without necessarily being concerned with the SWR.

7) the importance of thinking in terms of *resistive and reactive components of impedance instead of SWR alone*, and why SWR by itself is ambiguous, especially from the viewpoint of the selection and adjustment of the coupling and matching circuitry of an external line-matching network.

SECOND, we need to become aware that with moderate lengths of low-loss coax, such as we commonly use for feed lines, loss of power because of reflected power in the HF bands can be insignificant, no matter how high the SWR. For example, if the line SWR is 3, 4, or even 5 to 1 and the line attenuation is low enough to ignore the reflected power, reducing the SWR yields no significant improvement in radiated power because all the power being fed into the line is already being absorbed in the load. This point has especial significance for center-loaded mobile whip antennas, because of the extremely low attenuation of the short feed line, which is explained in Chapter 6.

THIRD, we should become more familiar with the universally known, predictable behavior of off-resonance antenna-terminal impedance and its correlation with SWR (*Ref 2, Fig 2-7; Ref 71, p 2-6*). This knowledge provides a scientific basis for evaluating SWR-indicator readings in determining whether the behavior of our system is normal or abnormal, instead of blindly accepting low SWR as good, or rejecting high SWR as bad. The following two examples emphasize the importance of this point by showing how easily one may be misled by a low SWR reading.

1) A ground system having 100 properly installed radials has negligible loss resistance (*Ref 20*). Many AM broadcast stations use either 120 or 240 radials, while the FCC requires a minimum of 90. With such a ground system, the terminal impedance of a ¼-λ vertical is 36.5 + *j*22 Ω, and becomes approximately 32 Ω when the antenna is shortened to resonance. When fed with a 50-Ω line, the SWR at resonance will be close to 1.6, rising predictably on either side of resonance. But a ground system having 15 radials has approximately 16 Ω of ground-loss resistance with this antenna. If we remove a few radials at a time from the 100-radial system, the increasing ground resistance adds to the radiation resistance and increases the total resistance terminating the feed line. Hence, as each radial is removed, the terminating resistance comes closer and closer to 50 Ω, reducing the SWR. When enough radials have been removed for the ground-loss resistance to reach 18 Ω, the terminating resistance will be 18 + 32 = 50 Ω, for a perfect one-to-one match! But while the SWR went down, so did the radiated power, because now the power is dividing between 32 Ω of radiation resistance and 18 Ω of ground resistance!

Ground systems having from two to four radials may have a loss resistance as high as 30 to 36 Ω, so now the SWR at the resonant frequency will be around 1.4 or 1.5. But when operating at other frequencies, instead of rising from this low value of SWR as it should at frequencies away from resonance, the ground-loss resistance holds the off-resonant SWR to lower values than would result with a good ground. The low SWR simply indicates that the line is well matched, but it offers no clue that approximately half the power is heating the ground.

2) Some amateurs who employ a one-to-one balun believe that "one-to-one" means it provides an impedance match between the feed line and the antenna. This is an erroneous concept, because "one-to-one" only specifies the output-to-input impedance ratio of the balun. No matter what antenna impedance terminates the output of the balun, approximately the same impedance is seen at the input, depending on the quality of the balun. Nevertheless, these amateurs are convinced the baluns are matching the feed line to the antenna, because the SWR sometimes goes down dramatically when the balun is inserted. When using some baluns having ferrite cores, the SWR is less than 2:1 over the entire 75-80 meter band, where somewhat over 5:1 is normal at the ends of the band when the antenna is cut to resonate at the center of the band. Off-resonance SWR is sometimes reduced with these baluns because the ferrite core saturates while attempting to handle the reactive current which exceeds the maximum core-current level. Thus, the full excursion of the reactive component of antenna impedance is prevented from appearing at the input of the balun. All power above the saturation level is lost in heating the balun, while the low SWR is deceiving the unsuspecting amateur. The true SWR will be unchanged by a 1-to-1 balun if it has a core capable of handling the current without saturating and if it has no significant leakage reactance. However, most transformer-type baluns having a ferrite core do have significant leakage reactance, and less than perfect coupling. Hence, these baluns cannot provide a true 1-to-1 impedance transfer, and the resulting SWR will not be the same as it would if the balun did have a true impedance transfer ratio. This is because when the leakage reactance is inserted between the antenna and the feed line, this reactance can either improve or worsen the match, depending on the magnitudes and signs of both the leakage and antenna-terminal reactances. In addition, an SWR indicator may not show the true SWR without a balun if antenna current on the outside of the coax is present at the SWR indicator (*Ref 36*). These

aspects of the balun problem, and how to avoid them, are discussed in detail in Chapter 21.

So it is important to know approximately what SWR to expect—if it is low, determine whether it should be low. Don't assume that a low SWR indicates success, or that it guarantees a great system! Be especially suspicious if the SWR remains low or relatively constant over a moderate frequency range, unless specific broadbanding steps have been performed on the radiating system. This knowledge is elementary and routine for an antenna engineer, but too little information in this area has been available for the amateur, considering the degree of his involvement with antennas. The variation of antenna-terminal impedance versus frequency is shown in *The ARRL Antenna Book* (*Ref 2, Fig 2-7, and Ref 71, p 2-6*). In addition, correlation of the impedance change with SWR is covered in detail in Chapter 12 in this book, to enable us to predict normal SWR, within limits with a nonresonant antenna terminating the feed line.

FOURTH, we need to re-examine the use of open-wire feed lines as tuned lines (*Ref 3, Part III, p 20; Ref 10; Ref 21, p 23*), to discover that the principles used there are exactly what we have been discussing. Remember, with tuned lines we completely ignore the mismatch at the junction of the feed line and the antenna. Then we compensate for the mismatch with the tuner at the input of the line, over the entire frequency range of the band. The SWR may run as high as 10, 15, or even 20 to 1, but the power reflected from the mismatch is re-reflected back to the antenna by the tuner. Adjusting the tuner to obtain maximum line current simply adjusts the phase of the reflected wave to re-reflect down the line in phase with the forward wave, again reaching the antenna. Thus, the reflection *loss* from the mismatch is canceled by the reflection *gain* of the tuner. The phenomenon of reflection gain is explained in detail in Chapters 4 and 7.

Many of us older amateurs know from age-old practice that a 600-Ω line made of two no. 12 wires with six-inch spacing would work every time. We had little incentive in the earlier days to learn how they worked and why they transferred power *efficiently* with such high reflected power and SWR. Nor did we particularly care that tuning the plate tank for a dip in plate current was really canceling the reactance, or that the plate-current dip was just another way of viewing the phase adjustment of the reflected wave. And we didn't realize that, at the current dip, the reflected wave would re-reflect in phase with the forward wave to obtain maximum feed-line and antenna current. This is probably one of the reasons for our misunderstanding of the similarity between open-wire and coaxial-line operation with mismatched loads. The principle is the same in both; only the degree of mismatch is different. In other words, for many applications, *coax can be used as a tuned line in precisely the same manner as open wire*. The spacecraft systems mentioned earlier are typical examples.

Thus, coax connected directly into the antenna may be operated successfully with substantial mismatch, in which case the SWR limits while operating away from the self-resonant frequency of the antenna are determined entirely by power lost because of line attenuation. Voltage breakdown and current heating should not be a problem at our legal power limit with RG-8 or RG-11, or with RG-58 or RG-59 at lower powers. This is because voltage at an SWR maximum is only the *square root* of the SWR times the voltage appearing on the line when the line is matched. The impedance at the input of the line will no longer be 50 Ω, but we can determine whether the output tank of the transmitter has sufficient impedance-matching range to permit feeding the line directly. This depends on the magnitude of the mismatch and the length of the line. The range is surprisingly high in some rigs, none in others (*Ref 4, Part III*). If the range of the rig is insufficient, an external matching device (antenna tuner, Transmatch, ATU) will be required to obtain an impedance match and the correct coupling between the input of the feed line and the transmitter (*Refs 9 through 12 and 22*). The point I am emphasizing here is that within the limits mentioned, *all required line matching can be transferred back to the operating position* instead of forcing the match to occur at the antenna feed point, without suffering any *significant* loss in radiated power. The use of this technique, which may come as a surprise to many, does not contradict any theory. It is actually an embodiment of the fundamental principle of network theory called conjugate matching (*Ref 17, p 243; Ref 19, p 38; Ref 35, p 49; Ref 69*) which is the basis for all antenna tuner, or Transmatch operation with either open-wire or coaxial lines.

After learning of the benefits obtained with line-input matching in the two spacecraft examples described earlier, it is interesting to compare the results using this same input-matching technique in typical 80- and 40-meter situations. The 80-meter amateur band is the widest in terms of percent of center frequency, and thus suffers the greatest SWR increase with frequency excursion to the ends of the band. A dipole cut for resonance at 3.75 MHz yields an SWR in a 50-Ω feed line somewhat above 6:1 at 3.5 MHz and about 5:1 at 4.0 MHz. As shown in Fig 1-2, in a 100-foot length of nonfoam RG-8, an SWR of 5:1 adds only 0.46 dB loss to the matched (that is, flat) line loss of 0.32 dB at 4.0 MHz. So out almost to the ends of the band, less than 1⁄12 of an S unit is lost because of the SWR, an imperceptible amount. This further verifies the principle and proves that full-band, coax-fed dipole operation on 80 meters also is practical. Even with the high SWR at the ends of the band, the loss cannot be distinguished from what it would have been had the SWR been a perfect one-to-one! At 40 meters, with the dipole resonated at 7.15 MHz, something is amiss if the SWR exceeds 2.5 at the band ends. And from Fig 1-2 it may be seen that this SWR adds only 0.18 dB to the matched loss, which at 7 MHz is 0.44 dB for 100 feet of RG-8 coax.

Sec 1.5 Nonreflective Load Versus Line-Input Matching

Now is a good time to contemplate the distinction between the no-reflection, perfectly matched load conditions and the line-input matching conditions. From the standpoint of good engineering, as long as the SWR does not exceed the value above which one cannot afford to compromise further power in exchange for improved operating flexibility, the convenience and increased bandwidth afforded by matching at the line input is obvious.

But line-input matching also presents a real challenge to learning more about complex impedance, because in the presence of reflections the line-input impedance has resistive and reactive components, both of which vary with changes in line length and with frequency. Thus, we need to understand complex impedance in order to choose and adjust corrected external conjugate-matching circuitry to couple the transmitter to the line, or to adjust the transmitter directly to the line if sufficient matching range is available. Practically all problems encountered while attempting to obtain proper coupling or loading to a line with reflections can be traced simply to not understanding the correlation of line length and relative phase of the forward and reflected waves with the resulting complex impedance seen at the input terminals of the line.

A detailed discussion of reflection mechanics and feed-line propagation is presented in subsequent chapters. Included is a novel means for explaining impedance transformation along the line in direct relation to forward and reflected waves, which simplifies the understanding of

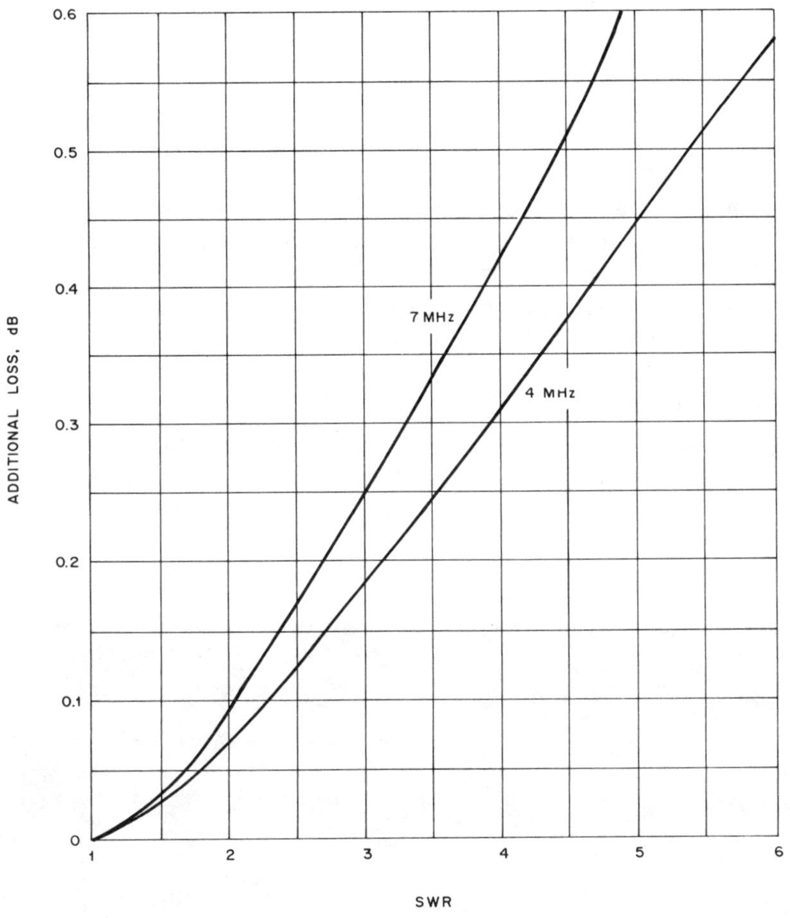

Fig 1-2—Effect of standing-wave ratio on line loss at 4 and 7 MHz. The ordinates give the additional loss in decibels over those for a perfectly matched 100-foot length of RG-8 line for the SWR values shown on the horizontal scale.

what does and what does not happen when a line length is changed, and how to select the correct length for given conditions. The relation of line attenuation to permissible SWR while using conjugate matching techniques, along with details on how to obtain proper coupling and loading of a transmitter to a line for which the input impedance has changed because of reflections, is also presented.

Antennas in Space

The TV dish antenna in Photo 1 was used by the Apollo astronauts during their moon ride in one of the moon buggies. It transmitted TV signals directly from the moon buggy to Earth and into the TV networks from a camera and S-band transmitter mounted on the buggy. When the buggy was stopped for a TV broadcast, the astronauts aimed the dish toward Earth, using an optical telescope attached to the dish mounting. To facilitate aiming with the high accuracy required, the optics in the telescope were designed so that Earth just filled the viewing area. The dish is 28 inches in diameter, constructed of gold-plated fine-mesh wire, folds up like an umbrella, and has a gain of 26 dBic at 3600 MHz.

The dish is shown at RCA on a test fixture designed by the author, and mounted atop the vertical stanchion of an antenna-positioner turntable (not shown). Here the dish received its final RF measurements of radiation pattern and gain. It is situated in an anechoic chamber, or RF

Photo 1—The strange device shown here that looks like a gossamer umbrella is actually a TV dish antenna. It is presently mounted on a lunar rover vehicle parked in one of NASA's used-car lots on the moon.

darkroom. The room eliminates reflections that would affect the accuracy of the measurements. For the pattern and gain measurements, the dish is illuminated with RF from a transmitting antenna positioned downrange 900 feet away. The pattern measurement is made by rotating the dish in azimuth (through the rotation of the turntable supporting the stanchion), and recording the signal level received by the dish during the rotation. A standard-gain reference horn is visible in the lower right in Photo 1. The horn is mounted on the same shaft as the dish, so as the dish rotates in azimuth, so does the horn. As the dish-horn combination rotates through the 180° position in azimuth, the open end of the horn becomes illuminated by the downrange transmitting antenna. As this occurs a coaxial switch just to the left of the horn is turned to switch the receiver from the dish to the horn to record the pattern of the horn as it rotates through the RF illumination. The gain of the dish antenna is then determined by comparing the recorded level of the signal received by the dish with that by the reference horn. The uniformity of the radiation pattern around the axis of the dish is also measured. These measurements are performed by setting the dish at selected angles in azimuth relative to the downrange illumination, and then rotating the dish on its own axis, again recording the signal level. Measurements on all the antennas that went to the moon on lunar rovers were performed by the author. In the event that you saw TV pictures sent from the lunar rover by the Apollo astronauts, the antenna in Photo 1 is the one that transmitted the TV signals from the moon.

Photo 2 shows author Maxwell at the RCA Laboratories working on the four crossed-dipole transmitting antenna system of the world's first weather satellite, TIROS-1. The spacecraft mockup is a full-scale electrical model constructed of sheet aluminum over a wooden frame, used for testing all of the spacecraft antenna systems during the design phase. The model was constructed to be electrically identical to the flight models to obtain valid measurements of impedance and radiation patterns as various antenna configurations emerged during the design. The makeshift mounting for the model was used temporarily while the antenna test laboratory was being constructed at RCA for the TIROS project. The matching and transmitter coupling network that Maxwell designed (not shown) comprised ring diplexers, baluns, frequency-selective filters and many stubs and phasing lines, all fabricated in 25, 50, and 70-Ω printed-circuit, sandwich-form stripline transmission line. The network provided four outputs with a progressive phase of 0, 90, 180, and 270° for feeding the four crossed dipoles to obtain circular polarization in the 108-MHz telemetry band and the 235-MHz TV band. The dipoles were fed from four transmitters operating simultaneously on different frequencies within the two bands. The first pictures of Earth

Photo 2—This 1958 picture shows an 18-sided full-scale spacecraft mockup of TIROS-1, the world's first weather satellite. Author Maxwell is working on the four crossed-dipole transmitting antenna system.

and clouds ever taken by cameras in an orbiting weather satellite were transmitted back to Earth on Maxwell antennas during the first orbit after the launch of TIROS-1 from Cape Canaveral on April 1, 1960. All systems were GO! An unused flight model of TIROS-1 with the Maxwell antennas attached is on display in the spacecraft museum of the Smithsonian Institution in Washington.

A multifrequency sleeve-dipole antenna was used in the four crossed-dipole transmitting antenna system of TIROS-1. The author used the arrangement shown in Photo 3 to make slotted-line impedance measurements during its empirical development. The half-shell scale model of the TIROS spacecraft and single dipole over a ground plane simplified the taking of many impedance measurements on various sleeve configurations at different frequencies, because it avoided the difficult obstacles involved in measuring through a complex feed system that would be required if two or four dipoles had to be fed simultaneously. Only one dipole was required for obtaining valid impedance measurements, because its image, and that of the half-shell of the spacecraft were both formed in the four-foot-square aluminum ground plane shown behind the model. Using the techniques of scale modeling and ground-plane imaging achieves the same results as with a full-scale model. However, scale modeling was preferred because it permitted using smaller components and a smaller work space for conducting the experiments. The model was ¼ size. Hence, the test frequencies were

Photo 3—The author, right, and Technician John Luzansky at the RCA Laboratories in 1958, viewing the half-shell 4:1 scale model of TIROS-1 mounted on a ground plane.

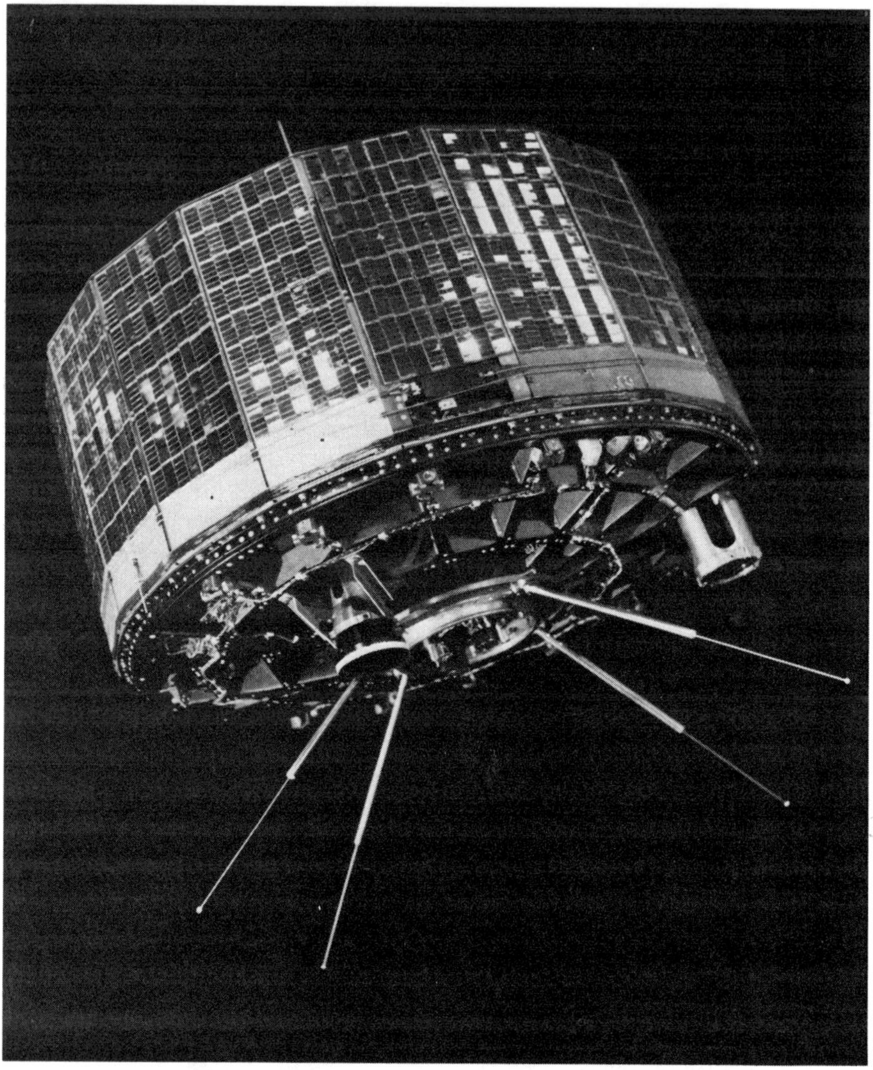

Photo 4—A flight model of the TIROS-1 weather satellite.

430-440 MHz for the telemetry band, and 930-950 MHz for the TV band, four times the actual operating frequencies. The sleeve dipole is visible projecting upward from the top of the model. The equipment used in the impedance measurements were Hewlett-Packard HP-608C and HP-612A signal generators, an HP-805A slotted line, and an HP-415B SWR indicator, located on the bench at the left of the ground plane.

Photo 4 portrays a flight model of the TIROS-1 weather satellite showing solar cells, cameras, and the Maxwell multifrequency crossed dipoles.

Chapter 2

Countdown for a Journey from Mythology to Reality

(Adapted from *QST*, June 1973)

Sec 2.1 Introduction

In Chapter 1 we saw that obtaining a low SWR is relatively unimportant for an efficient transfer of power when line attenuation is low. Four steps to assist in understanding the operation of lines with reflection were suggested, and the concept of matching the complex impedance at the line input in the presence of reflections, based on the principles of conjugate matching, was introduced. The paragraphs which follow present for consideration some of the basic principles involving efficient power transfer through any line terminated in a mismatch.

A match exists throughout the entire system when the optimum load resistance of the source is made equal to the resistive component of the line-input impedance (or vice versa), and when all residual reactance components in the source and line-input impedances are canceled to zero. In this condition the *system* is resonant. All available power from the source enters the line, and reflections from any terminating mismatch or other line discontinuities are compensated by a complementary reflection. Such a reflection is obtained by introducing a nondissipative mismatch at the match point. This nondissipative mismatch is one which, if placed in the system by itself, would produce the same magnitude of reflection, or SWR, as is produced by the mismatched line termination. The result is a precise and total reflection of the arriving reflected wave. Andrew Alford makes an elegant presentation of this concept *(Ref 39, pp 10-15)*.[1]

Although this sounds very complicated, the entire set of conditions is automatically fulfilled simply by completing a correct tuning and loading procedure. It matters not whether a transmitter having sufficient matching range feeds the line directly, or whether an external matching network (Transmatch) is used where additional range is required. If the transmitter, or source generator is now replaced by a passive impedance equal to its optimum load impedance, the line can be opened at any point. Then, from this open point, and looking in either direction, one will see the conjugate of the impedance seen in the opposite direction—whatever $R + jX$ value is seen in one direction, $R - jX$ is seen in the other.

Contrary to what many believe, it is not true that when a transmitter delivers power into a line with reflections, a returning reflected wave sees the internal generator resistance as a dissipative load. Nor is the reflected wave converted to heat and lost, while at the same time damaging the final amplifier. When a matched RF power amplifier is actively supplying power when the reflected wave returns, the reflected wave encounters total reflection at the match point appearing at the input terminals of the pi-network tank circuit, and the reflected power is entirely conserved because it never sees the source resistance of the amplifier as a dissipative terminating load. This is because the source and reflected voltages and currents superpose, or add at the match point, just as if the reflected power had been supplied by a separate generator in series with the source. The phasor sum of their voltages yields a net current flow which is always in the forward direction. The reflected power adds to the source power, deriving reflection gain which compensates for the reflection loss suffered at the mismatched termination. More details on the reflection phenomenon are presented in later chapters.

Sec 2.2 Line Losses

All reflected power reaching the source is returned to the load, as part of the forward or incident wave. The only reflected power lost is that from line attenuation during its return to the source and once again

---

1 All references appear in Chapter 24.

during its return to the load. The higher the line attenuation, the less reflected power reaches the source to add to the forward power. Thus, the lower the line attenuation, the higher the allowable SWR for a given loss because of SWR. No reflected power is lost in a lossless line, no matter how high the SWR, because it all ultimately arrives at the load. This is why open-wire line functions so efficiently as a tuned line with any reasonable mismatch value—its attenuation is almost negligible. Since the attenuation is higher in coax, that imposes lower limits on the mismatch and may require calculation of the loss penalty for a given SWR. Line attenuation and SWR must both be quite high to incur any substantial additional loss over and above the matched-line loss. The additional loss due to SWR and line attenuation may be determined graphically from Figs 1-1 and 6-1, or calculated using Eq 6-1.

Coax has higher RF losses than open wire at HF for two reasons:

1) It has a lower impedance, causing higher current flow at lower voltage for the same power. This results in higher I^2R loss for the same effective conductor size. (Electric power distribution lines minimize I^2R loss by using high voltage and low current). Skin effect also increases the loss with rising frequency because of decreased effective conductor size.

2) The increased amount of dielectric material separating the conductors in coaxial line is a substantial contributor to the attenuation loss factor. The attenuation increases with frequency, especially at VHF and UHF. Hence, it is understandable why RG-8, especially the foam type with its larger center conductor (*Ref 23*) allows higher SWR (more bandwidth) than RG-58 for the same additional loss penalty. And for any cable, the shorter it is, the less loss is added for a given SWR.

A fifth step in improving your understanding of the reflected-power problem is to view the situation objectively, asking yourself, "Have I fallen prey to any of the erroneous teachings? Can I spot the wrong dope when I hear it discussed? Do I understand the principles well enough to convince others of the correct version if the opportunity arises?" Several pertinent short statements follow which may be used as self-test material. They highlight and summarize many reflection-related concepts known to be generally confused among amateurs. In the interest of brevity they are not intended to be completely self-explanatory, but sufficient material for obtaining a complete understanding of each point appears later, or is available in the bibliography references. Support for nearly every statement can be found in *The ARRL Antenna Book* alone.

Sec 2.3 True or False?

1) Reflected power does not represent lost power except for an increase in line attenuation over the matched-line attenuation. In a lossless line, no power is lost because of reflection. Only when the matched-line attenuation and SWR are both high is there significant power lost from reflection. On all HF bands with low-loss coax, the reflected power loss is generally insignificant, though at VHF it becomes significant, and at UHF it is of extreme importance.

2) Reflected power does not flow back into the transmitter and cause dissipation and other damage. Damage blamed on reflections is really caused by improper output-coupling adjustment—not by SWR. Tube overheating is caused by either or both overcoupling and reactive (mistuned) loading. Tank-coil heating and arc-overs result from a rise in loaded Q caused by undercoupling. With some manipulation, proper output coupling indicated by a normal resonant plate-current dip at the correct loading level can be attained no matter how high the SWR. The transmitter doesn't "see" an SWR at all—only an impedance resulting from the SWR. And the impedances are matchable without concern for the SWR. This is one of the most important points of confusion at issue.

3) Any effort to reduce an SWR of 2:1 on any coaxial line is completely wasted from the standpoint of any significant increase in power transfer. See Figs 1-1 and 6-1.

4) A low SWR is not proof of a good-quality antenna system or an indication that it is working efficiently. On the contrary, lower than normal SWR values exhibited over a frequency range by a straight dipole or a vertical over ground is a clue to trouble in the form of undesired loss resistance. Such resistance can be from poor connections, poor ground system, lossy cable, etc.

5) The radiator of an antenna system need not be of a self-resonant length for maximum resonant current flow, the feed line need not be of any particular length, and a substantial mismatch at the junction of the line and the antenna does not prevent the radiator from absorbing all of the power available at the junction (*Ref 3, Part III, p 20; Ref 24*).

6) If a suitable matching network cancels all the reactance developed by a nonresonant-length radiator and a random-length feed line which is mismatched at the antenna feed point, the antenna system is

resonant, the mismatch effect is canceled, maximum current flows in the radiator, and all the power available at the feed point is absorbed by the radiator.

7) The majority of tower radiators used in the standard AM broadcast band (from 540 to 1600 kHz) are of heights which are not resonant lengths at the frequency of operation.

8) The SWR on the transmission line between the antenna and a matching network at the input to the line is determined *only* by the mismatch conditions at the load, and is not changed or "brought down" by the matching network. "Low SWR" obtained by using the device indicates only the mismatch remaining between the input impedance of the network and the impedance of the line from the transmitter.

9) Adjusting the matching network, or antenna tuner, for maximum line current creates a perfect mirror termination for the reflected wave, causing it to be totally re-reflected on arrival at the input end of the line. The tuner provides the proper reactance to cancel the equal but opposite reactance between the source and reflected wave at the input. This causes the reflected wave to add *in phase* to the source wave to derive the total incident, or forward power, which is the sum of the source and reflected power.

10) Total re-reflection of the reflected power at the line input is the reason for its not being dissipated in the transmitter, and why it is conserved, rather than lost.

11) With a good antenna tuner and a well-constructed open-wire feeder, a 130-foot center-fed dipole does not radiate significantly more power on 80 meters than one 80 feet long for the same power fed from the transmitter (*Ref 3, Part III, p 20; Ref 7, pp 50 and 124 or pp 41 and 42 in reprint version; Ref 10; Ref 21*).

12) A dipole cut to be self-resonant at 3.75 MHz and fed with either RG-8 or RG-11 coax does not radiate significantly more on 3.75 MHz than on 3.5 or 4.0 MHz with any feeder length up to 150 or 200 feet.

13) With a 3.75-MHz dipole, the SWR on a 50-Ω feed line rises to around 6.5 at 3.5 MHz, and around 5.0 at 4.0 MHz, thus utilizing the coax as a tuned feeder, but with *insignificant* loss in radiated power across the entire 80-meter band.

14) With the use of a Transmatch or a simple L network at the line input, proper coupling between the transmitter and the tuned-coax

feeder can be attained over the entire 80-meter band *with any random coax length.*

15) From the standpoint of line loss because of SWR resulting from the change in quality of the impedance match between the line and antenna, changing the height of the dipole above ground or lowering the ends of a horizontal dipole to make an inverted-V dipole has an *insignificant* effect on the amount of power it absorbs from the transmitter.

16) As a tuned line at 4.0 MHz, RG-8 can handle 700 watts CW *continuously* within ratings, at an SWR of 5:1. With the duty cycle of SSB, it is far below maximum ratings at 2 kW PEP. With a 100-foot length, the *total* attenuation (SWR = 5) is just 0.78 dB (0.46 dB because of SWR), which is insignificant in terms of received signal strength.

17) If the line length is critical in order to satisfy a particular matching condition, the same input impedance can be obtained with any length of line, shorter or longer, by adding a simple L network of only two components: either two capacitors, two inductors, or one of each, determined by the specific impedance change required of it. This statement is pertinent to coiled-up coax in mobiles (*Ref 19, pp 118-128; Ref 24; Ref 30, p 48; Ref 31*).

18) High SWR in a coaxial transmission line caused by a severe load mismatch does not produce antenna currents on the line, nor does it cause the line to radiate (*Ref 2, p 101; Ref 32*).

19) High SWR in an open-wire line at HF caused by a severe load mismatch does not produce antenna currents on the line, nor cause the line to radiate if the feed currents in each wire are balanced, and if the spacing is small at the wavelength of operation (also true at VHF if sharp bends are avoided (*Ref 2, pp 101, 106*).

20) Both coax and open-wire feed lines may radiate (*Ref 32*), though not at a significant level, by reradiating energy coupled into the line from the antenna because of asymmetrical positioning with respect to the antenna. The energy coupled from the antenna results in antenna currents flowing on the *outside* of the outer coax conductor, or *in-phase* currents flowing in the wires of the open-wire line. But this condition has no relation to the level of the SWR on the line in either case (*Ref 2, pp 101, 106*).

21) SWR indicators need not be placed at the junction of the feed line and the antenna to obtain a more accurate measurement. Within its own accuracy limits, the indicator reads the SWR wherever it is located in the line. The SWR at any other point on the line may be determined

by a simple calculation involving only the SWR at the point of measurement, the line attenuation per unit length, and the distance from the measured point to the point where the SWR is desired. In any case, the primary reason for using the SWR indicator is to assist in proper loading of the transmitter.

22) The SWR in a feed line cannot be changed, adjusted or controlled in any practical manner by varying the line length (*Ref 7, p 51*). (Also see Chapter 21.)

23) If SWR readings change significantly when moving the SWR bridge a few feet one way or the other in the line, it indicates either "antenna" current flowing on the *outside* of the coax, or else an unreliable instrument, or both, or even a reliable bridge incorrectly adjusted to the line impedance, but it is not because the SWR is varying with line length. Some writers insist the bridge must be placed at a ½-λ interval from the load to obtain a correct reading. This is *incorrect*. All readings are invalid if they change significantly along the line, even though they may repeat at ½-λ intervals (*Ref 2, pp 101, 106, and 132*). (Also see Chapter 21.)

24) Any reactance added to an already resonant (resistive) load of any value for the purpose of compensation to reduce the reflection on the line feeding the load will, instead, only increase or worsen the reflection. It is for this reason, although contrary to the teaching of several writers, that the lowest feed-line SWR occurs at the self-resonant frequency of the radiating element it feeds, completely independent of feed-line length. Any measurements which contradict this indicate that either the measuring equipment or the technique (or both) are in error.

25) Of the several types of dipoles such as the thin wire, folded, fan, sleeve, trap, or coaxial, none radiates more field than another, providing each has insignificant ohmic losses and is fed the same amount of power (*Ref 3, Part III*).

26) If coax at least the size of RG-8 is used in mobile installations (80 through 10 meters), any matching required to load the transmitter may be done at the input end of the coax without significant power loss compared to matching at the antenna terminals, and with improvement in operating bandwidth.

27) With center-loaded mobile whips of equal size having no matching arrangement at the input terminals, the best radiating efficiency is obtained on models having the lowest measured terminal

resistance (highest resonant SWR, model for model). Models having lowest SWR are wasting power in the loading coil, because of either a low value of coil Q or excessive distributed coil capacitance, or both. (See Chapter 6 for more details.)

28) The resonant frequency of an antenna cannot be determined by probing the input terminals of a feed line for resonance with a grid-dip oscillator or a noise bridge that has no capability for measuring reactance. Resonances measured at the feed-line input are resonances of the combination of the antenna and the feed line, not the resonance of the antenna alone. A change in length of the feed line results in a different resonant frequency of the combination for every different length of feed line. A ½-λ feed line is not the answer, because, since the resonant frequency of the antenna is unknown, the frequency at which the feed line should be ½ λ is also unknown. If the feed line is ½ λ at some frequency other than the resonant frequency of the antenna, the resonant frequency measured at the input of the feed line will be different than that of either the antenna or the feed line.

Sec 2.4 How Did You Do?

Did you mark all of the statements correctly? You did if you marked each one TRUE. Yes, *all* of the above statements are true. These examples have been centered around 80-meter operation because bandwidth and dipole length on this band present the maximum SWR problem. Of all the amateur bands, 80 m has the largest bandwidth, 13.3% of center frequency, as compared to 4.2% on 40 m, 2.5% on 20 m, 2.1% on 15 m, and 5.9% on 10 m. Having the longest wavelength (excepting 160 m, of course), 80 m poses the greatest problem with respect to physical construction of radiating systems on existing real estate. Some antenna sites just won't permit an entire ½ λ on 80 m. So these examples should have a special interest if you wish to work 80 meters but are forced to use a short antenna. Since the bandwidth and antenna-length problem are really one and the same, the 80-meter examples have maximum practical value. But practices recommended at the 80-meter level are also valid on the higher frequencies. Interestingly enough, as we go to the higher bands, the line losses increase but the percentage bandwidth of the amateur bands decreases. This means inherently lower maximum SWR values will be obtained during frequency excursions from the design center to the ends of the band.

Chapter 3

Going Around in Circles to Get to the Point

(Adapted from *QST*, August 1973)

Sec 3.1 Basic Reflection Mechanics

In this chapter I discuss reflections, how they are developed, and how they effect operations on transmission lines. In describing the direction of wave travel, the terms "forward" and "incident" are used interchangeably.

It is generally well understood that the size or magnitude of a reflection arising from a mismatched line termination is determined by the degree of the mismatch, or how much power in the forward, or incident wave is unabsorbed by the load terminating the line. Following ASA Standard Y10.9-1953, the symbol ρ (rho) represents the magnitude of the reflection coefficient, which is expressed as the ratio of the voltage reflected by the mismatch relative to the forward voltage, the voltage incident on the load. The *complex* reflection coefficient, $\overline{\rho}$ is determined quantitatively from the line and load impedances by the expression

$$\overline{\rho} = \frac{Z_L - Z_C}{Z_L + Z_C} \qquad \text{(Eq 3-1)}$$

where

Z_L = the complex load impedance, R + *j*X

Z_C = the characteristic impedance of the transmission line

This expression shows that $\rho = 0$ (no reflection) when $Z_L = Z_C$. However, the load must be purely resistive (R + *j*0) for zero reflection, because we are considering only lossless and low-loss lines with a characteristic impedance Z_C that is purely resistive. Perhaps somewhat less appreciated than reflection magnitude ρ is reflection phase θ (theta), which describes the character of the mismatch expressed as an angle, the phase angle of the reflected voltage wave relative to the phase of the forward wave. Hence, the magnitude ratio, ρ, and the relative phase, θ together comprise the *complex* reflection coefficient $\overline{\rho} = \rho\underline{/\theta}$, which appears later in discussions of transmission-line propagation and matching techniques.

An open circuit (infinite impedance), a short circuit (zero impedance), or a pure reactance as a terminating load on a transmission line is incapable of absorbing any power from a forward wave, and therefore causes total reflection of forward waves of both voltage and current. The magnitude, ρ, of the reflection coefficient at such a load impedance is therefore 1.0 (unity) for both voltage and current. The step-by-step process by which the reflections arise on low-loss and lossless transmission lines, both coaxial and open wire, is as follows. The incident or forward electromagnetic wave sees the characteristic impedance, Z_C, of the line as a resistive load as it leaves the generator in its forward travel down the line. Half of the energy in the wave is stored in the magnetic field, because of forward current, and the other half is stored in the electric field, because of forward voltage. The voltage and current travel forward in phase with each other because Z_C is resistive.

On reaching an open-circuit load termination, the magnetic field collapses because the current goes to zero. The changing magnetic field produces an electric field equal in energy to the original magnetic field. (Remember that motor-generator action results from mutual motion between a field and a conductor. Here the field is changing while the conductor is stationary, as in a transformer.) The new electric field resulting from the changing magnetic field adds *in phase* to the existing electric field, resulting in a corresponding increase of voltage at the open circuit to twice the voltage of the forward wave. At this instant, a standing wave is developing, because now there is a current minimum and a voltage maximum at the open-circuit termination, where an instant before, current and voltage were constant all along the line.

The increased voltage, along with its new electric field, now starts the reflected voltage wave traveling in the opposite direction, as if it had been launched by a separate generator at the open-circuit point. (Remember motor-generator action?) Since no energy was absorbed by the open-circuit load, the returning reflected wave is of the same magnitude as the original forward wave, which is why $\rho = 1$. As the new electric field starts its rearward journey, its motion sets up a new magnetic field in opposite phase to the original, and as in the forward electromagnetic wave, the reflected energy also divides equally between the electric and magnetic fields. The motion of the new magnetic field causes the current to build up again to the same magnitude as before, to be relaunched into the line as the reflected current wave, but in the opposite polarity and in the opposite direction relative to the forward current (*Ref 17, p 139; Ref 19, p 4; Ref 35, p 21; Ref 43; Ref 70*).[1]

The total voltage (or current) at the load at any instant is the sum of the voltages (or currents) of the forward and reflected waves. The in-phase reflected voltage wave is verified because the sum of the two voltages at the load is double that of the forward voltage. And since the two current waves add to zero at the open-circuit load, the generation of the reversed-polarity reflected-current wave is also verified. The phase angles, θ, of the reflection coefficients at the open-circuit load are therefore 0° for voltage and 180° for current.

When the load impedance is a short circuit, the reflection-generation process is the same as in the open-circuit case, except that the electric- and magnetic-field actions and the polarities of the reflected-wave components are reversed. This is expected when we recall that while current goes to zero in an open circuit, voltage must be zero in a short circuit. For the voltage to be zero, the forward and reflected voltage waves must cancel one another at the load, thus verifying the reversed polarity of the reflected voltage wave with the short-circuit termination. The corresponding currents add to double the forward value, as the voltages did when the load was an open circuit. The phase angles θ, of the reflection coefficients at the short-circuit load are therefore 180° for voltage and 0° for current. When the load impedance is a pure capacitance, it is equivalent to an additional length of open-circuited line, while a purely inductive load is equivalent to an additional length of short-circuited line.

---

[1] All references appear in Chapter 24.

We have observed that during the reflection process, either the voltage or current changes polarity, but not both. Repeating for clarity, when the load is an open circuit the current changes polarity, but not the voltage; when the load is a short circuit the voltage changes polarity but not the current. Hence, it is evident that while the forward voltage and current travel in phase, the reflected voltage and current are 180° out of phase.

When the load impedance contains resistance, the reflection is generated in a manner similar to that described earlier. However, the reflection is less than total, the amount depending on how much power is absorbed in the resistance. The reflected wave is again generated by the changing electric and magnetic fields at the mismatch point, caused by the change in voltage and current when the forward wave encounters a change in load impedance relative to the characteristic impedance Z_C of the transmission line. Hence, the reflection coefficient, ρ, is dependent on the difference between forward-wave voltage on the line and the voltage measured across the load. If the resistance in the load is greater than Z_C, the reflection angles θ are the same as for the open-circuit termination; if the load resistance is less than Z_C, the θ angles are the same as for the short-circuit termination.

No reflection arises when the load is a pure resistance equal to the line Z_C, because all of the forward energy is absorbed in the load, and there is no voltage or current variation when the energy passes from the line to the load. Thus, there is no change in the electric and magnetic fields, no new voltage or current generated, and hence no reflected wave.

The reflected wave travels back up the line as a new electromagnetic traveling wave, separate and distinct from the forward-traveling incident wave. As the reflected wave travels, it encounters only the same resistive line impedance Z_C encountered by the forward wave in its forward travel to the load. Hence, the magnitudes of both the reflected voltage and current waves remain constant as they plow rearward, having the same values as when leaving the reflection generator. (This is true on lossless lines, but there is a *gradual* diminishing effect because of line attenuation in real lines, as discussed later.) They are completely unaffected by the standing wave that is developing as the reflected and forward waves slide past one another going in opposite directions. The forward voltage and current waves are similarly unaffected, continuing in their forward travel with constant magnitude until reaching the load.

Also, as in the forward wave, both the reflected voltage and current pass through zero simultaneously, and reach their maximum values ¼ cycle later, because the line Z_C is resistive.

Are not the reflected voltage and current then also in phase with each other, like the forward voltage and current? No! As explained earlier, only the voltage *or* the current changes polarity during reflection, but not both. The maximum of the reflected voltage is positive when the current maximum is negative, and vice versa, which means, as stated earlier, the reflected voltage and current are always 180° out of phase with each other. But does the phase really matter here? Indeed it does—there is no other relationship more important to wave propagation on a transmission line! The 180° out-of-phase relationship between the reflected voltage and current is *directly responsible* for the development of the standing wave, as well as for the change in the input impedance of the transmission line from Z_C to a complex value when the load terminating the line is mismatched.

Even though the forward and reflected waves travel separately in opposite directions, they are inescapably related to each other through their common line and load characteristics. Their respective voltages and currents add *vectorially* at every point along the line as the two waves slide past each other, thus forming the standing wave. In other words, the phasor sum of the forward and reflected voltages (or currents) at every point along the line produces the amplitude of the standing wave at the corresponding point on the line. In addition, the value of the resultant phasor voltage divided by the resultant phasor current gives the complex impedance on the line at the corresponding point. Hence, the polarity relationships between the voltages and currents in the forward and reflected waves determine both the characteristic of the standing wave and the input impedance of the transmission line. We explore the development of the standing wave in more detail later in this chapter, and also in Chapter 9.

Sec 3.2 Wave-Travel Analysis

Some people have difficulty in understanding the polarity relationships between voltage and current waves on transmission lines, so now let's examine polarity and the establishment of a polarity reference for convenience. Consider a single wave traveling on a two-conductor line. By following conventional current flow we can select the appropriate conductor as the voltage-polarity reference for a given direction

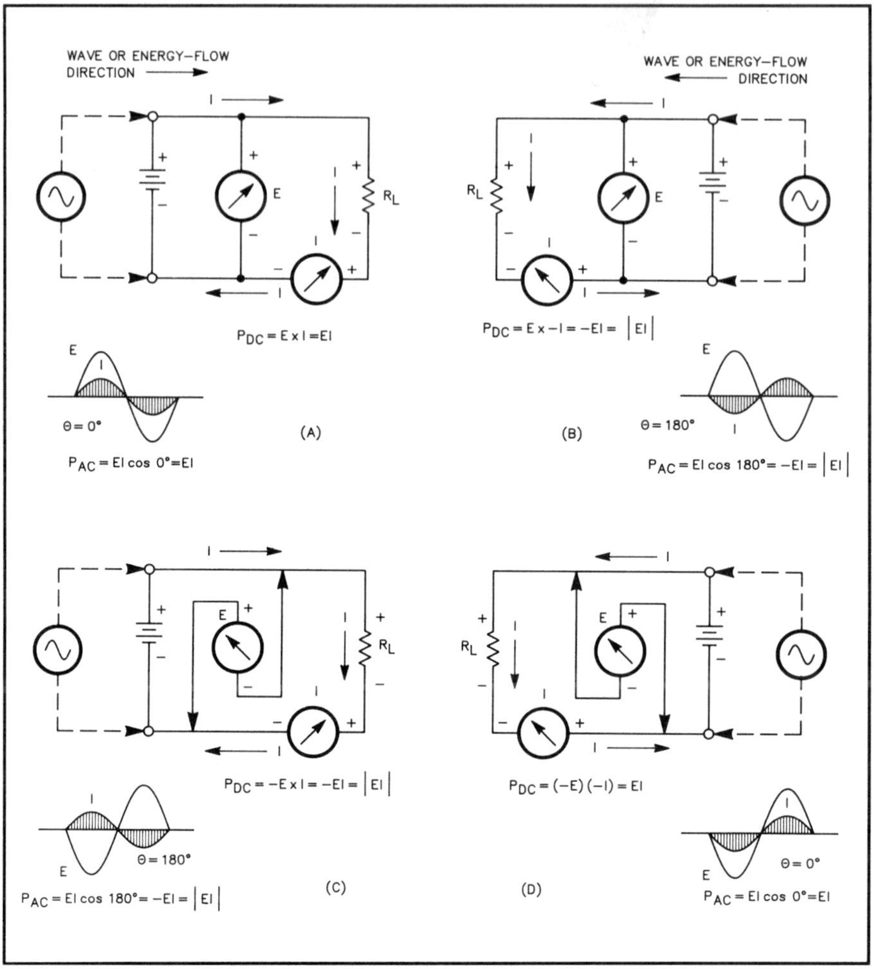

Fig 3-1—Illustrating relationships between current and various combinations of voltage reference and energy-flow direction.

of wave travel which causes the voltage and current maxima to occur with the same polarity. This polarity relationship may be reversed simply, either by selecting the opposite conductor for the voltage reference or by reversing the direction of wave travel. Obtaining the opposite polarity or phase relationship by reversing the conductors is a simple enough concept, but obtaining it by reversing the wave-travel direction has been a point of confusion for many people.

To reduce the confusion, a set of simple current-flow diagrams showing both ac and dc treatment is presented in Fig 3-1. Use of dc with zero-center meters as indicators makes the explanation of polarity easy. Conventional needle movement is to the left for negative polarity and to the right for positive. Once polarity is clear, the battery may be replaced with an ac generator, and phase will also become clear using the waveforms as indicators. The circuits in A and B of Fig 3-1 are the same as in C and D, respectively, except that the voltmeter terminals have been reversed. Observe that the wave or energy-flow direction and voltage-polarity reference selected in A cause line voltage and current flow to be in the same polarity. Now notice that reversing either the wave direction (as in B) or the voltage reference (as in C) both result in *opposite* voltage and current polarities, as stated previously. Observe also that reversing both direction and voltmeter reference polarity (as in D) again results in line voltage and current flow with the same relative polarity, though reversed from A. It may be helpful at this point to perceive that changing the wave or energy-flow direction is simply equivalent to reversing the terminal connections of the current meter, because the source changes sides. This is the key to understanding the polarity-reversal problem, because for a given voltage-polarity reference, the current-flow direction must reverse when the wave-flow direction reverses.

On the basis of these conditions, assume a generator is now placed at each end of a single two-conductor transmission line. Then, a reference that is selected to make the voltage and current in phase on the line for one generator will result in 180° out-of-phase voltage and current for the other generator. This is the situation which exists with the mismatched RF transmission line—a source generator at one end of the line and the reflection generator at the other end. By selecting the conventional reference that makes forward voltage and current in phase with each other, it verifies that reflected voltage and current are 180° out of phase with each other (*Ref 35, p 23*).

It is of interest at this point to be concerned with the nature of the power in the forward and reflected waves. Some writers contend erroneously that the voltage-current phase relationship in the reflected wave is 90°. If this were true, then the reflected wave would contain only reactive volt-amperes, and no real power. The evidence presented above disproves this contention since we have seen that the voltage-current phase relationship in the reflected wave is 180°, not 90°. And certainly we will agree that if real power is conveyed in A of Fig 3-1, it is also real power in B, or C, even with reversed current-meter or voltmeter terminals. We will agree that real power P equals EI cosine θ, in which cosine θ is the power factor. It matters not whether the phase angle is 0° or 180°, because cos 0° = 1, and cos 180° = –1. This simply denotes the polarity difference discussed above.

When conductor spacing is restricted to the near field, the fundamental principles governing transmission-line propagation are the same as those which govern all general ac circuit relationships, including electric power transmission. From these principles we know that real power flows for every value of theta in all four quadrants, except at 90 and 270° where the cosine is zero, yielding zero power factor. Wherever the phase is other than 0, 90, 180, or 270°, both real power and reactive volt-amperes are present. But at 0 or 180°, only real power exists because the absolute value of the power factor is 1.0 in either case. This clearly proves that reflected power and forward power are both real power, and that no fictitious power, or reactive volt-amperes, exists in either one. This is indeed true because, as I have shown earlier, the voltage and current in the forward wave are always in phase, and the voltage and current in the reflected wave are always 180° out of phase.

The problem concerning real versus reactive power in the reflected wave arises in part from confusion between traveling and standing waves, because of insufficient familiarity with both types. To broaden the familiarity, I have concentrated first on the *traveling* forward and reflected waves. From the physical viewpoint, as stated earlier, standing waves are derived from the resultant interaction between the two traveling waves. And thus, sufficient knowledge of traveling waves is essential before one can correctly understand the formation of standing waves and other correlated phenomena occurring on the transmission line. These others will become apparent as we proceed.

Now that you have a reasonably enlightened background concerning the ingredients of standing waves, we'll explore the details of their development. After that it is appropriate to return to the real versus reactive power problem for a few brief comments to clear away any remaining confusion.

Sec 3.3 Vector Graph Explanation and Standing-Wave Development

The newly launched reflected voltage and current waves, in their rearward travel toward the generator, combine with their respective

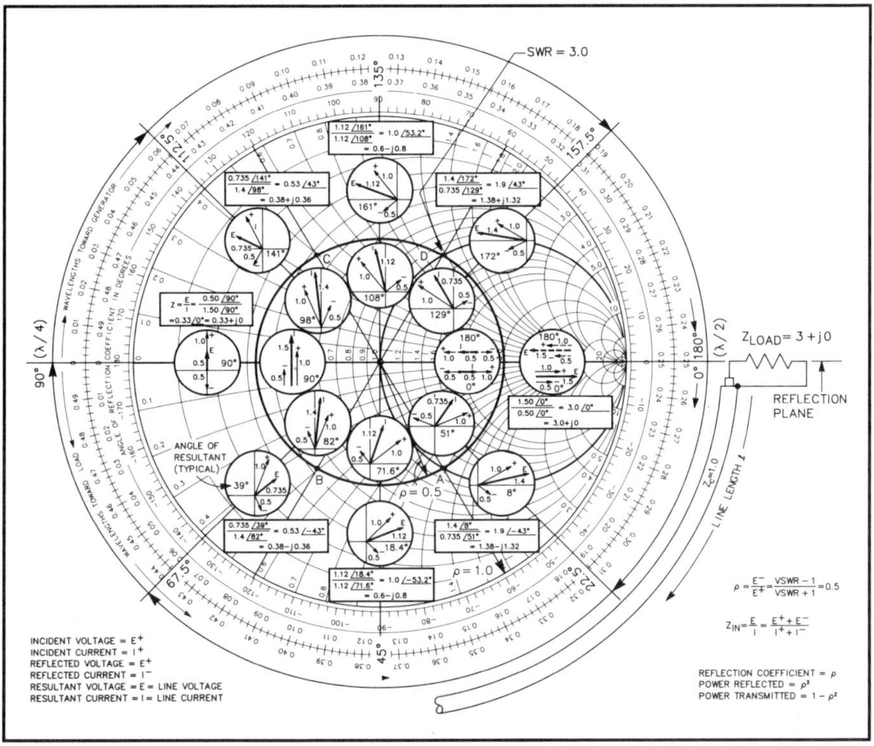

Fig 3-2—The W2DU Reflection Mechanics Vector Graph. For greater readability, this graph is printed in larger size on a foldout page near the back of this book.

forward waves at every point on the line. Their continuously changing relative phase differences along the line cause alternate cancellation and reinforcement of the voltage and current distribution on the line. This results in the formation of the standing wave and a change in the input terminal impedance, Z_{IN}, from the initial value of the line characteristic impedance, Z_C.

A physical picture of this complex relationship enhances the understanding of the phenomenon. Accordingly, the W2DU Vector Graph of Fig 3-2 graphically illustrates the progressive phase relationships. (A larger version of Fig 3-2 appears on a foldout page near the back of this book.) Circular insets every 22.5° along the line (1⁄16 λ) show accurately scaled vector plots of the forward and reflected wave for visual comparison. The reference for the angular distance along the line begins at the termination point (the reflection plane, drawn at the right), with angular movement going clockwise toward the generator. These vector plots superimposed circularly around the Smith Chart present certain symmetrical phase-angle relationships with respect to line length that are not obvious in previously published displays (*Ref 2, p 72; Ref 17, p 146; Ref 18, p 110*). In addition to providing a new way to show the formation of the standing wave, Fig 3-2 provides a new visual basis to explain many other aspects of transmission-line operation. Included are the development of capacitive and inductive components of complex impedance, the understanding of ¼-λ impedance inverting action, the impedance-repeating phase-inverting action of the ½-λ line, and the reciprocal relationship between impedance and admittance.

For this illustration we are letting the line impedance $Z_C = 1.0$, and we have terminated the line with a pure resistance of 3 Ω, three times the line impedance Z_C. Hence, a circle representing an SWR of 3.0 is shown. The line length len°, designated on the drawing by a script letter L, is measured from 0° at the reflection plane, and is represented by clockwise rotation around the chart.

Vectors representing voltage are displayed in the insets outside the 3:1 SWR circle. Vectors marked "+" represent incident, or forward voltage (magnitude 1.0), vectors marked "–" represent reflected voltage (magnitude 0.5), and the vectors marked "E" represent the resultants of the forward and reflected vectors, both as to length and angle. (The resultant voltage, E, is also the *line voltage*. The corresponding current vectors appear in the insets directly opposite and inside the SWR circle. The resultant vectors marked "I" represent *line current*. The number in degrees shown with each vector set indicates the angle of the resultant in that set. The vector lengths are proportional to the amplitudes of their respective voltage and current waves at each point along the line. By setting the lengths of the forward vectors equal to 1.0, the length of the reflected vectors automatically equals the magnitude of the reflection coefficient, ρ. Hence, the lengths of the reflected vectors are 0.5 because ρ = 0.5 for an SWR of 3.0 from the relationship

$$\rho = \frac{\text{SWR} - 1}{\text{SWR} + 1} \qquad \text{(Eq 3-2)}$$

While the vector lengths are proportional to each other, the lengths are not scaled to the chart dimensions. However, the various vector plots on the graph contain the necessary amplitude and phase information to define both the standing wave and the impedance at each corresponding point on the transmission line.

The SWR = 3.0 (or ρ = 0.5) circle is based on the chart scales, and may be observed to have a radius of half the chart radius. The perimeter of the chart has a radius of 1.0, representing total reflection, that is, an infinite SWR. The radius of the ρ circle is thus directly proportional to the reflection coefficient, ρ. Thus, an SWR or ρ circle may be constructed for any value or reflection by making the radius equal to either SWR or ρ.

In order to make valid phase comparisons between points on the line, wave motion has been frozen at an arbitrary point in time so that all vectors are shown in their true positions relative to each other. It makes no difference when the motion is stopped, but the symmetry of the presentation has been enhanced by stopping the motion when the forward vectors at the reflection plane are pointing in the zero direction in the standard polar coordinate system (to the right).

Observe that at any point on the line, the vectors representing forward voltage and current are always pointing in the same direction, indicating that they are in phase. In contrast, the vectors representing reflected voltage and current are always pointing in opposite directions, indicating that they are always 180° out of phase. Note that at the reflection plane (len° = 0°), all components are in phase except the reflected current, which is 180° from all of the others. This is as it should be when the terminal impedance lies between Z_C and an open circuit. These vector relationships graphically illustrate the concluding comments of Sec 3.2.

With clockwise travel from the reflection plane toward the generator, each forward and each reflected vector rotates the same number of degrees as those in the angular change in position along the line. But observe carefully, for this is very important: The forward-wave vectors (+) rotate counterclockwise (phase leading), while the reflected-wave vectors (–) rotate clockwise (phase lagging). For example, at 45° from the reflection plane, the forward voltage vector is at +45°, while the reflected voltage vector is at –45°, for a total phase difference of 90°. Thus, for every degree of motion along the line, the relative phase angle between the forward and reflected voltage changes two degrees. (The relative phase angle between the forward and reflected voltage is θ, the *angle* of the reflection coefficient. Also observe on the graph that the degree scale labeled "Angle of Reflection Coefficient" increases at twice the rate of increase of the line-length scale, len°.) This can be readily understood when you consider that in the distance from the reflection plane to our observation point at 45° along the line from the termination, the reflected wave has traveled twice as far as the forward wave. From the observation point, the forward wave travels only to the reflection plane, while the portion of the forward wave that is reflected travels an equal additional distance in returning to the observation point.

Now let's see what happens at len° = 90°, or after ¼ λ of travel from the load. At the load, where len° = 0°, the forward and reflected voltages are exactly in phase with each other. This gives the reinforcement mentioned earlier, and shows a resultant length of 1.5. But at this point, 90° toward the generator, the two voltage vectors have each rotated 90° in opposite directions. They are now 180° out of phase and opposing, showing a resultant length of 0.5. Continuing on to len° = 180°, or ½ λ from the load, we see that each voltage vector has now rotated 180°, but in opposite directions of rotation. Hence the vectors have rotated 360° relatively. Once again, they are exactly in phase with each other, and are again reinforcing. (On the chart, the dashed vector lines indicate the vector positions after 180° of travel.)

At the points between len° = 0° and len° = 90°, the resultant voltage vector, E, is seen to diminish gradually from the maximum of 1.5 to a minimum of 0.5 and then increase back to the 1.5 maximum at len° = 180°. In Fig 3-3 the values of these resultant vector magnitudes at each 22.5° point have been plotted on the more familiar rectangular coordinate graph. The smooth curves connecting the plotted values generate the familiar standing-wave patterns of voltage and current. The curves verify the relationship

$$\text{SWR} = \frac{1+\rho}{1-\rho} = 3.0 \qquad \text{(Eq 3-3)}$$

With $\rho = 0.5$, the curves of Fig 3-3 show first adding ρ to and then subtracting ρ from the forward voltage, 1.0. Also bear in mind that the amplitude of each curve at any given point indicates the line voltage and line current, respectively, at the corresponding points on the transmission line.

Line lengths greater than ½ λ are accommodated simply by continuing on around the circle again (Fig 3-2), repeating the same values encountered 180° earlier, and thus establishing the periodicity of the standing wave. Only lossless line is being considered here; correction factors for attenuation, which change the circle into an inward spiral, are presented later. The basis for the phase- or polarity-reversing characteristic of the half-wave or 180° line may be observed on the Vector Graph: Note that the specific phase of the voltage vectors at the 180° point on the line is 180° from their phase orientation at the reflection plane (len° = 0°).

The significance of the constant 180° phase difference between the reflected voltage and current emerges if we now compare the phase and magnitude of the current vectors with the voltage vectors just discussed. We see that at the reflection plane where the forward and reflected voltage waves are in phase and adding to create a voltage maximum, the corresponding currents are out of phase and opposing to create a current minimum. And at len° = 90°, where the voltage waves are out of phase and opposing, resulting in a minimum, the currents are *in phase*, resulting in a maximum. This is the reason, graphically illustrated, why the maxima and minima of the voltage standing wave are *always* separated by 90° from the corresponding maxima and minima of the current standing wave. This phenomenon is caused directly by the constant 180° phase difference between the reflected voltage and current, as stated earlier and as shown here on the vector display.

You can make a visual comparison of the voltage and current vector resultants in Fig 3-2 (both angular positions and magnitudes) with the corresponding positions along the plot of Fig 3-3. This comparison should enhance your understanding of the important concept of

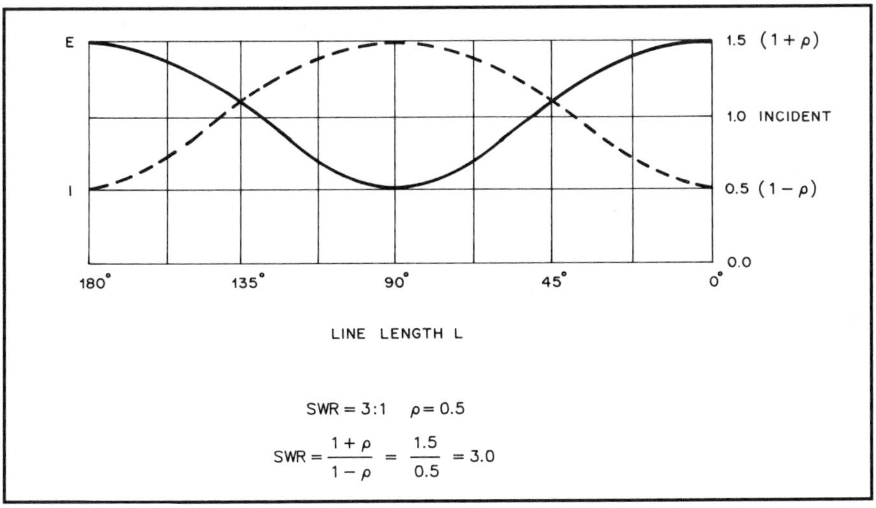

Fig 3-3—Illustrating how the voltage and current standing waves relate to the magnitude of the reflection coefficient, ρ.

a 180° voltage-current phase difference in the development of the standing wave. The concept is of further importance because, as we proceed, you'll see that this phase difference is the basis for the impedance-transforming properties of the transmission line, including the ¼- and ½-λ sections (which are only two specific conditions of the general case).

As we discuss the concept of impedance in Chapter 4 and later solve matching problems, conjure up a mental image of the action occurring on the line. This will at times be more helpful in understanding the difficult points than through logical reasoning alone. It is especially important to have a clear image of the formation of the reflected wave. Remember, the reflected wave must be considered as a separate traveling wave. It is identical to the forward wave except for the direction of travel and, usually, the magnitude. This point is important, because it helps us keep in mind that the reflected voltage and current waves travel the line 180° out of phase with each other, and thus transfer actual, real power during their travel. It is essential to the process of reinforcement and cancellation of voltage and current along the line in the formation of the standing wave that real power be conveyed in the reflected wave, as if it had been developed by many tiny little generators at every point all along the line. It will also become clear in the chapter to follow why the impedance along a line changes in the presence of reflections only because real power is flowing in *both* directions (*Ref 2, p 70; Ref 35, p 24; Ref 42*).

I stress this point because, as mentioned previously, some writers have presented the erroneous viewpoint that the reflected wave conveys no real power, with the argument that the reflected voltage and current are 90° out of phase with each other and are therefore wattless. The reflected wave would indeed convey zero power if its voltage and current were 90° out of phase, but the argument is incorrect because, as stated previously and shown in the Vector Graph, they actually travel 180° out of phase. At least three writers, Woods (*Refs 67 and 97*), DeLaMatyr (*Ref 98*), and Drumeller (*Ref 68*) would have us believe otherwise. It is easy to reach the wrong conclusion, however, because of the lack of a clear image of the reflection process and the somewhat complicated wave mechanics on the line. Woods and Drumeller have confused the *reflected wave* with the *standing wave*, while all three writers may have confused *reflected* voltage and current (E^- and I^-) with *line* voltage E and *line* current I. Drumeller says ". . . application of logic has demonstrated that reflected power is a vicious fiction." The true nature of the reflected wave as a separate electromagnetic traveling wave must be appreciated. It is interesting to note that Woods states, ". . . re-reflection of power at the input end (of the line) is impossible to accept since the necessary conditions of impedance mismatch are not present." However, his Fig 2 shows a circuit which, when properly tuned, forms an impedance-matching network that re-reflects all reflected power reaching it. This is also true in his Fig 1, since he implies that his final amplifier is loaded and tuned properly. Now the very essence of this impedance-matching network is its totally reflecting mismatch for waves traveling toward the generator, while presenting a perfect match for waves traveling toward the load. Misunderstanding of this important basic concept is widespread among the amateurs. This concept is basic to the operation of passive frequency-selective filters, and the mechanism for developing the match simply constitutes a filter of this type. The mechanism behind its operation involves wave interference and reflections, which are described in detail in Chapter 4, again using the Vector Graph as a visual aid.

Mistaking the difference in phase between the reflected voltage and current as the 90° difference of position on the line that exists between the voltage and current maxima (or minima) of the standing wave (described a few paragraphs earlier and shown in Fig 3-3) is understandable. And the similarity between *line* voltage and *line* current behavior with the voltage-current relationship in conventional ac circuitry also makes it easy to understand why line voltage E and current I are being confused with reflected voltage E^- and current I^-. This is because, in addition to the in-phase line voltage and current components, which convey only the net power flow, line voltage and current do contain reactive components which are 90° out of phase with each other when reflections exist on the line. Obviously, these reactive components convey no real power. Unfortunately, some people who are well versed in ac circuitry using lumped constants, but who are less familiar with the principles of transmission lines, sometimes make the error of assuming that the two circuit types are identical in electrical performance. Hence, we should be wary of making unwarranted comparisons which can bring about disastrous consequences of the type we are attempting to straighten out here.

Another erroneous argument set forth is that reflected power cannot be real power because it cannot perform work. I'll prove this argument erroneous by showing how power in the reflected wave *can* do work. Now a simple RF voltmeter connected across the line indicates line voltage, and an RF ammeter in the line indicates line current. As stated earlier, line voltage and current are the *resultants* of the combined forward and reflected waves (E and I on the Vector Graph), the product of which, when multiplied by the cosine of the phase angle between them, yields only the net power that is absorbed by the load. But there are several devices in common everyday use which selectively extract either the reflected or the forward wave from the line, independent of the standing wave. These devices permit separate measurement of the power associated with the waves traveling in either direction. One such device is the directional coupler. Another is the circulator. The circulator is a three-port directional device. The wave reflected from a mismatched load that terminates the second port is completely diverted away from the input feed line, and emerges from the third port. The reflected wave cannot get back onto the feed line to interact with the forward wave to develop a standing wave, or to change the line-input impedance at the source feeding port one. However, current flow through a resistor terminating port three develops I^2R heat equal to the amount of the power reflected from the mismatched load that terminates port two.

A four-port hybrid coupler can be connected to perform in the same manner as the circulator. Directional RF devices most familiar to amateurs are the simple reflectometer SWR indicator and the directional wattmeter (*Ref 18, p 188, and Refs 38, 40, and 42*). The meter in either instrument is actuated by RF power absorbed from one of the traveling waves on the line—either the forward or the reflected, as selected. If the reflected wave were wattless reactive power, no power would be available to actuate the meter movement in the SWR indicators, or to produce heat from the current flow in the resistor on port three of the circulator. Furthermore, for power to become wattless on being reflected would violate the most general and fundamental of all physical laws, namely the law of conservation of energy (*Ref 35, p 25*). On the basis of this law, if all the energy flowing in the line toward the load cannot be absorbed in or dissipated by the load, then that portion which is not absorbed must appear somewhere else. It cannot just disappear or cease to exist as if by magic. The reflected power recovered as heat in the circulator is a typical proof of this fact.

Here is another way to express net power flow through a transmission line, one that enables us to break power down into its separate forward and reflected *power* components. The final expression is obtained from the transmission-line power formulas found in many engineering textbooks on transmission lines. The formulas which follow are taken from Johnson (*Ref 18*).

$$P = EI \qquad \text{(Eq 3-4)}$$

$$P = \frac{E^2}{Z} \qquad \text{(Eq 3-5)}$$

On RF lines,

$$\text{Power} = P = E_{MAX} \times I_{MIN} \qquad \text{(Eq 3-6)}$$

or

$$P = \frac{E_{MAX} \times E_{MIN}}{Z_C} \qquad \text{(Eq 3-7)}$$

Now E_{MAX}, produced by $E^+ + E^-$, occurs as shown in Fig 3-2, where the forward and reflected voltages are in phase at len° = 0° and 180°, and E_{MIN}, produced by $E^+ - E^-$, occurs where they are 180° out of phase, ¼ λ away at len° = 90°. Later we shall see that both the resultant voltage E and resultant current I are nonreactive at these points on the line, while being reactive everywhere else along the line between these points. However, because E_{MAX} and E_{MIN} are also nonreactive at these points on the line, respectively, the product of these voltages divided by the line impedance, Z_C, yields the net power flow, *exactly*. But recalling that E^+ and E^- are always nonreactive, we can replace the term E_{MAX} by $E^+ + E^-$, and E_{MIN} by $E^+ - E^-$, and thus

$$P = \frac{E_{MAX} \times E_{MIN}}{Z_C} = \frac{\left(|E^+| + |E^-|\right) \times \left(|E^+| - |E^-|\right)}{Z_C} \qquad \text{(Eq 3-8)}$$

Multiplying out the numerator terms gives the desired forward and reflected components of power.

$$P = \frac{|E^+|^2}{Z_C} - \frac{|E^-|^2}{Z_C} = \text{net power flow} \qquad \text{(Eq 3-9)}$$

The first term on the right of the equal sign in Eq 3-9 expresses the power associated with the forward wave and the second term expresses the reflected power. This simple separation of power into two components, each associated with one of the traveling waves, can be done on a lossless or low-loss line, where Z_C is resistive. If the line has appreciable loss, the interaction of the two waves gives rise to a third component of power which we can disregard, because lines normally used by amateurs are usually in the low-loss category (*Ref 18, p 150; Ref 37, p 129*).

This separability of the forward and reflected powers forms the physical basis for the operation of reflectometers and directional wattmeters. These devices sense either the forward or reflected component by taking advantage of the 180° out-of-phase relationship of the reflected components of voltage and current, while the forward voltage and current components are in phase with each other.

Bruene's explanation of directional wattmeter operation is a classic paper (*Ref 38*). In these instruments, a sample of the voltage across the line is added to a sample of voltage derived from the current in the line. When the amplitudes of the two samples are adjusted to the correct relationship determined by line impedance Z_C, the two components of the reflected waves cancel, so that the sum represents only the forward components. By reversing the phase of the current sample, the forward components cancel and the resulting sum represents only the reflected components. A meter connected to indicate these voltage sums can now be calibrated in *power*, because the voltage squared is proportional to power.

The beautiful aspect of the directional wattmeter is that, when properly calibrated, it indicates the *true* power in the transmission line regardless of the load impedance terminating the line; the load can be either a match or a mismatch, and it can be reactive or nonreactive. The meter achieves this because the forward power value is always equal to the sum of the line-input power and the reflected power. Hence, the meter indicates the true power that is incident on the load. In the reflected-power position the meter indicates the amount of the forward power which was not absorbed by the load, but which adds to the line-input power from the transmitter at the line input, or at whatever point in the line the match is performed (*Ref 18, p 191*). Hence, the difference between the forward and reflected readings is the net power flow in the line at whatever point the wattmeter is inserted. In a lossless line the net power flow indicates the line-input power, which is the absorbed power exactly; the two are identical anywhere on the line. In a line with attenuation, the meter indicates the line-input power if it is placed at the line input, or it reads the absorbed power if it is placed immediately ahead of the load. The difference between these readings is related to the line attenuation. Of course, for practical reasons there may be errors in the actual results from SWR measurements—for one example, diode nonlinearity at various power levels (*Ref 40*).

A further explanation of why the points mentioned above occur, and the beginning of a discussion on the wave mechanics of impedance matching appears in Chapter 4.

Chapter 4

A View Into the Conjugate Mirror

(Adapted from *QST*, October 1973)

Sec 4.1 Introduction

Chapter 3 presents some basic concepts concerning the generation of reflections, wave propagation along the line, and the development of standing waves. Then I showed mathematically how net power is separated into its forward and reflected components. After that it followed logically to explain how the power separation is physically realized by directional devices, such as SWR indicators and directional wattmeters. While learning about wattmeter operation and how to interpret the indications, we saw that the incident or forward power in the line between the matching point and the terminating load impedance is greater than the power supplied by the source generator when the line is terminated in a mismatched load. In this chapter I explore this transmission-line phenomenon in detail, because it is of considerable importance to the amateur. It relates directly to the operational flexibility of the antenna system. Appreciation of the fundamentals involved in this seemingly anomalous situation will free you from the prevalent opinion that we are restricted to operating with little or no mismatch at the junction of the transmission line and the antenna terminals.

The explanation of directional wattmeter operation in Chapter 3 should help you understand why the forward power appearing on the line between a Transmatch and a mismatched load is higher than what the transmitter delivers. This is a normal condition that enables a mismatched load to absorb all the power delivered by the transmitter, while at the same time reflecting a portion of the total power it receives. To do this, the load must receive enough more forward power than what is delivered by the transmitter so it can absorb the power delivered and still reflect the portion demanded by the mismatch. The basis for understanding this rather subtle concept lies in the wave mechanics behind the principles of impedance matching introduced in Chapter 1 and defined in Chapter 2. As far as I know, the *wave* aspect of this subject has been presented in the literature only by Slater and Alford (*Refs 35 and 39*).[1] Perhaps this restricted exposure may account for some of the confusion in this area among engineers and amateurs alike. For example, the many "cook book" recipes and graphic directions for stub-matching a mismatched line tell "how" to do it, but offer little insight concerning the wave mechanics through which the match is accomplished. However, this insight goes to the heart of the transmitter-to-line coupling problem. It clarifies how the reflected wave becomes re-reflected at the matching point. If the matching did not produce this effect, the reflected wave would travel back to the generator, and reduce the amount of power made available by the generator (*Ref 19, p 37*).

Sec 4.2 The Normalizing of Impedances

I will shortly be discussing the reflection mechanics involved in impedance matching with the transmission-line stub technique. In this discussion I use normalized impedances represented graphically on the Smith Chart W2DU Vector Graph, Fig 3-2. Thus, I now digress briefly to explain impedance normalization. We are rather accustomed to thinking of 50 Ω as a standard *system* impedance because of the preponderance of RF components and coaxial lines for that impedance. However, many calculations are greatly simplified by using *normalized* impedances, in which all impedance values have been divided by the system impedance. The system impedance is usually taken as the characteristic impedance of the transmission line, Z_C. Normalizing an impedance by dividing it by Z_C amounts to a change of scale such that the unit of impedance is 1 Ω, rather than Z_C Ω. The W2DU Vector Graph uses the normalized system to take advantage of the simplification in the calculations. To obtain normalized values occurring in a system of any impedance, simply divide all impedances by the system impedance. For example, in a 50-Ω system 150 Ω becomes 3.0 Ω. To convert back to the 50-Ω system, simply reverse the process and

---

[1] All references appear in Chapter 24.

multiply the normalized values by 50. For example, the normalized impedance $z = 0.6 - j0.8$, found at len° = 45°, becomes $Z = 30 - j40$. Normalized values of impedance components are customarily shown in lower case, and nonnormalized values are capitalized.

Sec 4.3 Reflection Mechanics of Stub Matching

An introduction to the reflection mechanics involved in matching concerns concepts of line-input impedance and the angle of the reflection coefficient, θ. This subject is discussed in detail in Chapter 9. The reflection coefficient angle was introduced briefly in Chapter 3. As stated with the definition of the conjugate match in Chapter 2, the matching is accomplished by inserting a nondissipative mismatch at the match point, which produces a complementary reflection that compensates and cancels the waves reflected from the mismatched load. It was also stated that matching can be achieved by a correct adjustment of either the final tank tuning circuit (*Ref 4, Part III*), or of a line-matching network if one is used. Because stub matching uses the same principles as a matching network and is easier to visualize, I will use the stub technique to demonstrate the wave mechanics.

In stub matching, the stub provides what may seem like an anomaly—a nondissipative discontinuity, or mismatch. While we usually think of the stub as providing a *match*, rather than a mismatch, we will discover that the match achieved by the stub results from the mutual cancellation of two complementary reflected waves generated by two complementary mismatches. One wave is that reflected from the terminating load mismatch, and the other is a new reflected wave generated by the stub mismatch. This new wave is equal to the load-reflected wave in both magnitude and phase *but of opposite phase sign*. Wave interference between these two complementary waves at the stub point causes a cancellation of energy flow in the direction toward the generator. This cancellation results from the *difference* between the two reflected waves. The wave interference also produces an energy maximum in the direction toward the load, resulting from the *sum* of the two reflected waves and the source wave.

The effect of the energy-cancellation wave interference presents a virtual one-way open circuit to waves traveling toward the generator. This open circuit blocks both the load-reflected waves and the stub-reflected waves at the stub point from any further rearward travel, and causes these waves to be totally re-reflected toward the load, in phase with the forward waves. Hence, the power in the re-reflected waves adds directly to the forward waves, causing the forward power to be greater than that delivered by the transmitter, as stated earlier.

This wave-interference mechanism which accomplishes the matching will become more clear as we investigate the reflection coefficients of the load and stub mismatches with the aid of an example using the Vector Graph in Fig 3-2. In this example we are using *normalized* values of impedance, resistance and reactance. Here we show a terminating load of $3 + j0\ \Omega$, which produces a 3:1 standing-wave ratio along the whole line. The SWR of 3 is shown by the dark concentric circle, which is the locus of impedances anywhere on the line when the terminating load is $3 + j0$. This circle intersects the resistance circle marked 1.0 at the two points A and D, as shown. At these corresponding points on the transmission line, the resistive component of the line impedance equals 1.0 times the characteristic impedance Z_C of the line. Hence, points A and D are appropriate positions for attaching a series matching stub.[2] Points A and D on the graph are also intersected by reactance circles marked 1.15. At point A the reactance is negative, 1.15 times Z_C, and at point D the reactance is positive, also 1.15 times Z_C. Thus, the line reactance is $-1.15\ \Omega$ at stub point A, and $+1.15\ \Omega$ at stub point B. The conditions for obtaining reflected-wave cancellation by wave interference are the well-known stub-matching requirements, as follows:

1) A series stub is placed where the resistance component on the line equals Z_C, such as at points A or D, and

2) The stub is tailored to have a reactance equal in magnitude and opposite in sign to the line reactance at the stub point (that reactance resulting from the phase relationship between the forward and reflected waves at the stub point), so that the stub and line reactances cancel to zero. (See *Ref 2, p 116* and *Ref 19, p 97*).

This sounds almost like the conjugate-match definition itself, doesn't it? The correct point for inserting a stub in series with the line nearest the load is at point A on the SWR circle. Alternatively, the stub may also be placed at point D, or at any ½-λ interval from points A or

---

[2] For this example we are using *series* stubs to permit *impedance* treatment for clarity. Parallel or shunt stubs are used more prevalently in practice, but using the shunt form would also require an explanation of *admittance* treatment, which is beyond the scope of this example.

D. This is possible because the impedances are repeated every ½ λ on the diagram, and on the line.

Using the complex coefficients of reflection, ρ and θ, we now examine how reflections add at a stub matching point to produce the matching effect. We'll see why a directional wattmeter gives a true reading of forward power between the matching point and the load—power that is greater than that supplied by the transmitter when the line is terminated with a mismatched load. A little later we'll also see how these principles apply to practical feed-line matching networks.

At point A, which is 30° from the load (at len° = 30°), the unmatched voltage reflection coefficient is $\bar{\rho}_E = 0.5 \angle -60°$. This means that the phase of the reflected voltage wave lags the forward wave by 60° at point A. The line impedance E/I at this point is $1 - j1.15$. A match can be effected by connecting an inductive reactance, such as a stub or a lumped inductance, of $0 + j1.15$ in series with the line at point A. The reactance-cancellation effect of the positive-reactance stub on the equally negative reactance of the line impedance is generally understood, but several points are not always clear regarding the effect on the component wave. What characteristics of the stub cause it to counteract the reflections from the load? Also, why does the stub cause the forward power to rise between the matching point and load?

In answer to these questions, let's first determine the reflection coefficient produced by the stub if it were inserted at point A in a perfectly matched line. In this condition we may analyze the reflection generated by the stub in the absence of any other reflection on the line. If we station ourselves just on the load side of stub point A with the stub attached and look into the line toward the matched termination, we will see a pure resistance equal to the characteristic impedance of the line, Z_C.

We know from matched-line theory that if we remove the line portion extending from the stub to the load and insert the terminating resistance $R = Z_C$ directly in series with the stub across the open-ended line, we may look into the line toward the stub from the generator and see the same conditions of reflection as were present before the line section was removed. Thus, the series circuit comprising the matching resistor and the stub performs as a $1 + j1.15$-Ω mismatched load terminating the line. Precisely this same reflection is produced no matter where the stub is inserted in a matched line. The Vector Graph shows

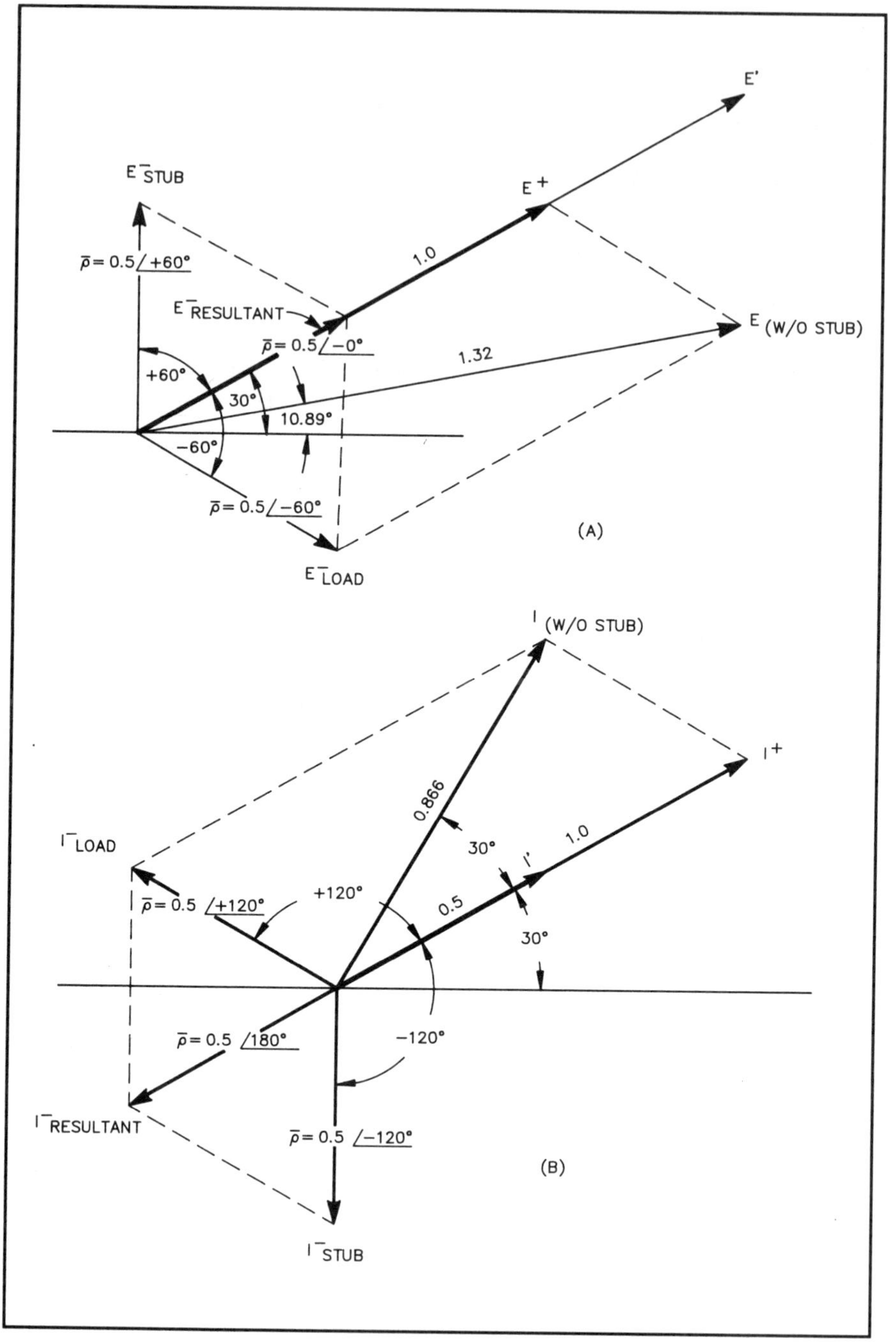

Fig 4-1—At A, various voltage-vector relationships, and at B, various current-vector relationships at len° = 30°.

this impedance of $1 + j1.15$ to appear at point D, for which the voltage reflection coefficient is $\bar{\rho}_E = 0.5 \underline{/+60^\circ}$. Note that this is the same magnitude and phase *but of opposite phase sign* of the reflection coefficient appearing at point A resulting from the load mismatch of $3 + j0$.

Thus, the stub mismatch produces the same magnitude of reflection (and the same SWR) as was produced by the load mismatch, but the stub-reflected voltage wave *leads* the forward wave by 60°, while the load-mismatch wave *lags* by 60°. If the stub is now attached at matching point A (corresponding to len° = 30° on the Vector Graph) with the $3 + j0$ load terminating the line, both the stub- and load-mismatch reflections will be produced simultaneously. As a result of their opposite-sign phase relationship, the *leading* stub-reflected wave and the *lagging* load-reflected wave cancel each other at the match point.

The voltage reflection coefficients of the load and stub thus add vectorially to zero degrees, $\bar{\rho}_E = 0.5 \underline{/0^\circ}$, which tells us that the resultant of the two voltage reflections is exactly in phase with the forward voltage wave at match point A, as shown in Fig 4-1A. The amplitude resulting from this trigonometric addition is considered later, but knowledge of these angular relationships should add to the understanding of the mechanics of line-reactance cancellation by the stub.

Now that we know what is happening with the voltage wave, we also need to investigate the current wave to learn about the impedance relationship at the matching point. As defined earlier, reflected current is 180° out of phase with reflected voltage. Hence, the reflection coefficient for current at point A on the line is found on the Vector Graph 180° away, or diametrically opposite the corresponding voltage-coefficient point. Thus, we find the current reflection coefficient for the load mismatch at point C with $\bar{\rho}_I = 0.5 \underline{/+120^\circ}$. Similarly, the reflection coefficient for current of the stub mismatch is found opposite point D, at point B, and is $\bar{\rho}_I = 0.5 \underline{/-120^\circ}$. Note that, similar to the case with voltage, the current coefficients of the load and stub are also of equal magnitude and phase, but of opposite phase sign. But while the voltage angles added vectorially to 0°, the current angles add to 180°, $\bar{\rho}_I = 0.5 \underline{/180^\circ}$. Hence, the resultant of the two reflected *current* waves is 180° out of phase with the forward current, as shown in Fig 4-1B. So we now have forward and reflected voltages in phase, and forward and reflected currents out of phase; hence, the wave arriving at the match point from the generator sees a perfect match.

These specific phasor conditions on the line at the stub matching point send a significant message. In the opening paragraphs of Chapter 3, I described the wave mechanics involved in a line terminated by an open circuit. There we learned that angle theta (θ) of the voltage reflection coefficient of an open circuit is 0° and that of the current is 180°. With an open-circuit termination, the unabsorbed voltage wave which is incident on the termination is reflected with no change in phase, while the unabsorbed current wave is reflected with a phase change of 180°, a complete reversal of current polarity.

It is interesting to note here that the phase relationships of the reflected wave at the stub-match point which we just established are precisely the same as those which prevail in a line terminated in an open circuit, so far as the reflected wave is concerned (but not the forward wave). Therefore, the effect of the two reflected waves arriving simultaneously at the match point is to establish an open circuit at the match point to the waves generated by the load and stub mismatches. Thus, the waves from the two mismatches become totally re-reflected at the match point and undergo open-circuit phase changes as described above. Hence, the resultant re-reflected voltage wave does not change phase during this re-reflection. (Remember, it was already in phase with the forward wave prior to its re-reflection.) However, the resultant *current wave changes phase* by 180° on re-reflection. And because it was 180° out of phase with the forward current wave just prior to its re-reflection, the present 180° reversal now places it also in phase with the forward current wave. Now that both voltage and current re-reflected waves are in phase with their corresponding forward waves, addition of the voltages and currents of the forward and reflected waves occurs at the stub-matching reflection point. Thus the conclusion: *The power contained in the reflected wave adds directly to the forward power*, and enlarges it between the matching point and the load in the manner stated earlier.

Now let us consider that the above conclusion assumes an open-circuit condition was established at the match point for the reflected wave traveling toward the generator. The assumption was based on the identical relationship between the reflection coefficients which we established at the match point by wave interference and those known to exist at an open-circuit termination. We can verify this assumption by an alternative method. The method is based on the discussion in Chapter 3 concerning the electric- and magnetic field relationships that occur with an open-circuit termination.

Let us first observe the net value of all currents flowing at the match point at the instant the two reflected currents form their resultant angle $\theta = 180°$. At this instant we will see an initial sudden drop in resultant line current, I, because of wave cancellation as the reflected-wave resultant becomes aligned exactly 180° out of phase with the forward current. This drop is shown graphically in Fig 4-1B, where the original resultant current, I (as with no matching stub present) suddenly drops to the new instantaneous resultant value I' from the effect of the stub discontinuity.

Now recall the open-circuit field theory presented in the early part of Chapter 3. We know that when the current drops, the magnetic field also drops. The changing magnetic field produces an electric field equal to the energy reduction in the magnetic field. The new electric field then adds in phase to the existing electric field, producing a voltage increase at the match point. This voltage increase now starts a wave traveling in the opposite direction, which is now in the same direction as the forward wave, and thus adds to it. The increased electric field (now an enlarged *forward* electric field), as it moves toward the load, produces a new magnetic field. The new field is equal in magnitude but of opposite polarity to that of the original field. The new magnetic field now causes current to build up again to the same magnitude as the original reflected current, but of opposite polarity and direction. Thus, the new current wave is now also traveling in the same direction, and with the same polarity as the forward current wave. The new wave adds to the forward wave, enlarging it just as the re-reflected voltage wave added to and enlarged the forward voltage wave.

By following these field-current-voltage reactions through their natural sequence of events, you can see that we have obtained the same conclusions as those previously obtained on the basis of reflection coefficients. This justifies the assumption that the resultant reflection coefficients at the stub-matching point have defined an open circuit to the reflected waves, thus preventing any further travel of the reflected waves toward the source. The existence of the reflectance (ratio of reflected power to forward power) at the matching point is therefore verified, with the result that both the reflected voltage and current have indeed been re-reflected, and the power associated with them has thus been effectively added to the power contained in the source wave. Thus, when the line is terminated in a mismatch causing reflected power to exist on the line, the *sum* of the source and re-reflected powers (traveling only toward the load) must be greater than the power delivered by the generator alone. And since we have shown how the stub acts to counteract reflections from the load, our original questions concerning the stub characteristics have been answered.

Sec 4.4 Effect of Line Attenuation on Reflected Power

Now we'll consider the effects of line attenuation, α (alpha), on the increased loss of power delivered to a load when the load is mismatched to the transmission line impedance, Z_C. We'll assume a 3:1 SWR, where $\rho = 0.5$, and where 100 watts supplied by the transmitter will yield 133.33 watts of forward and 33.33 watts of reflected power. We get 133.33 watts of forward power because forward power is $1/(1 - \rho^2)$ times the supplied power, so 1/0.75 = 1.3333, and 1.3333 times 100 watts equals 133.33 watts. If we initially neglect the losses, 100 watts will also be absorbed in the load, because reflected power $\rho^2 = 0.25$, or 25% of the forward power, leaving 75% absorbed by the load ($1 - \rho^2 = 0.75$), and 75% of 133.33 is exactly 100.

Consider a typical realistic case. In 175 feet of RG-8 at 4 MHz, the matched-line attenuation is 0.5 dB. If the load were perfectly matched to the line (1.0 SWR), the 100 watts delivered by the transmitter would be attenuated to 89.13 watts during travel to the load. But with a 3:1 mismatched load the *additional* one-way line attenuation (because of the SWR) is 0.288 dB. Hence, the forward power at the conjugate-match point would then be 124.78 watts (0.288 dB below 133.33 watts), and 111.21 watts of power (0.5 dB below 124.78 watts) reach the load. Of that power, 27.80 watts (25%) are reflected, leaving 83.41 watts to be absorbed. Of the 27.80 watts reflected, 24.78 watts arrive back at the input to join the 100 watts of source power to develop the 124.78 watts of forward power. The 5.72 watts difference between the power absorbed in the *matched* load and the 3:1 *mismatched* load (0.288 dB) is insignificant. It actually amounts to less than 1⁄12 of an S unit. Information on calculating these values is presented in Chapter 6. These values are typical of data obtained during actual routine measurements in a professional laboratory. They provide additional evidence that reflected power is real, not fictitious. If it were fictitious power, no more than 66.85 watts (75% of 89.13 watts) would be available to the 3:1 mismatched load. But the 83.41 watts actually absorbed in the mismatched load is 93.58% of the amount absorbed in the matched load, the loss of 6.42% being completely accounted for in line attenuation alone.

Sec 4.5 Matching Networks and Reflection Mechanics

Now I shall delve further into the wave-interference principles demonstrated with the stub technique in Sec 4.3. The principles also apply to both resonant ¼-λ series matching-transformer operation and to the typical amateur "antenna tuner," line-matching network, or Transmatch. In order to visualize the inherent generality of these principles, we need to develop some additional concepts concerning stub matching and embark on a somewhat different line of reasoning. As you can surmise from the example presented above, the fundamental principle behind the elimination of reflections is this: Have each reflected wave canceled at the point where the elimination of the reflection is desired, by interference from another wave of equal magnitude and phase but opposite phase sign (*Ref 35, p 58*). A transmission line of the appropriate length which has one end effectively open circuited and the other end short circuited possesses the reflection-producing characteristics required to develop canceling waves of the correct phase in relation to the wave to be canceled.

Canceling waves can be developed by using other line arrangements, but for the purpose of demonstrating the principle, I will use the arrangement just stated. Fig 4-2 shows how a stub of this arrangement performs the matching function in practice. Fig 4-2A is the conventional representation of a typical series-stub circuit (using the values of our previous SWR = 3 example), in which section F is called the *feed line*, section S is the *stub*, and section T is an impedance-transforming section which I will call the *transformer*. We'll now examine these in greater detail. To clarify the approach, Fig 4-2A is redrawn in Fig 4-2B, with the stub and transformer shown as one continuous straight-line section. This straight-line section will presently come to life as the heart of the wave-interference-producing mechanism found in all stub-matching operations. This is because its physical length will be adjusted specifically so that waves reflected at each end will return to the feed point with equal magnitude and phase, but with opposite phase sign.

We have seen earlier that a voltage wave is reflected with zero phase change from an open circuit (or from any resistive termination greater than a matched load), and is reflected with 180° of phase change from a short circuit (or from any resistive termination less than a matched load). With a current wave, the opposite is true. So with the $3 + j0$ load terminating the transformer, from the viewpoint of reflection

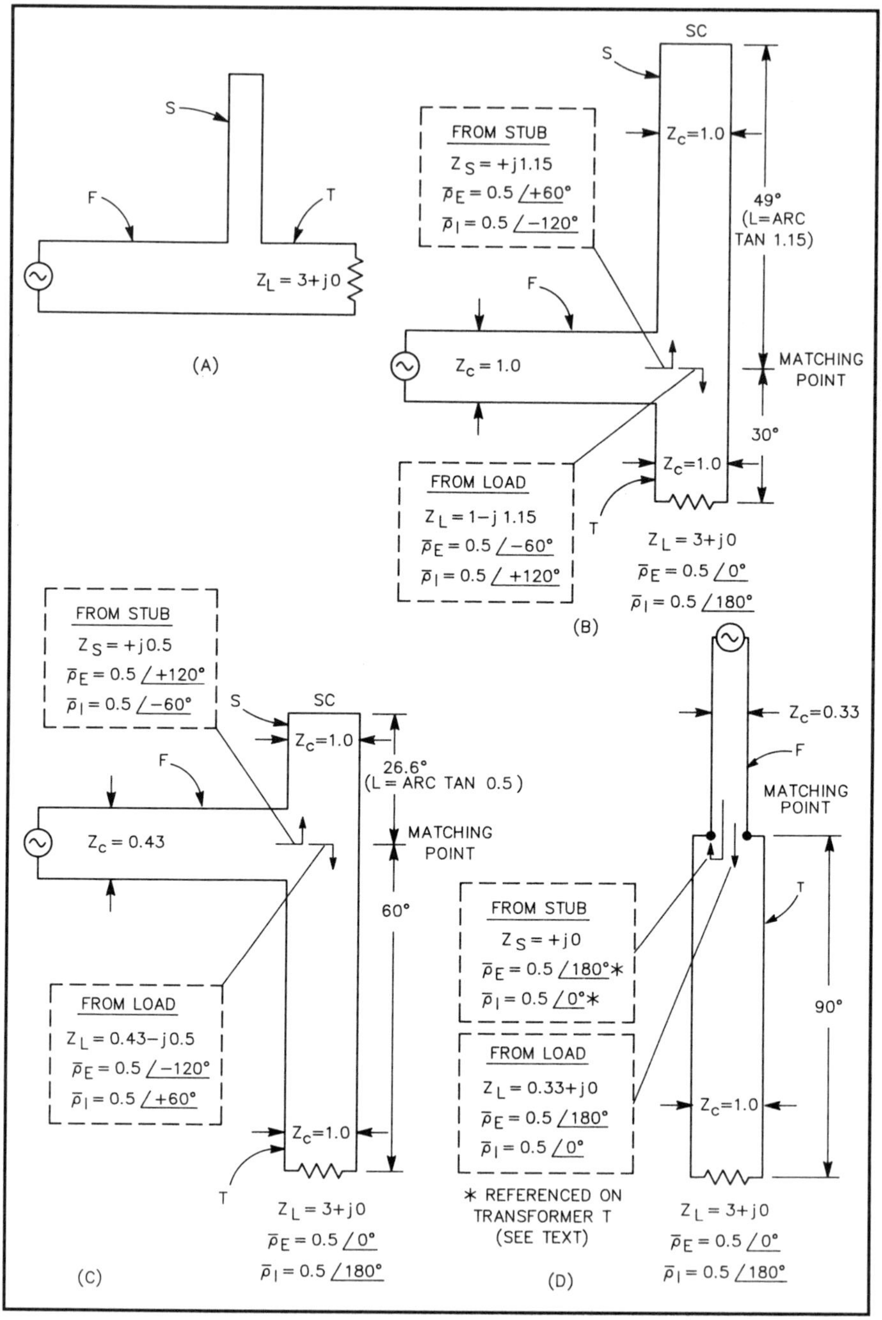

Fig 4-2—At A, the concept of stub matching with a series-connected stub. At B, the same matching arrangement redrawn. At C, a similar matching arrangement but with different stub and transformer line lengths. At D, a ¼-λ line-transformer section evolves as the stub length goes to zero. SC = short circuit.

behavior, the load end of the straight-line section is considered as being open circuited and the other end short circuited. As can be seen, which end is open and which end is short circuited depends on the character of the load. In our example in Fig 4-2B, the load end behaves as an open circuit as far as wave reflection is concerned because R is greater than Z_C, while the opposite end (the stub) is short circuited. The action occurring in this line section in the process of developing the interfering wave-canceling relationship is as follows. Either a voltage or current wave is assumed to enter the line section at the feed-line entry point. The energy divides, one portion of the wave traveling toward one end of the section, and the other wave portion traveling toward the opposite end. After each wave portion encounters one reflection, the returning waves will each have the same absolute value of phase but opposite sign, or polarity, on return to the point of entry.

The opposite phase polarity between the two reflected waves arriving from opposite directions results because reflection at one end is accompanied by a 180° phase reversal, while reflection from the other end is not. As stated above, the phase reversal of one wave but not the other is caused by the opposite conditions of reflection at the two ends of the line, one end open-circuited and the other end short-circuited. Note in the dashed-line boxes in Fig 4-2B that in each case, the phase of the reflected wave (of both voltage and current) is of opposite polarity on opposite sides of the feed line as they return from the stub and load directions. The wave-entry point, where the feed line is attached, is the matching point, and divides the line section into its two complementary portions: the *stub* portion, S, and the *impedance-transformer* portion, T. Electrically, each portion is the complement of the other, because the waves reflected from the end of each portion returning to the match point are complementary in their phase relationship and equal in magnitude. Herein lies the basis for the term *complementary mismatches* as used earlier, because each portion presents a complementary mismatch to the feed line.

Later I show that this complementary mismatch concept is of great importance to matching in general, because the complementary relationship holds no matter where the feed-line entry point is positioned on the stub-transformer line section. The importance prevails because the canceling wave and the reflected wave to be canceled will be of the same magnitude and phase, but opposite phase sign, at whatever point on the matching line section the feed line is attached.

This is true with two provisions: (1) the characteristic impedance Z_C of the feed-line section F must be the same as the resistive component of the transformed impedance appearing on the transformer line at the feed point, and (2) the length of the stub portion S must be adjusted to produce a reactance equal and opposite in polarity to the line reactance appearing at the feed point. The length may be found from the expression

$$S_L = \arctan \frac{jX}{Z_C} \qquad \text{(Eq 4-1)}$$

where

S_L = stub length in electrical degrees
jX = line reactance (obtained from the Vector Graph)
Z_C = characteristic impedance of the stub section; $Z_C = 1.0$ here because we are using normalized impedances (discussed in Sec 4.2)

Transformer section T transforms the load impedance to varying values of both resistance and reactance along its length. Hence, for a proper match, the magnitude of the feed-line section F impedance, Z_C, depends on the location of the feed point, and vice versa. This is because the feed-line impedance, Z_C, must equal the resistance appearing on the transformer line at whatever point the feed line is attached to the transformer. This concept is not generally appreciated, and it is certainly not readily apparent from the usual stub-length and position-indicating graphs appearing in many publications.

Sec 4.6 Stub Matching Versus Network Matching

We are now getting closer to seeing how stub-matching principles extend to line-matching-network operation. If we look further into the reflection characteristics of what have generally been considered to be different techniques of matching, a fascinating revelation of the similarity among all of these various techniques emerges. Stub, hairpin, ¼-λ series transformer, Transmatch, L network, pi network, and so on, are all in this category. And there is a logical reason for this similarity—these techniques all have one essential ingredient in common—reflections! Reflection and matching are applied in transformers that rely on reflections from the end terminals, where a change in impedance level exists. As I discussed in Chapter 3, any abrupt change in impedance level appears as a discontinuity to the smooth flow of the electromagnetic wave, and results in producing a reflection. The transformer accomplishes the task of matching its input and output impedances by controlling the phase and magnitude of the waves produced by reflection at its boundaries, or end terminals, so that all the reflections produced at either end are canceled by those arriving from the other end (*Ref 35, p 58*). This is what was meant in the reference to "controlled reflections" early in Chapter 1. A corollary to the seeming anomaly of the stub producing a mismatch, instead of a match, is that we match to avoid reflections, but we can't match without reflections when different impedance levels are involved.

From Eq 3-1, we know that the magnitude of a load reflection is determined by the ratio between the load impedance Z_L and the impedance Z_C of the line transformer. So in furthering the understanding of the role played by reflections in the process of impedance matching, it is interesting to make two additional observations on the Vector Graph, Fig 3-2. First, the reflection magnitude, ρ (or SWR) determines the position on transformer section T where the resistance component of the transformed line impedance E/I equals the line impedance Z_C. This position, we recall, is the matching point, and fixes the length of transformer section T. In making this observation, remember that the diameter of the SWR circle is proportional to the SWR. By tracing along the R = 1.0 circle we can see that as the diameter of the SWR circle changes, the point where the R = 1.0 circle and the SWR circle intersect moves accordingly. A radial line drawn through this intersection point, and extending to the line-length scale len°, thus indicates the angular distance (T) from the load to the matching point for a given SWR. We recognize this observation as simply the conventional method of using the Smith Chart for determining the stub position when the transformer and feed-line impedance Z_C are equal. But second, using a radial indicating line in a similar fashion while tracing along the SWR = 3.0 circle as it intersects the various other resistance circles, we see that for a given SWR the resistance component of the line impedance E/I changes with position along the transformer. These two observations together reveal a flexibility available in the approach to a matching design that provides a step toward visualizing the fundamental similarity of the different matching techniques. This flexibility includes the following three conditions, which are explained in more detail later.

1) There is no restriction on the characteristic impedance Z_C of

the transformer section T that requires it to be of the same value as the section F feed-line impedance Z_C—it can range from low (coax) to high (open-wire line).

2) The *length* of the transformer section having a given Z_C can be found to yield the resistance component of the transformer-line impedance that matches the feed-line section F impedance Z_C, which differs from the transformer Z_C. However, the transformer section can have a length which is not limited to the distance from the terminating load to the first point at which the resistance component is seen to equal the feed-line Z_C. The transformer can extend from the load to either of the two points where the resistance component is seen to equal the feed-line Z_C on the SWR circle, or any electrical length extending beyond these points by an integral multiple of ½ λ. Later I show how the use of matching networks assists in obtaining the required electrical length. This removes *all restrictions* from any specified physical transformer length, that is, from the load all the way to the operating position.

3) The action of the stub portion can be performed by any reactance of the proper value. This can be either by a lumped-constant component, or by a separate line section of any reasonable Z_C value having the proper length to present the required value of reactance. The *electrical* length of the stub is always directly related to its reactance. Now we again return to wave or reflection mechanics. We will see in those terms how matching, obtained by the various techniques recited previously, is described by three parameters: transformer impedance, transformer length, and stub reactive elements.

In our earlier example using the stub technique, the magnitude ρ of the reflections appearing at each end of the *transformer* section was the same (0.5, or an SWR of 3:1 for the load at one end and the stub at the other). In other words, the magnitudes of each complementary mismatch were identical. For the present, we will retain the characteristic impedance $Z_C = 1.0$ for the entire stub-transformer line section. But based on conditions 1 and 2 above, we may change the feed-line impedance as conditions dictate. Consider now the effect of increasing the length of the transformer section and shortening the stub section in accordance with Eq 4-1. For example, while referring to Fig 4-2C and the Vector Graph, Fig 3-2, let us move the feed-line entry point farther away from the load, from len° = 30° to len° = 60° This increases the transformer length to 60° and reduces the stub length to 26.6°. Now on the Vector Graph observe the radial line extending from the len° = 60°

point (where $\theta = -120°$) through point B, where the SWR = 3.0 and the R = 0.43 circle intersect. See that the corresponding movement along the SWR circle results in a change in the resistance component at the feed point, from R = 1.0 to a new resistance, R = 0.43.

Recall from the earlier statement that the complementary mismatch relation holds constant wherever the feed line is attached. A feed line having a characteristic impedance Z_C = 0.43 will be perfectly matched when attached at the len° = 60° point. The load-mismatch voltage reflection coefficient is now read as $\overline{\rho}_E = 0.5 \underline{/-120°}$ at point B. The magnitude is the same as before, but the phase angle is larger because we are farther from the load. And applying the complementary-mismatch principle, we see that the voltage reflection from the stub mismatch becomes $\overline{\rho}_E = 0.5 \underline{/+120°}$, as read at point C. So we ask the question, how does this new combination produce a canceling wave having the same magnitude as before? Recall that previously, when the line-characteristic impedance of sections F, S and T were all Z_C = 1.0 as in Fig 4-2B, the canceling reflection was generated by the stub alone. This happened because no line-junction mismatch existed between the feed-line section F and the transformer T. But now that the impedance of the feed line differs from the impedance of the transformer, we have an additional discontinuity at the feed point, which also generates a reflection. And the *shorter* (series) stub portion now generates a reflection which is smaller than when all the line sections had a Z_C = 1.0. The stub-reflection magnitude is reduced by the amount of the reflection presently being generated by the feed-line-to-transformer mismatch. Thus, by the complementary mismatch principle, the resulting canceling wave still retains the correct magnitude and phase to cancel the load-mismatch reflected wave at this new feed point on the transformer. This canceling wave is evidently generated by the *combined discontinuities* of both the differing line impedances at the feed-line junction, and of the stub with the corrected length. We have thus matched a feed line of Z_C = 0.43 to a load of $Z_L = 3 + j0$ through a transformer of Z_C = 1.0.

Using this same type of reasoning, to obtain a match for feed lines of higher impedance, we may conversely shorten the transformer section and change the stub section, according to the tangent relation in Eq 4-1. We need merely to position the feed-line entry point where the resistance component of impedance in the transformer T has been transformed to the value of the feed-line impedance Z_C that we wish to use, and then adjust the stub length to cancel the line reactance. As explained above, the resistance circle which is intersected by the SWR circle for a given transformer length indicates the resistance component of the feed point. This is also the Z_C value of the feed line which will be perfectly matched when attached at the feed point. The data presented in Table 4-1, taken from points along the SWR = 3.0 circle on the Vector Graph, show a few selected transformer-length examples, and pinpoint some of the pertinent information for clarity. Notice especially that the feed-point resistance component decreases as the transformer length increases. It is interesting to discover that when the feed point goes beyond the len° = 45° position, the θ angles of the voltage- and current-reflection coefficients pass through 90° from opposite directions. The result of this is that their respective resultants shift 180°. Thus, the resultant θ angles of reflection coefficient interchange. The voltage-coefficient resultant angle θ now becomes 180° and the current θ angle becomes 0°. This means that the effective reflecting termination at the match point shifts from an open circuit to a short circuit when the feed line is attached more than 45° away from the E_{MAX} position (at len° = 0° on the transformer).

Consider now the effect of increasing the length of the transformer section until the reactance component of the line impedance disappears by itself without requiring a stub to cancel it. The Vector Graph shows this condition to occur at len° = 90°. At this point the reflected voltage wave from the load mismatch is exactly 180° out of phase with the forward wave, and therefore no reactance component is developed. The resistance component of the line impedance at this point is $0.33 \times Z_C$, as shown on the chart. Hence, we may now connect a feed-line section F having an impedance $Z_C = 0.33$ at this point and obtain a perfect match, as shown in Fig 4-2D. No reflections appear on the 0.33-Ω feed line. How come? Again, because of the canceling reflections in the transformer section! Note the present length of the transformer section—90°, or ¼ λ. The transformer section alone is now using the *entire length* of the line section, and the stub section has disappeared.

Note that simply by moving the feed point along the transformer to the point where the line reactance vanishes, the resistance component becomes Z_C/SWR = ⅓, and we slip smoothly from the stub form into the series ¼-λ transformer form of matching. You can see what's happening by observing the numbers in Table 4-1. Remember, the

Table 4-1

Matching Characteristics with Various Transformer- and Stub-Section Lengths

Transformer Length len°	Stub Length S_L°	Resistance Component R	Stub Reactance jX	Angle of Reflection Coefficient — Voltage — Load Mismatch θ°	Voltage — Stub&Line Mismatch θ°	Current — Load Mismatch θ°	Current — Stub&Line Mismatch θ°
0	0	3.0	0.0	0	0	180	180
10	48.2	2.42	+1.12	−20	+20	+160	−160
22.5	52.9	1.38	+1.32	−45	+45	+135	−135
30	49.0	1.00	+1.15	−60	+60	+120	−120
45	38.7	0.60	+0.80	−90	+90	+90	−90
52	33.0	0.50	+0.65	−104	+104	+76	−76
60	26.6	0.43	+0.50	−120	+120	+60	−60
67.5	19.8	0.38	+0.36	−135	+135	+45	−45
90	0	0.333	0.0	180	180	0	0

characteristic impedance of the transformer is still $Z_C = 1.0$, which has become the geometric mean between the input and output impedances that it is matching, $\sqrt{0.33 \times 3.0} = 1.0$. From inside the transformer, the impedance level looking toward the input terminals is stepped down 3:1 from 1.0, giving us short-circuit reflection behavior, just as the output impedance is stepped up 1:3 from 1.0, for open-circuit reflection behavior at the output terminals. It is therefore evident that a ¼-λ transformer section of line having a Z_C equal to the geometric mean of its two end-terminal impedances has equal mismatches at both ends, and thus produces reflections of equal magnitude at both ends. These reflections from each end cancel each other at the junction of the feed line and the transformer input terminals, as waves reflected at the output load mismatch return to the input junction exactly 180° out of phase with the wave reflected at the input mismatch. This is because the wave reflected from the load mismatch has traveled 90° from the input point to the load mismatch, and an additional 90° in returning to the input.

To clarify further what is happening here, recall that previously when the feed-line section F and transformer T were of equal impedance $Z_C = 1.0$ (Fig 4-2B), the canceling reflected wave was generated entirely by the stub mismatch. In the present case where the stub length is zero and the transformer length is ¼ λ, the canceling reflected wave is generated entirely by the 3:1 mismatch at the junction of the feed line and the transformer. The voltage reflection coefficient angle of the wave reflected from this feed-line-junction mismatch is $\theta = 0°$, referenced to the feed-line Z_C, because the Z_C of the transformer is greater than the Z_C of the feed line. However, after both the load- and input-junction-mismatch reflected waves have joined at the input of the transformer to cancel one another, the input-junction reflection no longer travels toward the generator, but is re-reflected into the transformer toward the load in the same manner as with the previous stub-reflected wave. Therefore, referenced from within the transformer, the reflection coefficient angle of the feed-line-junction reflection is $\theta = 180°$, as indicated in Table 4-1. The matching point with this configuration is at the input terminals of the ¼-λ transformer section. Hence, it is evident that the above discussion has explained the reflection mechanics involved in the general impedance-matching action of all series ¼-λ transformer matching sections.

Chapter 12 explains why the ¼-λ series transformer is an impedance *inverter* for any complex terminating load. It is not restricted to purely resistive loads.

Chapter 5

Low SWR for the Wrong Reasons

(Adapted from *QST*, April 1974)

Sec 5.1 Introduction

I made the statement in Chapter 1 that misconceptions concerning SWR and reflections are prevalent among amateurs, both in print and on the air. So I reiterate, this book was written with one primary goal in view—to identify some of the misconceptions and to clarify some of the confusion resulting from the misconceptions.

One outstanding area of confusion concerns the nature of reflected power and how it is accounted for in the circuit. In short, is it *real* or is it *fictitious*, and where does it go? The nature of reflected power is discussed in Chapter 3, where it is shown why reflected power is real power. However, the understanding of this subject is so vital that an entire chapter, "The Reality of Reflected Power," Chapter 8, was written to present additional proof for those who still believe that reflected power is fictitious. And Chapter 4, in discussing the role of reflections in conjugate matching, delved into the question of where the reflected power goes. I used the stub form of matching to illustrate the wave action which accomplishes the matching function, and which also derives the total forward power from the combined source and reflected power. Recall that learning of this wave action stripped away the mystery of how a mismatched load can absorb all of the power delivered by the source. We learned this as we saw how the reflected power adds to the source power at the conjugate match point so that the reflected power can be subtracted from the total, enlarged, forward power at the mismatch point to leave a net power in the load equal to the source power.

Now that we have established this relationship between the

source, reflected, and forward powers in terms of the wave mechanics of the conjugate match, we have the necessary background for identifying some of the improper usage of SWR. We can also clarify in greater detail the reasons for the misunderstanding that still prevails concerning what happens to the power reflected from an antenna that is mismatched to its feed line. Further clarification of the misconceptions will enhance the appreciation of the mismatched feed line as simply a tuned impedance-transforming device, particularly as we see somewhat later how the Transmatch type of feed-line matching network and the pi-network tank circuit of the transmitter perform the matching function in the same manner as the stub. Additional perspective in relating the discussion to practical feed-line operation will be gained as some of the thoughts presented in Chapters 1 and 2 are now expanded.

If it appeared that the importance of SWR was overly minimized or downgraded in the treatment accorded it in Chapter 1, I did not so intend. The intent there was to focus attention on the importance of understanding the subject of reflection and SWR correctly and in such depth that we may retain complete control over them in our antenna system design engineering. Thus, instead of letting SWR become *king* to take control and deprive us of a breadth and flexibility, we may use SWR in the system design choices in ways which many are unaware exist.

How many of us have acquiesced to "King SWR" in pruning an 80-meter dipole, taking great pains to obtain the best possible match to a ½-λ feed line at a specific frequency, and fearing to operate more than a few kilohertz from that frequency without worrying about the king's ominous threats of damage to our equipment? But how many are aware that King SWR can be outwitted and his consequences averted without pruning either the dipole or the feed line? And how many have been aware that the matching operation can be performed at the transmitter end of the line at any frequency within the entire 75-80 meter band without suffering any significant loss in power in spite of the SWR remaining on the feed line? Although it contradicts the word published in many articles during the past two decades, this revelation is true. This revelation indicates the flexibility or freedom that really is available in our choice of antenna systems designed for all the HF bands, simply by having a better understanding of SWR and reflection.

Sec 5.2 Valid Reasons for Low SWR

There are good and valid reasons for being concerned with SWR and reflection from both the amateur and commercial viewpoints—with this there can be no argument. As we well know, these reasons are concerned basically with voltage breakdown and power-handling capability, efficiency and losses, and with line-input impedance as it relates to transmitter output coupling. However, in amateur practice, power-handling capability and voltage breakdown don't become serious problems unless we try to shove the legal limit of power through RG-58 or RG-59 at a high SWR. Losses and efficiency concern us, but to a much smaller degree than is generally realized, and for a different reason than many are aware, as we'll see very shortly.

The chief reason why the amateur should be concerned but not alarmed with SWR is in its relation to line-input impedance and transmitter coupling. This is discussed in great detail in Chapters 6, 7 and

13. There we see how to tame impedance and coupling for any reasonable value of SWR, and in those discussions the relative *unimportance* of having a *self-resonant* antenna also becomes evident. But it is of great importance that we first clarify some of the prevalent misunderstandings of SWR and reflected power, because they are causing many amateurs to strive for a low SWR for the wrong reasons, and often needlessly. Probably the most serious and widespread misconception concerning SWR prevailing throughout the amateur fraternity is the erroneous notion that there is a direct one-for-one relationship between reduction in reflected power and a resulting increase in radiated power. In other words, many believe that for every decreased watt of reflected power there is an additional watt of increased output. *Not so*, but the number of amateurs who have been misled to believe this invalid and unscientific premise is unbelievable.

Another related concept, popular, but also erroneous, is that when terminated in a mismatch, the coaxial feed line becomes part of the radiator, causing radiation from the feed line because of the standing wave. (See Chapter 2, statement 18.) This is untrue because the line voltages and currents, and the standing wave resulting from the mismatch, are entirely contained in and between the outer and inner conductors, inside the coax. *No standing wave develops on the outside of the coax because of impedance mismatch.* However, feed-line radiation may result from standing waves on the outside of the coax because of current flowing on the outside, if a balanced dipole is fed without using a balun. This feed-line radiation may or may not be of any consequence, but the topic is covered well by McCoy (*Ref 45*).[1] Also see Chapter 21. There is also no radiation from tuned open-wire feeders, because the current is flowing in opposite directions in each wire; hence the fields developed by the current in each wire are canceled.

Misunderstanding of how the benefits accrue from a low SWR, and of just how little benefit is obtained, is driving many of us to attain SWR values far lower than where the benefits continue to be significant in relation to the efforts expended to attain them. It is for this reason that we often set an unrealistically low limit on allowable SWR that needlessly restricts the operating bandwidth—the range of usable frequencies on either side of the resonant frequency of the antenna—to a

---

1 All references appear in Chapter 24.

far more limited range than is necessary. In rectifying a misunderstanding such as this, it often helps to first learn how the misunderstanding originated.

Sec 5.3 "Impedance" Bridges

One aspect of the misunderstanding has been created to some extent by narrow, and often erroneous interpretations of matching principles found in various instructions for instruments such as noise bridges (*Ref 100*) and the antenna-scope for determining the terminal "impedance" of an antenna. Contrary to what is stated in some of the instruction manuals, these devices cannot measure *impedance—they can measure resistance only—and then only in the absence of reactance.*[2] Look up and compare the definitions of impedance and resistance; the term impedance is often misused when the correct term should be resistance (*Ref 46*). Consequently, in using these devices we have been coerced into finding only the resistance component of the antenna terminal impedance, and only at the resonant frequency of the antenna, because this is the only frequency where the impedance has zero reactance, or R + *j*0.

In following this tack, erroneous emphasis has been given to requiring the antenna radiator itself to be resonant, thus nurturing the misconception that it needs to be self-resonant to radiate all the power being supplied to it (*Refs 100, 101*). Thus, many have been misled to believe that the antenna just won't perform properly at any frequency except the self-resonant frequency. (See Chapter 2, statements 5, 6 and 7; also *Refs 20, 21, and 24.*) In addition, emphasis on the further necessity for obtaining an antenna-terminal resistance component equal to the line impedance Z_C has in many cases caused us to go to extreme measures, such as adjusting the antenna height above ground in small increments to achieve that precise resistance reading in quest of the perfect 1.0 match (*Ref 101*). See Chapter 2, statement 15. Adjusting antenna heights in large increments to obtain control of radiation in the vertical plane is realistic. But controlling radiation *resistance* by adjust-

---

2 [AUTHOR'S NOTE: The original *QST* text from which this chapter is adapted was written before noise bridges were constructed with the capability of measuring reactance.]

ing the height is neither necessary nor practical, because the efficiency thought to be gained through this action is illusory.

The truth of this will become evident somewhat later as we see why there is no justification for expending any matching effort at the load, or antenna, to improve a mismatch of 2:1 or less, simply to remove the standing wave with the expectation of improving efficiency. (See Fig 6-1.) Furthermore, because of the reactance that appears as we depart from the resonant frequency of the antenna, the sacred but overrated perfect match found at some carefully adjusted height can be obtained at only one frequency without retrimming the radiator length, thus continuing the vicious cycle. However, the widespread practice of this philosophy in antenna-system operation has conditioned us to think only in terms of using a ½–λ transmission line with no reflections, and to obtain its perfect 50-Ω nonreactive input impedance by operating only at the resonant frequency. So we have, in effect, been deterred from learning of the real effect of reactance in antenna impedance, and how the transmission line transforms any antenna impedance in a straightforward and predictable manner for use at the line input.

In becoming so conditioned, many of us have forgotten that we can obtain the desired 50-Ω nonreactive input impedance from the line-transformed antenna impedance with a simple line-input matching network in the shack, much more easily than it can be obtained at the antenna. In fact, with most tube transmitters the impedance seen by the transmitter at the input of the feed line for SWR values in excess of 2:1 can be matched for optimum loading by adjustment of the transmitter tank circuit itself. If a transmitter does not contain sufficient matching range, a separate line-matching network between the transmitter and line input offers a more judicious matching arrangement than playing games out at the antenna. Chapter 6 explains why there are many situations where this same matching approach should be considered when the load mismatch yields SWR values of even 5:1 or higher, as one departs from the self-resonant frequency of the antenna.

One further misconception exists that has also resulted in needless and unwarranted reliance on the ½–λ feed line to repeat the resonant antenna resistance at the transmitter. This one concerns the effect of line-input reactance on tank-circuit resonance when the line with reflections is fed directly by the pi network. Consider a tank circuit which is first loaded and tuned to resonance with a resistive load, and then when the load is changed to one containing reactance. If the tank components have sufficient retuning range to compensate for the reflected reactance to return the circuit to resonance at the proper load level, all is well; the tubes still see a proper resistive load as before. The misconception about this point has been generated by some writers who apparently don't understand resonant circuits, for they proclaim that the retuning *introduces* reactance that detunes the circuit, causing improper loading and increasing plate current and dissipation. Chapters 7 and 13 contain much more detail on this point.

Sec 5.4 Low SWR for the Wrong Reasons

We have discussed "low SWR for the wrong reason," as practiced (often unwittingly) in using the perfectly matched antenna operated only at the self-resonant frequency of the radiating element. But another wrong reason for desiring a low SWR is interpreting feed-line SWR as the sole indicator of the quality of an antenna's radiating performance across a band of frequencies, with low SWR across the band getting the raves and high SWR getting the boos. This is a definite misuse of SWR information, because there are cases where the low and high SWR values occur in just the opposite relation, with respect to indicating antenna efficiency over a given bandwidth. Reasons for this are given shortly. As a result of this misuse of SWR values, *good* antennas are too frequently rejected as "bad" because the feed-line SWR swings relatively high, and *poor* antennas are accepted as "good" when the SWR remains relatively low.

In most cases the use of feed-line SWR alone to indicate antenna efficiency is completely invalid, because SWR indicates only *the degree of mismatch*, not efficiency. However, we will see presently how a relative change in SWR, to a value either lower or higher than a previous value known to be correct in a given antenna system, can indicate that a possible undesirable change has occurred somewhere in the system. That change may affect its radiating efficiency. The popular vertical antenna having from two to four ground radials (an insufficient number for efficient operation), or perhaps having only a buried water pipe or a driven rod for a ground terminal, is one case where a lower-than-normal SWR obtained over a frequency range indicates a poor quality of radiating efficiency, rather than a good one. But conversely, improving the ground system by adding a sufficient number of radials can increase the radiating efficiency to nearly 100%, and this improvement is accompanied by a *significant increase in SWR readings* over the same frequency range to higher values, which are the normal or expected values.

With an adequate ground system, the SWR is predictable over the frequency range, because a load impedance of any specific $R + jX$ value yields an exact value of SWR on a given feed line, and because we can determine approximately what the antenna impedance should be at whatever frequency we may wish to use (*Ref 2, Fig 2-7; Refs 47, 48, 49 and 58; Ref 71, p 2-6*). But when the ground system is inadequate there is an unknown ground-loss resistance added to the known antenna impedance, which changes the SWR to some lower and unpredictable value. Yet, without being aware of these facts, we often tend to be happier in the discovery of an unsubstantiated low SWR than we are in determining whether we have SWR values that *should* be obtained with the existing configuration. This is a very important concept that requires a clear understanding if we are to avoid misinterpretation of SWR data in our effort to optimize radiated power.

It will help in understanding this concept if we have a clear physical picture of how the ground-loss resistance develops. Still another misconception exists here, this one concerning the current and field behavior in the vertical-over-ground antenna system. Most of us know that conventional grounding techniques used for lightning protection, such as rods or pipes driven deeply into the ground, provide an excellent low-resistance current path for the lightning current (dc). Many are unaware, however, that these techniques are totally inadequate for conducting the entirely different pattern of current flow of the vertical antenna system.

Sec 5.5 Vertical Radiator over Earth

Let us digress a moment for a brief look into the field and current behavior of the vertical antenna system, to see what type of ground system it takes to meet the requirements of a proper current-flow pattern. Consider a base-fed vertical antenna as shown in Fig 5-1. One terminal of a generator is connected to the base of the vertical radiator and the other generator terminal is connected to ground, just below the base of the radiator. During the half cycle in which the conduction current in the antenna flows upward, all the current returns to ground through displacement currents, which follow the lines of force in the RF electric field through the radiator-to-ground capacitance, as shown in Fig 5-1.

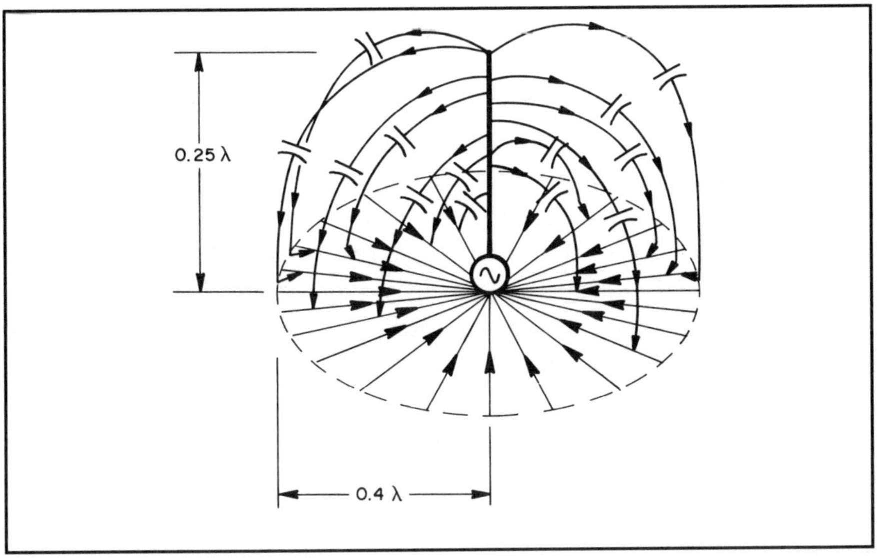

Fig 5-1—The hemisphere of displacement current that flows as a result of the capacitance of a ¼-λ vertical radiator to the earth or a radial system. At frequencies above 3 MHz, RF currents flow primarily in the top few inches of soil, as explained in the text. Ground rods are of little value at these frequencies, and spikes or large nails are sufficient to secure the outside end of each radial wire. With a sufficient number of radials, annular wires interconnecting the radials offer no improvement in antenna efficiency, as the current path is radial in nature.

The electric field surrounding the antenna, which excites the displacement currents, fills the entire volume of space surrounding the antenna in the shape of an oblate or somewhat squashed hemisphere. This hemisphere intersects the ground to form an imaginary circle having a radius of slightly over 0.4 λ for radiators of ¼ λ in electrical height. This radius defines the distance from the antenna at which the returning displacement currents become insignificant, and don't justify radials of greater length. (The radius decreases as the height of the radiator decreases, because the size of the hemisphere surrounding the antenna deceases. Also, radials in or on the ground have no resonance-versus-length characteristic; when outward-flowing current reaches the end of the radial, it continues flowing in the ground instead of going to zero and becoming reflected to flow in the opposite direction, as it does in elevated radials.) The displacement currents enter the ground *every-where over the entire surface within the circle* and then flow back *radially* to reach the grounded generator terminal. Although some of the current penetrates somewhat more deeply, most of the current flow at frequencies above 3 MHz is restricted by skin effect to the upper few inches of the ground.

Now a ground system comprising only a simple water pipe or a driven rod or two is simply a terminal—the ground-feed terminal of the antenna system. So all the returning currents using this arrangement must flow entirely through the poorly conducting ground from all directions everywhere within the circle to reach the terminal. This ground system is often measured to have an "acceptably low" resistance *at dc* (which may be satisfactory for lightning protection), but it injects a loss resistance in series with the antenna *at RF*. The RF ground resistance often exceeds the radiation resistance of the antenna itself! Adding a few wire radials to the system provides good conductivity toward the ground terminal for the currents which reach those radials. However, only a tiny amount of the total current entering the earth's surface inside the circle is intercepted by these few radials. Thus, all the remaining currents still flow only through the lossy earth, and the result is that we still have a high loss resistance.

If a sufficient number of equally spaced radials (90 to 100) extending out to 0.4 λ is present to intercept all the current, all the returning displacement currents find highly conductive paths everywhere within the circle, which lead the currents through negligible loss resistance directly back to the ground terminal of the generator. This can be visualized by examining Fig 5-1. Currents which enter the ground between the closely spaced radials quickly diffract to a radial wire, and thus travel only a short distance through lossy earth before reaching a good conductive path. Thus, with a sufficient number of radials, we have a nearly perfect ground system, which adds only a negligible amount of resistance to the true antenna impedance measurable between the base of the radiator and the ground terminals (*Refs 20, 50, 51 and 57*). From this we can see why the lightning-type ground system, although in prevalent use, is unsatisfactory for an efficient antenna system (*Ref 57, p 82*).

I am not suggesting that ¼-λ antennas with less than ideal ground systems should not be used, nor that fair results cannot be obtained without their use. But the difference between no radials, or only 3 or 4 compared to 100, can amount to over 3 dB. This is far in excess of the loss resulting from an SWR of 4:1 or 5:1 on the average coaxial feed line used by amateurs. The point made here is that the value of ground resistance is unknown and unpredictable in systems using less than an adequate number of radials. This makes the resulting SWR readings unpredictable and therefore useless for the purpose of evaluating the quality of the system, unless some means is available for determining what the change in SWR would be if the loss resistance could be switched in or out.

In practical amateur installations, the ground resistance is sufficiently low if only 40 to 50 radials are used *with a ¼-λ radiator*. The small improvement in radiated power for the addition of still another 40 to 50 radials with the ¼-λ radiator probably does not justify the extra cost and effort. However, if a *short* vertical antenna (from ⅛ λ or less to ¼ λ is contemplated, remember that the radiation resistance decreases as the radiator is shortened. The ground resistance then becomes a larger part of the total resistance, decreasing the efficiency. Thus, the ground resistance should be kept as low as possible for the full capability of the short antenna to be realized (*Refs 51, 56 and 57*). There is practically no difference between the radiation capabilities of the ¼-λ antenna and a radiator even shorter than ⅛ λ, except for the effect of ground resistance and the loss in the resistance of the inductance coil used to cancel the capacitive reactance in the terminal impedance of the shortened antenna. The professional literature is replete with references confirming this point (*Refs 20, 52 and 53*).

Sec 5.6 Resistive Losses and SWR

In this section I discuss how *any* additional resistive losses that are separable from the true antenna impedance affect the true load SWR. By separable I mean such losses as ground-loss resistance, corroded connectors and other poor connections, cold-solder joints, and so on. These all contribute loss resistances that we can control or reduce. In contrast is the resistive component of the antenna terminal impedance, which comprises both the radiation resistance and the inherent conductor-loss resistance in the radiating element. In most cases the conductor-loss resistance in practical radiating elements is negligible, unless excessively small wire is used.

There are several useful relationships between load impedance $Z_L = R + jX$, line impedance, Z_C, and SWR. For example, it is well known that when the load impedance is a pure resistance R, equal to the line impedance Z_C, the reflection coefficient ρ is zero, and the standing-wave ratio is thus one to one. But the reflection is no longer zero and the SWR becomes equal to the ratio R/Z_C when the resistance is larger than Z_C, or Z_C/R when the resistance is smaller than Z_C. It is also well known that ρ and SWR increase with the addition of any reactance component in the load impedance that increases the total reactance, whatever the resistive component may be. This relationship may be verified by observing Eq 3-1. And as noted previously, any combination of $R + jX$ yields an exact value of SWR when terminating a line of given impedance Z_C. We also know that the reactance, X, appearing in the impedance at the terminals of an antenna contributes more to the rise in SWR at frequencies away from antenna resonance than the change in antenna resistance. This is because the reactance changes more rapidly than the resistance during the change in frequency, and because no power is absorbed in the reactance.

However, there is an interesting relationship between the resistance and reactance components of a load impedance which is not generally well known. This relationship sheds light on how these two impedance components affect mismatch reflection and SWR, and it also explains why the unknown ground resistance and other losses mentioned above reduce the usefulness of SWR readings. When reactance is present in the load impedance, the minimum possible SWR occurs when the resistance R is greater than Z_C. The value of the resistance that yields the lowest SWR in combination with a given value of reactance in the load, which I call the minimum-SWR resistance, is dependent solely on the reactance present in the load. This value may be obtained from the relationship

$$r = \sqrt{x^2 + 1} \qquad \text{(Eq 5-1)}$$

where

r and x are normalized to the system Z_C
r = minimum-SWR resistance
x = reactance present in the load

(For more information concerning Eq 5-1 and its proof, see Chapter 11 and Appendix 1 in Chapter 23.) It can be seen from Eq 5-1 that when x becomes zero, r = 1, for an SWR of 1:1. But it is interesting to know that the resulting SWR *always equals exactly* the arithmetic sum of the minimum-SWR resistance value r, and the reactance value, x. This interesting relationship is shown in the following expression.

$$SWR_M = r + x \quad \text{(Eq 5-2)}$$

where

SWR_M = SWR obtained with the minimum-SWR resistance

r = the normalized minimum-SWR resistance

x = the normalized reactance

The relationship shown in Eq 5-2 can help us to understand how unwanted loss resistance, separable from the true antenna-load impedance, affects SWR. In the case of the vertical radiator over earth, the unpredictable ground losses change the SWR from a predictable value, based on known available antenna impedance data (*Ref 48, p 176*) to some unpredictable and usually lower value. A general application of the relationship is presented in the following statements. When the resistance component of the true load impedance is lower than the minimum-SWR resistance, as determined for any reactance component also present in the true load, adding of resistance separate from the true load impedance will cause the SWR to decrease from the value obtained with the true load resistance. This is true until the total resistance is equal to the minimum-SWR resistance. The further addition of resistance will cause the SWR to rise again. These statements apply especially to the vertical antenna of ¼–λ heights or less in proving why ground resistance which reduces the efficiency also reduces the SWR. This is because the true antenna resistance component, R, is generally less than the impedance, Z_C, of normally used feed lines, while the minimum-SWR resistance, R, is always equal to or greater than Z_C.

The effect of reactance in the antenna impedance raises an additional factor of importance in understanding the relationship between SWR values and antenna performance. As stated earlier, the rate at which the SWR rises as the operating frequency departs from the resonant frequency of the antenna depends on the resulting change in the impedance at the antenna terminals, which in turn is dependent on the Q of the antenna. One factor that has a primary influence on antenna Q is the amount of capacitance between the opposite halves of the dipole. (Although it is more commonly called a monopole, a vertical antenna over ground can also be considered as a dipole, because the lower half is simply the image of the upper half, with the opposite polarity.) This dipole capacitance is determined by the ratio of the radiator length, abbreviated len, to its diameter, abbreviated diam.

The len/diam ratio found in the usual simple thin-wire dipole is very high, resulting in a low dipole capacitance and high Q. When the frequency is changed with such an antenna, rapid changes in impedance, reflection, and SWR result. This is why a thin-wire dipole is considered a narrow-band device. However, specific broadbanding steps may be taken to increase the dipole capacitance and thus reduce the Q, thereby reducing the rate of change of SWR. One such step, for example, is decreasing the len/diam ratio by using a multiwire cage configuration for each dipole half, or by fanning out multiple wires from the feed point. Curves of SWR versus frequency are valid here in comparing bandwidths obtained while you experiment with different radiator configurations. However, any separable loss resistance must now be either minimized or held constant to prevent it from introducing unknown variables. Otherwise, the unknown variables can cause differing errors in the SWR readings obtained with different configurations, and thus render the results of the experiment invalid. But unless actual broadbanding steps have been taken to reduce the Q, the rate of change in SWR as frequency changes will not differ dramatically between various types of dipoles having roughly equivalent values of Q. These types include the inverted-V dipole. If a dramatic difference is noted with no *valid* broadbanding steps taken, troubleshooting is called for to determine the cause. More than likely some unwanted loss resistance will be flushed out, if this is the case.

I have seen SWR curves published with descriptions of quite simple antennas, where it would have been impossible for the SWR to remain as low as indicated over the frequency range shown—the Q of the antenna configuration shown would have simply been too high. Two possible explanations for this sort of contradiction are that (1) perhaps the readings were obtained from an inaccurate SWR indicator—many read on the very low side (*Refs 40, 54*) or (2) as suggested above, an unrecognized trouble existed somewhere in the antenna system which was lowering the Q by means of a separable loss resistance. Yet these articles were published because the antennas they described were purported to have "improved SWR characteristics." How many times have you heard someone praise his newly hung skywire by simply telling how low the SWR indicator reads across an entire band? It should now be clear, and it cannot be emphasized too strongly, that an unrecognized and unwanted loss resistance in an antenna system can cause a low SWR reading when *it should not be so low*. So in Chapter 6 we explore the relationship between antenna impedance and SWR in detail so that we may determine what is a proper SWR for given conditions.

Sec 5.7 Reflected Power and SWR

Let us now return to the subject of why we worship low SWR for the wrong reason. As stated earlier, the misunderstanding of this aspect of reflected power is based primarily on the prevalent, but erroneous, idea that any reduction in SWR or reflected power effected on a line feeding an antenna results in a direct one-for-one increase in radiated power. The erroneous reasoning in this idea is in the assumption that if the power is being reflected, it cannot be absorbed in the load or radiated, and that the power which is reflected returns to be lost by dissipation in the transmitter. The assumption is false on both points, for the truth is, because of the reflective conditions in the circuitry used in coupling the transmitter to the line, all power that enters the line is absorbed by the load (except that dissipated in the line itself because of line attenuation). This is true even when the load is not matched to the line impedance (*Ref 55*). *All* of the power delivered by the transmitter is completely absorbed in the mismatched load, because the power reflected from the mismatch is conserved and returned to the load by re-reflection from the line-coupling or matching circuitry, in accordance with the principles discussed in Chapter 4.

In light of the above statement, let us consider a lossless line for a moment. Here it is axiomatic that if all power delivered to the line is already being absorbed in the load (because none can be absorbed in a lossless line), a reduction of the reflected power cannot have any effect whatever on the amount of power taken by the load. And, obviously, there is no power left over to be dissipated in the transmitter.

Following this same reasoning in a real line having attenuation, *all losses in power* must be attributed to the basic I^2R and E^2/R losses arising from the line resistance or attenuation. These losses are unavoidable, even when the load is perfectly matched. The only *additional losses in power* which can be attributed to SWR or reflection occur because the same resistive attenuation is encountered by the reflected power as it travels back along the line from the load to the input. The amount of power lost in this manner is very small at frequencies in the HF range when good-quality, low-loss line is used, because during its return to the input, the reflected power suffers only the same *rate* of line-attenuation loss (in dB) as the forward power suffers in its forward travel toward the load. And as stated previously, all the reflected power which arrives back at the input now becomes part of the forward power.

Another way of explaining the relationship between SWR and lost power is to recall from Chapters 3 and 4 that, because the forward power is the sum of the source and reflected powers, the forward power is greater than the source power whenever the SWR is greater than 1.0. Thus, for a given source power, the resistive losses are somewhat higher in the portion of line where the forward power is higher than the source power, simply because the average line current I and voltage E are higher in that portion.

So from this discussion concerning improper usage of SWR data, we learn that from the viewpoint of efficiency, our concern for SWR involves only the loss from line attenuation. Hence, we can tolerate a higher SWR when the attenuation is low, but when attenuation is high the SWR limit must be lower for the same amount of additional power lost from SWR. The exact relationship between SWR and line losses for different values of line attenuation is shown graphically in Figs 1-1 and 6-1. From these graphs we can easily see that the amount of power actually lost is in sharp contrast to the amount mistakenly assumed to be lost in the improper concept of SWR, where it is thought that a reduction in SWR or reflected power results in a direct equivalent decrease in the amount of power lost in the system.

There is a certain twist of irony behind these various misunderstandings of reflections that have prompted the wrong interpretation or usage of SWR. The irony is that the correct reasons why SWR should be considered are frequently overlooked in the wrong usage, while the basis so generally accepted in support of the wrong usage doesn't even exist in the coupling methods used by amateurs to transfer power from the transmitter to the antenna. A part of this obtuse logic seems to have originated from the confusion among both amateurs and engineers in the meaning of "a matched generator." To some it means being matched in only one direction, and to others it means being matched in both directions. A "matched" signal generator is generally considered to be Z_0 matched, or matched to the line impedance in both directions. However, in transmitter operation, the match is in one direction only—forward—because of the total re-reflection of the load-reflected wave by the matching circuitry at the input to the transmission line. This subject is treated in great detail in Chapters 6, 7, 16, 17 and 19.

Chapter 6

Low SWR for the Right Reasons

(Adapted from *QST*, December 1974)

Sec 6.1 Introduction

Chapter 5 concludes with the statement that in transmitter operation, where conjugate coupling is normally used to deliver optimum power to a load through a transmission line, the match is in one direction only—forward. The transmitter (or generator) output is matched to the line, but looking back into the generator, the line is totally mismatched during the time the generator is actively supplying power to the line through the conjugate coupling of the pi-network tank. If the correct procedure is followed, the conjugate relationship may be demonstrated by making impedance measurements in either direction from any point on the line. These measurements will show an impedance $R + jX$ looking in one direction, and the equal but opposite-sign impedance $R - jX$ in the opposite direction. The net reactance of zero obtained from these two impedances proves that the system is resonant! However, these measurements cannot be performed while the generator is active; it must be *turned off* and replaced with a passive impedance equal to its optimum load impedance. When this is done the impedance terminating the generator end of the line will be seen as a dissipative load when measuring the impedance in the direction of the generator.

One of the prevalent misconceptions is that when reflected power returns to the generator, it sees the plate resistance of the amplifier tube as a dissipative terminating resistance. The fact that dissipation occurs in the impedance that replaces the generator impedance during the measurements described above is partly responsible for the erroneous inference that power reflected toward the generator is dissipated in a similar manner in the internal impedance of the generator. However, when the generator is supplying power to the line, its internal impedance is never seen as a terminating load for power reflected from a mismatched load terminating the line. This is true because of the wave interference between the source wave, the load-reflected wave, and the canceling wave supplied by the coupling network that eliminates the reactance and achieves the total re-reflection of the load-reflected wave. Hence, the line is totally mismatched looking in the direction of the generator. This entire phenomenon is described in detail in Chapter 4.

On the other hand, in laboratory work the generator is usually matched in *both* directions. Here the generator has a Z_C source impedance, usually 50 Ω, and it is also isolated from the line with a resistive pad or attenuator. The pad usually has an insertion loss of around 20 dB and the same impedance as the generator output and line Z_C. Thus, the generator sees a match looking into the pad, and the line also sees a match looking back into the pad. This is because the pad absorbs and dissipates both forward and reflected power like a lossy line, so that only about 1⁄100 of the source power reaches the load, and any power reflected from a mismatched load is also dissipated to 1⁄100 of its original value during its return to the generator. As a result, when the load is a totally reflecting short- or open-circuit termination, the reflected power reaching the source is about 40 dB below, or only 1⁄10,000 of the power delivered by the generator.

Reflected power reaching the source in this test setup is even less with practical terminations that are dissipative. Thus, for all practical purposes, the pad appears to the generator as either an infinitely long line, or a line having a perfect $R = Z_C$ termination. As a result, reflected power returning from a mismatched load termination on the line does not reach the generator with sufficient power to significantly affect its output power, or to modify the impedance which the generator sees as its load. Consequently the forward voltage entering the transmission line at the output of the pad is held constant, independent of the terminating load impedance (*Ref 19, p 48*).[1] Thus, it is understandable that confusion between these two forms of matching can be responsible for misleading us into thinking that reflected power in the transmitter case is dissipated and lost on return to the source.

---

[1] All references appear in Chapter 24.

Sec 6.2 Reflected Versus "Lost" Power

The erroneous conception that reflected power is lost is widespread, having been nurtured on the air for a long time, and supported in print in so many published articles it would be impossible to count them. Two such articles, one by Houghton (*Ref 102*) and the other an SWR-indicator review by Scherer (*Ref 103*), are especially pertinent, because they contain explicit statements supporting the erroneous concept, whereas statements in many other articles only support the error implicitly.

Let us now make a further analysis of the reflection mechanics involved in generator matching. In this analysis, two important ingredients that have been overlooked for a long time will be revealed. In so doing, we will see not only why statements concerning lost power published in the two articles mentioned above are incorrect, but also why it was so easy for these ingredients to be overlooked early in the amateur use of coaxial transmission line, with the result that many amateurs have been misled into seeking low SWR for the wrong reason.

Assume a lossless transmission line having a perfectly matched load termination. Assume also a matched condition between the generator or transmitter and the line characteristic impedance, Z_C. With these conditions, there is no reflected power in the line and therefore no *reflection loss*. The generator delivers what is defined as the *maximum available matched power*, and the load absorbs all the power delivered.

If the load termination is now changed, creating a mismatch between the line impedance Z_C and the terminating load, less power will be absorbed by the load. The amount of reduction in absorbed power resulting from the change in load impedance is the measure of the reflection loss. As the reflected power wave returns toward the generator, it causes a change in the line impedance from Z_C to $Z = E/I$ all along the line. This change is as described in Chapters 3 and 4, and as shown for an SWR = 3.0 in Fig 3-2. When the reflected wave reaches the input terminals of the line, the generator is presented with a change in line-input impedance from the original Z_C value to some new value determined by the E/I vector relationship at the line-input terminals. This new impedance at the line input has exactly the same degree of mismatch to the line Z_C as the mismatched terminating load that generated the reflection. Thus, the line is also now mismatched to the generator in the same degree, and in this condition the generator will automatically make less power available to the line.

The reduction of power delivered to the line is exactly the same amount as the power reflected at the load. In other words, the reflection loss at the load is referred back along the line to the generator. Thus, reflection loss is simply a *nondissipative* type of loss representing only the *unavailability* of power to the load because the generator makes less power available as a result of the impedance mismatch.

It will now become evident that reflection loss represents only the *unavailability* of power to the load, as we see that the load absorbs all the power the generator makes available to the line. On reaching the generator terminals and causing the mismatch to the generator, the reflected power adds to the reduced source power exactly the same amount of power as the reduction in power made available by the generator. Since forward power equals source power plus reflected power, the forward power reaching the mismatched load remains the same as before the generator made less power available. Hence, the reflection loss equals the amount of reduction in power made available by the generator. But because the power reflected by the load is now a part of the forward power reaching the load, the load continues to reflect the same amount of power and therefore absorbs the exact amount of power delivered by the generator.

If an impedance match is now provided anywhere along the line, even at the input terminals, the reflected power is prevented from traveling past the match point toward the generator, as explained in Chapter 4. Thus, the line impedance between the match point and the generator is now unaffected by the reflected wave, and remains at its Z_C value. Also, the generator no longer sees a mismatch, and again delivers its *maximum available* matched power to the line. The impedance match has thus provided a negative reflection, commonly called "reflection gain," which exactly equals and cancels the reflection loss. Consequently, all the power delivered by the generator is absorbed in the load in either case—with or without the reflection gain. The generator simply made less power available before the reflection gain restored the matched condition between the generator and the line (*Ref 19 p 37*).

So we now ask the question, how does this situation relate to the Houghton nomograph (*Ref 102*), where reflected power is stated to be "lost power," and to the "useful power" table from the Knight SWR indicator review (*Ref 103*)?[2,3] It is this: The nomograph simply converts SWR back to reflected power, ρ^2, which is what the SWR indicator actually measures but converts to SWR by means of its scale construction. As discussed in Chapter 3, ρ^2 is the measure of reflection loss or power *reflected*, which equals the reduction in power made available from the transmitter, calculated directly from the mismatch between the characteristic impedance Z_C of the line and the terminating load impedance Z_L. The reflected power is the square of the voltage (or current) reflection coefficient, ρ, from Eq 3-1 (also see Fig 3-2), but remember further that reflected power ρ^2 is a *nondissipative* power, because it all eventually reaches the load, as explained in Chapters 4 and 5.

The tabulated data in the SWR-indicator-review article correctly lists percentage of reflected power ρ^2 for corresponding values of SWR. However, the "useful-power" column is incorrectly labeled, and is therefore misleading because it is actually listing percentage values of $(1 - \rho^2)$, which is the portion of the maximum-available *matched* power the transmitter actually delivers, depending on the degree of mismatch it sees. In other words, the "useful power" column is simply specifying the amount of power the transmitter will deliver into the mismatch if first tuned to a line having a matched Z_C load, and is then switched to the mismatched load without the benefit of *retuning* or *rematching* to the new impedance at the line input. But we do not operate in this manner—we retune, thereby rematching the transmitter to the new load, and consequently establishing the reflection gain as

$$\frac{1}{1 - \rho^2} - 1$$

---

2 Houghton (*Ref 102*) states that reflected power is lost, accompanied by a nomograph for converting SWR into percent reflected power "to make it easier to determine just how much power is lost." The reader is invited to read an excellent rebuttal to this nomograph presentation by Anderson (*Ref 55*).

3 Scherer *(Ref 103)* erroneously shows data on page 90, where 100% minus the reflected power is given as "useful" power. The succeeding paragraph also states incorrectly that the SWR indicator must be placed at a multiple of 1/2 λ from the load to indicate the true SWR. This statement is simply *untrue*! The example demonstrates that the SWR indicator was either not properly adjusted to the impedance of the line, or that it was unreliable, or that no balun was being used to feed a dipole with an unbalanced feed line. See Chapter 2 (statements 21 and 23), Chapter 21, *Ref 38, pp 25-26*, and *Ref 59*.

which completely cancels the reflection loss ρ^2 and the effect of the load mismatch. The transmitter now returns to delivering *100%* of its matched available power to the line, whatever the SWR on the line may be! Thus, the two missing ingredients are

1) understanding the concept of reflection loss and reflection gain, and

2) the discovery that the reflected power is totally re-reflected at the transmitter output terminals, either with or *without* the reflection gain.

It is now evident that the information presented by Houghton and Scherer is not specifying "lost" power at all, but only the *reflection loss*—the amount of power made unavailable by the transmitter until an impedance match provides the reflection gain which cancels the loss and permits the transmitter to again deliver its maximum-available matched power. And as stated on several previous occasions, the impedance match is automatically attained (sometimes unknowingly) either by proper tuning of the transmitter tank circuit to the line-input impedance E/I, or (knowingly) by use of a line-matching network if the pi-network tank of the transmitter lacks sufficient range to obtain the match by itself. How the tank performs the impedance match, and the effects of undercoupling, overcoupling, and possible reactive loading of the tank which can result in the absence of the impedance match are explained in detail in Chapter 7.

Sec 6.3 Reflection Gain

Refer now to Fig 6-1. This figure was developed to illustrate the *reflection-gain* concept, in order to emphasize the effect of misinterpreting reflection loss to be "lost" or dissipated power. The impact of this single misunderstanding of transmission-line principles has been disastrous, because it is the principal cause of the prevalent "low-VSWR mania" (low-vis-war-ma'nya). It is the reason why so many amateurs wrongly believe that "getting the SWR down" is the most important factor in "getting the power into the antenna." They fail to realize that, *whatever the SWR with a low-loss feed line*, the reflection gain has canceled the effect of the load mismatch if the transmitter can be made to tune and load properly into the line, and that all the transmitter power is already being taken by the antenna. And therefore, as explained in Chapter 5, they have also been unaware that no more power of any significance reaches the antenna by achieving the lower SWR. They have also been unaware that line attenuation is the key to whether the SWR level has any *practical* impact on efficiency at all.

In Fig 6-1, the heavy curve marked ρ^2 and $1 - \rho^2$ is based on lossless-line conditions. It is also an exact replot of the Houghton "lost power" nomograph (*Ref 102*), and indicates both the reflected and so-called "useful power" columns from the Scherer's SWR-kit review article (*Ref 103*). Reading downward from the top of the chart, the curve represents reflected power ρ^2 versus SWR. Conversely, the power made available by the transmitter versus the mismatch it sees in terms of SWR $(1 - \rho^2)$ is found by reading upward from the bottom of the chart. Thus, in reading upward, the curve represents the power being made available with the transmitter tuned for a perfect Z_C match, but actually looking into an *uncorrected mismatch*. However, as explained earlier, when the reflection loss is canceled by the reflection gain of the impedance match obtained by retuning the transmitter, a new curve, $\alpha = 0.0$ dB, which now represents this matched condition, follows the heavy straight line across the top of the graph. This indicates that 100% of the power is being made available, and is also taken by the load *regardless of the value of the SWR*. Suddenly the "lost" power is found! Since the curves we've just been discussing represent conditions on *lossless* lines, it is evident that neither Houghton nor Scherer was even considering line attenuation in their presentations.

As stated earlier, power can be "lost" in a transmission line only through line attenuation, alpha (α). If the attenuation is zero, lost power is also zero, as shown along the straight-line $\alpha = 0.0$ dB curve at the top of Fig 6-1. When the line has attenuation, power is lost, as shown by the various loss curves marked $\alpha = 0.03$ dB, etc. Since no allowance for the attenuation factor was made in either case in the material presented by Houghton and Scherer, we have still another reason why the terms "lost power" in one case and "useful power" in the other are incorrect and misleading (*Ref 55*).

In Chapter 15 I deal in detail with attenuation effects and show how to perform some of the pertinent calculations based on Eqs 6-1 and 6-2, which are used as the basis for the α loss curves appearing in Fig 6-1. However, these loss curves are in a form practical for visualizing the correct relationship between the losses actually encountered in different feed lines of various lengths, values of attenuation, and values of SWR. From the curves we can determine the total loss encountered on the line arising from attenuation for the values of SWR indicated on

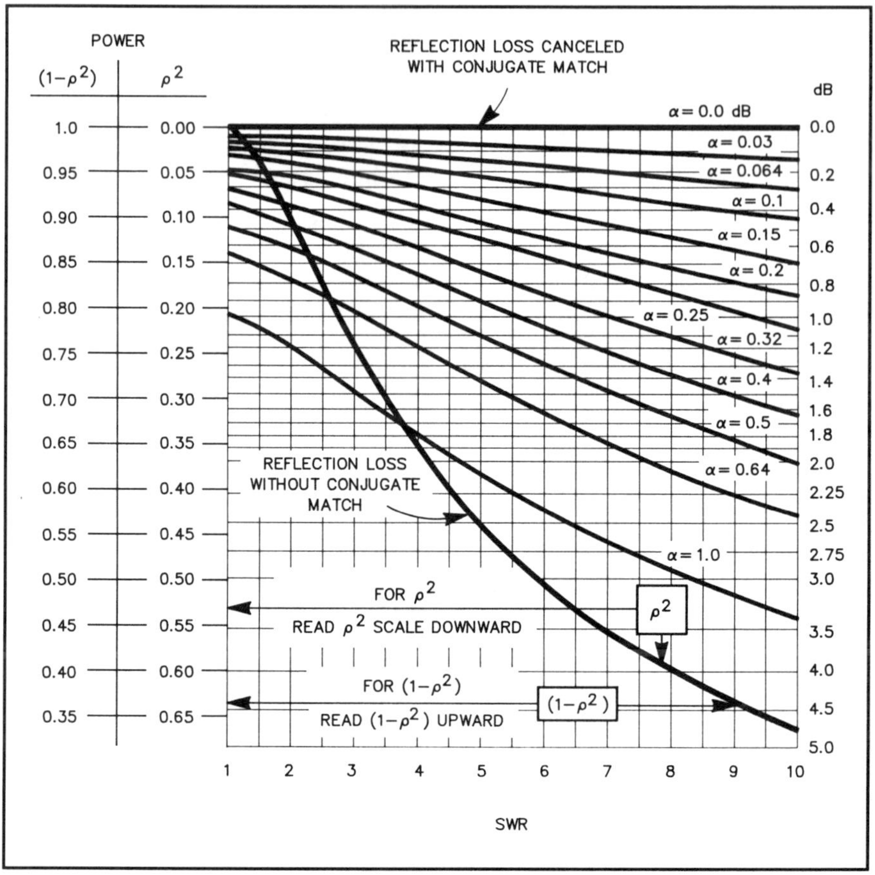

the scale at the bottom of the chart. The curves represent the condition in which the transmitter is matched to the input of the line, and therefore signify that the effect of the load mismatch is canceled in each case. Each curve starts at the left, where the SWR is 1.0, thus indicating the actual attenuation encountered by a particular line when terminated in a perfectly matched load.

The loss value along each curve is seen to increase logarithmically as the SWR on the line rises with increasing line termination mismatch. Thus, the difference between the loss incurred with a 1.0-SWR termination compared to that for any other given SWR value on the same line indicates the amount of additional loss that will be incurred for that SWR. Hence, the graph in Fig 6-1 presents further evidence that when the line attenuation is low, the additional loss from reflection is surprisingly small, even when the SWR is quite high.

←Fig 6-1—Reflection loss versus SWR and matched-line loss of RF transmission lines. Total attenuation in a line operating with SWR may be determined from the dB scale at the right of the chart. The calibration scales at the left are discussed in the text. The α curves in the body of the chart represent the matched-line loss for any given transmission line. For example, the following types and lengths of line would exhibit the α attenuation factors indicated at the frequency of 4 MHz: α = 0.03 dB for 100 ft of no. 12 open-wire line; α = 0.064 dB for 20 ft of RG-8; α = 0.1 dB, for 100 ft of Amphenol twin-lead, no. 214-022; α = 0.2 dB for 62.5 ft of RG-8; α = 0.32 dB for 50 ft of RG-59, 100 ft of RG-8, or 200 ft of RG-17; α = 0.5 dB for 87 ft of RG-59 or 175 ft of RG-8; α = 0.64 dB for 100 ft of RG-59 or 200 ft of RG-8; α = 1.0 dB for 119 ft of RG-58, 350 ft of RG-8, or 700 ft of RG-17. For mathematical derivations of Eqs 6-1 and 6-2 below, see Appendix 5 in Chapter 23. The curves are plots of the following expressions.

INCIDENT, OR FORWARD POWER (multiply by source power delivered) − POWER REFLECTED = POWER ABSORBED*

POWER AT MATCH POINT; POWER AT LOAD (after line attenuation)

$$\frac{1}{1-\rho^2\varepsilon^{-4\alpha}} \qquad \frac{\varepsilon^{-2\alpha}}{1-\rho^2\varepsilon^{-4\alpha}} - \frac{\rho^2\varepsilon^{-2\alpha}}{1-\rho^2\varepsilon^{-4\alpha}} = \frac{(1-\rho^2)\,\varepsilon^{-2\alpha}}{1-\rho^2\varepsilon^{-4\alpha}} \qquad \text{(Eq 6-1)}$$

With lossless line ($\alpha = 0$),

$$\frac{1}{1-\rho^2} \qquad \frac{1}{1-\rho^2} - \frac{\rho^2}{1-\rho^2} = 1 \qquad \text{(Eq 6-2)}$$

where

ρ= magnitude of voltage-reflection coefficient (see Chapter 3, Sec 3.1)

α = line attenuation in nepers = dB ÷ 8.6859

ε = 2.71828, the base of natural logarithms

$$\text{* POWER ABSORBED} = \frac{(1-\rho^2)\text{ less one–way line attenuation}}{1-(\rho^2\text{ less two–way line attenuation})} \times \text{source power}$$

Examine the SWR region between 1:1 and 2:1. Do you see enough difference in power level on any of these curves to justify any effort in reducing a 2:1 SWR to *any* lower value whatever? Do you still think you'll get out better by squeezing that SWR of 1.8 down to 1.2? A review of Chapter 1 is now appropriate to re-emphasize how the use of these concepts can broaden our design flexibility. It is also reassuring to check the efficiency values of both the TIROS and NAVSAT spacecraft feed-line examples from Chapter 1 in the Fig 6-1 graph. In the NAVSAT example, comparison of the 65% value of reflected power with only 1.04 dB actual line loss will be especially revealing.

Fig 6-2 provides additional line-attenuation data, permitting us to extend the use of the Fig 6-1 curves to other frequencies and transmission lines. Fig 6-2 may also be supplemented with further data available in *The ARRL Handbook* and *The ARRL Antenna Book* (*Refs 1, 2 and 71*).

Sec 6.4 Radiation Resistance

In Chapter 2, statement 26 says, in effect, that with mobile whip antennas in the 80- through 10-meter bands, no significant amount of power is saved by using matching circuitry between the feed line and the antenna terminals. And statement 27 goes on to say that in the absence of such matching circuitry, more power is radiated from center-loaded mobile antennas that have a high feed-line SWR at resonance than those that have low SWR. The concepts involved in those two statements are also widely misunderstood, so this is an appropriate time to clarify both of these statements. These concepts also fit the category of "low SWR for the wrong reasons."

It is well known that the radiation resistance of the short whip-type mobile antenna is very low. And of all the HF amateur bands, the radiation resistance is the lowest at 80 meters, because the electrical length of the radiating portion is the shortest on this band. Depending on the exact length of the antenna and other factors, the *radiation resistance* of the center-loaded antenna is approximately 1 Ω at 80 meters, as shown by Belrose (*Ref 60*). The capacitive-reactance component in the terminal impedance of this short antenna ranges from $-j3000$ to $-j3500$ Ω in typical 80-meter models, as shown by Belrose and also confirmed by my own measurements using a General Radio 1606-A RF impedance bridge. This capacitive reactance is canceled by the equal $+jX$ inductive reactance of the loading coil.

However, it is not well known that there are *two other resistances* which become important for consideration in antennas of this type. These resistances, from loading-coil loss and ground loss, add to the radiation resistance to comprise the total resistive component of the impedance appearing at the antenna terminals. Thus, by ignoring these two resistances, it is erroneously thought by many amateurs that the 1-Ω

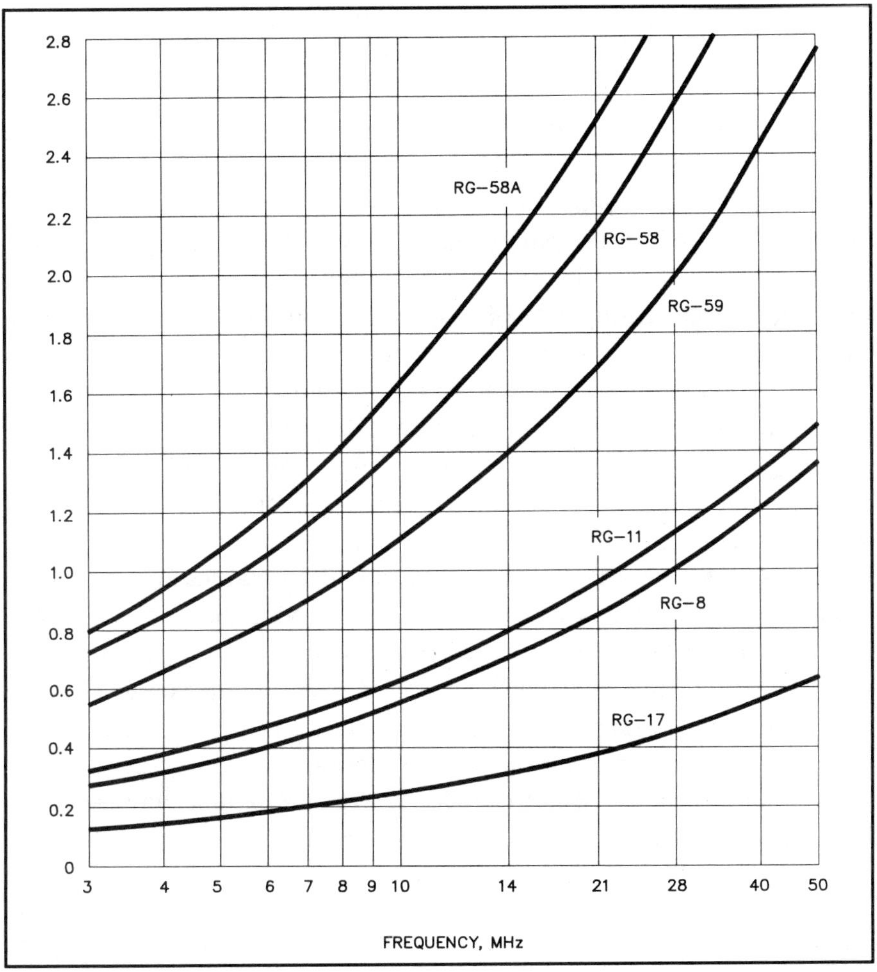

Fig 6-2—Attenuation in decibels per hundred feet for various coaxial cables.

radiation resistance alone comprises the entire antenna-terminal impedance, and also that it requires a matching device at the antenna to match what is thought would be about a 50:1 mismatch if fed directly with a 50-Ω feed line. Actually, the loss resistance in the loading coil and any ground-loss resistance both add to the radiation resistance, causing the terminal resistance of the antenna to be much higher than is commonly realized, although still lower than the 50-Ω feed lines that are normally used. Thus, the actual mismatch value is *much lower* than is usually realized.

While there are some who recognize that loading-coil loss appears as part of the total terminal impedance, only a few are aware that ground-resistance loss also exists, because, except for Belrose, most writers neglect to mention it or consider it in their system analysis. For example, see Swafford (*Ref 116*).

The Mobile Handbook (*Ref 117*) not only fails to recognize the existence of ground resistance, but mistakes what is actually the combined ground and radiation resistances as radiation resistance alone. For example, in subtracting 6 Ω of loading-coil and whip resistance from the 14 Ω of measured terminal resistance, the 8-Ω difference is simply taken to be radiation resistance, with no mention of any ground resistance. Hence, the ground resistance, which cannot be ignored, is unknowingly and improperly being included as a portion of the radiation resistance instead of as a loss resistance in the expression for efficiency. This oversight might have been avoided if an analysis had been made of the increase in radiation resistance actually obtained by raising the loading coil from the base to the center of the whip, because the 6 Ω of coil resistance plus a 1-Ω radiation resistance subtracted from the 14 Ω of measured terminal resistance leaves 7 Ω, which requires an explanation of where those remaining 7 Ω of resistance came from. Obviously, it is the ground resistance which was ignored.

From further study of the *Mobile Handbook* text, however, it is evident that an assumption of a greater amount of radiation from the coil than what is actually possible may have been the reason why such a high value of radiation resistance was considered plausible. But whatever the reason, we have been given the unrealistically high radiation resistance of 8 Ω and an impossible efficiency value of 8/14 = 58%. These values are fundamentally impossible to obtain in the center-loaded mobile antenna, and the true values have been obscured. In other words, a large portion of the power considered in the *Mobile Handbook* as being radiated is actually dissipated as heat in the ground. Belrose shows a proper analysis (*Ref 60*), which is supported by my own measurements.

It is practically impossible to obtain a mismatch of sufficient magnitude that requires *any* matching circuitry between the feed line and a properly resonated, conventional center-loaded mobile antenna *for the purpose of conserving any significant amount of power*. This statement is true, the many remotely controlled luggage-compartment tuning and matching arrangements notwithstanding. Now we will see why.

I have made measurements on loading coils and antennas using a Boonton 260-A Q meter and a General Radio 1606-A RF impedance bridge. Loading-coil loss resistances range from about 8 Ω for the better commercially available coils to as high as 31 Ω measured in poorer coils, depending on the Q and the self-resonant frequency of the coil. Ground-loss resistances encountered with conventional low-band mobile antenna installations range from about 5 Ω for low, wet ground to around 12 Ω for high, dry ground. Average ground yields about 7 Ω.

The ground-loss resistance in the mobile setup is less than that found in an antenna of a full ¼ λ in physical height with no radials, because the radius of the circle where the minimum space-displacement currents return to the ground is shorter with the shorter antenna. Therefore, the return currents travel a shorter distance through lossy earth (see Fig 5-1). The current-flow pattern in the mobile system is also described by Belrose.

Consequently, from all of this we can see that the total terminal resistance of the mobile antenna is nowhere near the 1-Ω radiation resistance alone (which would produce an SWR of 50:1), but it also includes the ground-loss resistance and the coil-loss resistance, all three appearing in series. The absolute minimum resistance is 1 + 5 + 8 = 14 Ω for an SWR of 3.5:1 at resonance when using the low-loss loading coils over good ground. The resistance can be as high as 1 + 12 + 31 = 44 Ω for an SWR of 1.1:1 when the higher ground loss and higher coil loss occur simultaneously. Hence, the actual resistance appearing at the antenna terminals lies in the range from 14 to 44 Ω, with corresponding SWR values of 3.5 and 1.1 to 1.

Thus, as strange as it may seem, the *higher* the minimum SWR attainable *at resonance*, the greater will be the power radiated for the same amount of power delivered to the feed line by the transmitter. It does not seem so strange, however, when we consider that the low radiation resistance of around 1 Ω is the only portion of the total terminal resistance that contributes to radiation. And this 1 Ω is constant, fixed by the radiator length. So by making the loss resistance lower through the use of a higher Q loading coil, less power is dissipated as heat in the coil, leaving more to be radiated. Conversely, if the lower Q coil is used simply to achieve a lower SWR, less power is radiated because more is now being spent in heating the coil.

Many amateurs unknowingly select loading coils of low Q because "they produce a lower SWR than coils having a higher Q." Some lower Q coils, such as the Hustler manufactured by Newtronics, are advertised as producing a lower SWR and greater bandwidth. They do, indeed, produce a lower SWR and greater bandwidth because of their higher loss resistance, which results in a substantial loss of power. However, when using the higher Q, lower loss loading coil, even though its lower loss resistance results in a larger load mismatch and higher feed-line SWR, the resulting increase in radiated power with the higher Q coil is still proportional to the decrease in total resistance. Any additional loss from the higher SWR is so small that it can be neglected, because line attenuation in the short feed lines used in mobiles is extremely low. Remember, line attenuation is the only cause of power loss in the feed line, regardless of the SWR level.

I don't know whether it is through unawareness or by intent in the design that Hustler loading coils have close-wound turns (no space between turns). Close-wound turns contribute to the high loss resistance that provides the low SWR and greater bandwidth of these coils. Let me digress here for a moment to examine the nature of coil resistance so that we may understand why some loading coils have high Q and low loss, while others have low Q and high loss, depending on their design.

The ac resistance loss in inductors is comprised of two separate resistance components, resistance from skin effect, related to wire size, and resistance from distributed capacitance, related to the spacing between turns. For a coil of given length, diameter and number of turns, there is an optimum wire size that yields minimum coil resistance and maximum Q. On one hand, the largest wire that will fit the given length (no space between turns) yields minimum resistance from skin effect. On the other hand, when the turns are close spaced, the resistance from distributed capacitance is maximum. Spacing the turns apart (by using smaller wire for the same number of turns over the same length) reduces the resistance from distributed capacitance. Hence, the optimum wire size is a trade-off between a large wire that minimizes the resistance from skin effect and a small wire that minimizes the resistance from distributed capacitance because of the increased spacing between the turns.

The decrease in skin-effect resistance from increasing the wire size is obvious. However, the decrease in resistance from lower distributed capacitance with the smaller wire size is not obvious, so I'll now

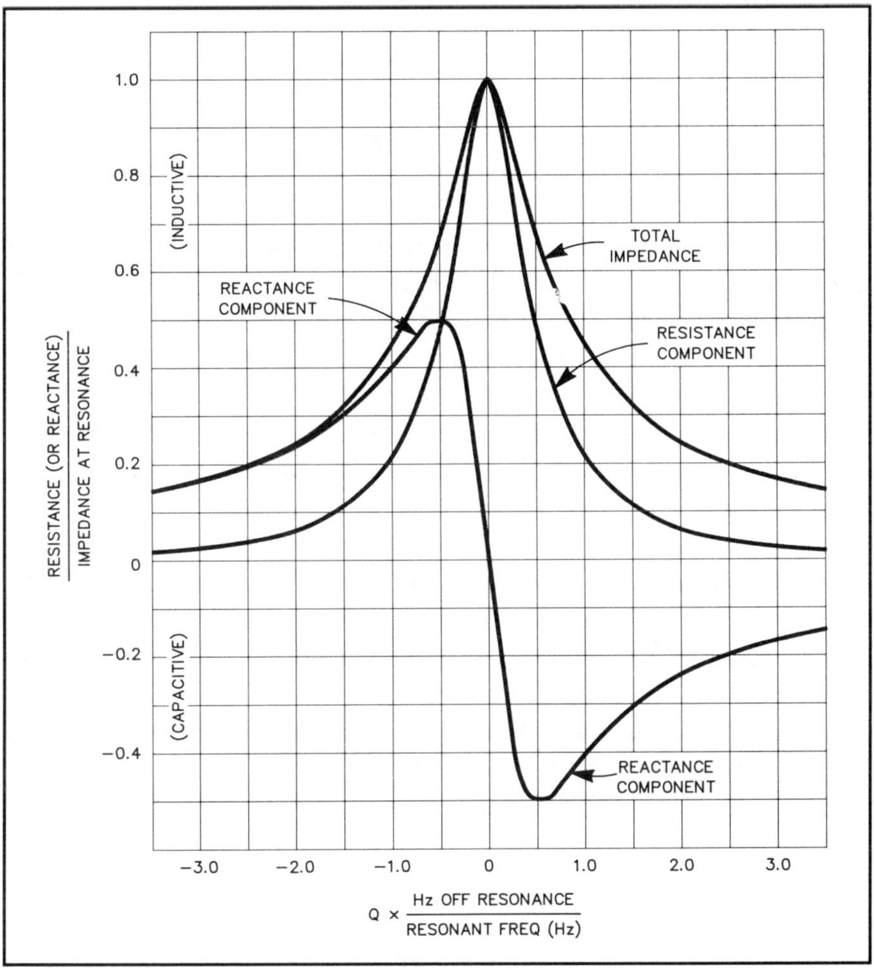

Fig 6-3—Universal curves giving the resistance and reactance components of the impedance of a high-Q parallel-resonant circuit as a function of frequency. *After Terman (Ref 126, p 145)*

explain why this is true. That there is distributed capacitance between turns of an inductor is well known. The closer the spacing between the turns, the greater the distributed capacitance, with maximum capacitance appearing when there is no spacing (except for insulation). The distributed capacitance appears inherently in parallel with the inductance of the coil. Consequently, there is a frequency (above the operating frequency of the coil) where the reactance of the distributed capacitance is equal to the inductive reactance of the coil, and the coil becomes self resonant.

At the coil self-resonant frequency, the impedance appearing at its terminals is a very high, pure resistance (which we don't want in a loading coil), and the natural inductive reactance disappears. At this frequency, the coil no longer supplies any inductive reactance to cancel the capacitive reactance of the antenna. Instead, the coil simply introduces its high resistance in series with the antenna terminals.

The curves of Fig 6-3 show the relative impedance, resistance and reactance components of the impedance versus frequency in parallel-resonant circuits, such as we have in the self-resonant inductor. For a given frequency, the resistance curve shows the amount of coil resistance relative to the maximum resistance that appears at the self-resonant frequency. We know that in a resonant circuit and with a given inductance, the smaller the capacitance, the higher the resonant frequency. Hence, in the loading coil, the lower the distributed capacitance (from increased turn spacing), the higher the self-resonant frequency and the lower the loss resistance for a given operating frequency. Certainly the design goal for obtaining minimum loss resistance in an inductor is to make its self-resonant frequency as high as possible relative to the operating frequency by optimizing the turn spacing. Because turn spacing (for a fixed winding length and number of turns) has no effect on the amount of inductance in the coil, there is an optimum spacing versus wire diameter that yields the lowest resistance, and hence the highest Q.

It is interesting to note that of the commercially available 80-meter mobile loading coils I have measured, the Hustler coils having the loss resistance of 31 Ω at 4.0 MHz were self-resonant at 6.0 MHz. The better Webster KW-80 coils having 8 Ω of resistance at 4.0 MHz were self-resonant at 14.0 MHz. The windings in the Hustler coils are close spaced, while the spacing between turns in the Webster coils is approximately equal to the wire diameter. Need I say more?

Before leaving the subject of mobile loading-coil loss relative to distributed capacitance, here is some important information and a warning concerning the use of top-hat capacitance loading. When using an adjustable loading coil, the increased capacitance to ground afforded by top loading can reduce the coil loss, and thus improve the efficiency of the loading coil substantially. This is because the extra capacitance in the top loading reduces the amount of coil inductance required to obtain resonance, and hence reduces coil loss. However, to obtain the maximum improvement in efficiency, the top hat must be placed at the top of the radiator, not part way down. If the hat is large enough to be effective, any portion of the radiator protruding above the hat is ineffective.

And now the warning. I have seen many top-loading devices placed directly above the loading-coil inductor, with the entire top portion of the radiator extending above the top hat. This placement is incorrect, an absolute no-no! This incorrect placement of the hat causes an undesirable increase in the hat-to-coil capacitance. That capacitance adds to the already undesirable distributed capacitance of the coil, lowering the Q and increasing the coil resistance as described earlier—just the opposite of the hat's intended purpose. Consequently, the coil performance is degraded by the hat, rather than enhanced. This incorrect placement of the hat does reduce the resonant SWR, as many have discovered, but this reduction in SWR is for the same wrong reason as explained earlier—it results from the increased resistance of the coil. For further information on capacitive loading, I refer you to a section I wrote for *The ARRL Handbook* on "Top-Loading Capacitance." It appears in Chapter 33 of all editions since 1986 (*Ref 1*).

In Fig 6-1, the loss curve α = 0.064 dB represents the loss characteristics of a typical mobile feed line—20 feet of RG-8 at 4.0 MHz. The curve shows a matched-line attenuation of 0.064 dB, plus an additional loss of 0.056 dB resulting from the 3.5:1 SWR, for a total loss of 0.12 dB. When this condition is accompanied by the 1:14 ratio of radiation resistance to total terminal resistance and a transmitter power of 100 watts, the difference in power radiated between matching at the antenna terminals and leaving the 3.5:1 SWR on the line and matching at the line input amounts to less than 0.1 watt!

It is also of interest to note that with the lower loss resistance ratio (1 to 14 Ω) while using the high-Q coil, the radiating efficiency is 7.14%, or 11.46 dB below the transmitter power delivered (excluding line loss). With the 1 to 44-Ω resistance ratio (with the higher loss coil and poor ground), the radiating efficiency is only 2.22%, or *16.53 dB* below the transmitter power. We thus have a 5.07 dB loss in efficiency in return for lowering the SWR from 3.5 down to 1:1 by using a "low-SWR" loading coil. Contrast these actual values of efficiency with the 58% recited in the *Mobile Handbook* (*Ref 117*).

So, contrary to statements found in numerous articles which have been insisting that we believe otherwise, no significant improvement in efficiency can be obtained on the lower frequency bands by performing the matching function between the feed line and the mobile antenna terminals when a low-loss loading coil is already in use. The matching can be performed equally well at the input of the feed line, either by the transmitter output tank itself, or by a separate matching network if the transmitter tank lacks sufficient range, *with the feed line connected directly to the antenna terminals!* See *Refs 4, 24* and *61*. Thus, as emphasized in the opening paragraphs of Chapter 5, the important point here, again, is that a flexible, open choice is available in our system design.

Whether the matching which is required to transfer maximum power from the transmitter into the line is to be performed at the input end or the load end of the feed line is a choice which should be determined according to personal preference of the operator. It should be based on convenience and accessibility to adjustment, and not on an arbitrary, low SWR dictated for the wrong reason by a decree of an SWR King who doesn't understand his Subject! But wherever the matching is performed during operation, the antenna that will radiate the strongest signal is the one with the loading coil that is capable of producing the *highest SWR at resonance* with no matching at the antenna terminals, for the reasons explained above.

Now I'll explain why loading coils of low Q have a high resistance that reduces the SWR, but also reduces the radiated power. A principal contributor to low Q in loading-coil inductors is the careless (or intentional?) lack of attention to coil design, particularly in relation to the spacing between turns. As is well known, there is distributed capacitance between every turn. The total distributed capacitance resonates with the coil inductance at some frequency, and at this frequency the coil becomes a self-resonant L-C tank circuit. As is also well known, the series ac resistance in a resonant tank circuit is very high. Insufficient spacing between turns results in a large distributed capacitance, which lowers the self-resonant frequency. Hence, the effective RF resistance of the coil will be high unless the operating frequency is substantially lower than the self-resonating frequency. On 80-meter coils, where I have measured 31 Ω of series resistance at 4.0 MHz, the resonant frequency of the coils was around 6 MHz. On the higher Q coils that measured 8 Ω, the resonant frequency was around 14 MHz—a considerable difference!

Finally, here is a suggestion which may be helpful in tuning a mobile antenna. The use of a dip oscillator or a noise bridge in an attempt to determine the resonant frequency of the coil and radiator combination can rarely give the correct answer when you measure at the input terminals of the feed line. The reason is that when using these measuring devices in such a way, they are measuring the resonant frequency of the entire system, including the length of the feed line. As the length of the feed line is changed, the resonant frequency of the antenna system will also change. However, an SWR indicator connected directly at the input of the feed line provides an accurate indication of antenna resonance at the frequency of lowest SWR, providing the instrument is reliable and accurately calibrated for the impedance of the line (*Ref 59*). This is because the minimum SWR for a given terminating load of the type we're discussing occurs at the resonant frequency of the load, *regardless of the length of the feed line.*

Sec 6.5 Low SWR for the Right Reasons

It is of interest to know that in TV broadcasting, where a long feed line is required to reach the antenna on a high tower, low SWR is an absolute necessity. However, the requirement here is primarily to avoid multiple, displaced, ghost images from appearing in the received picture images which would result from reflections on the feed line caused by a mismatch. Similarly, low SWR on the feed line is a necessity in FM stereo broadcasting to avoid cross contamination between the two audio modulation channels. However, in Amateur Radio we do not have the problems encountered in TV and FM broadcasting.

To summarize the discussions of SWR and reflections as they pertain to Amateur Radio operations, we have seen that *we do not need* a low SWR on the antenna feed line

1) to prevent reflected power from dissipating in the transmitter, because none dissipates in the properly coupled transmitter anyway, whatever the SWR;

2) to prevent feed-line radiation or TVI, because a mismatched load on the feed line doesn't cause feed-line radiation or TVI;

3) to attain proper coupling to the transmitter, because we can couple to or match the impedance at the input terminals of the line, whatever the SWR.

From examining Figs 6-1 and 6-2, it is evident that *we do not need an SWR lower than 2:1 on any feed line* to avoid any significant loss in efficiency, or with SWR values considerably higher than 2:1 when using feed lines having low attenuation. It seems that there aren't many reasons for needing a low SWR on amateur feed lines at HF, if an antenna tuner is used when required to obtain a satisfactory impedance match for the transmitter. (In fact, now would be an appropriate time to review statements 11 through 17 in Chapter 2.) So let's see how we can briefly develop realistic SWR limits in relation to the attenuation values found in practical feed lines. Here are a few time-tested rules of thumb to use as guidelines.

1) When operation is near the dipole-resonant frequency, either 50- or 75-Ω feed line may be used equally well. Depending on the height above ground, the antenna terminal resistance at resonance will fall somewhere between 40 and 80 Ω, so the resulting mismatch with either line impedance is so small as to be inconsequential, despite arguments to the contrary from those who are still afflicted with low-viswarmania. However, to obtain accurate readings with an SWR indicator, the impedance of the indicator must be compatible with the impedance of the line on which it's being used.

2) An impedance-matching device placed *anywhere* on the feed line between a mismatched load and the transmitter compensates for the mismatch at the load, with the resulting effect that a match now exists everywhere on the line (*Ref 17, p 243*). In other words, if a mismatched load impedance, $Z_L = R + jX$, is conjugately matched anywhere on the line, the reflection generated by the complementary mismatch at the matching point causes the impedance looking into the termination end of the line to change from Z_C to $Z = R - jX$. (See Chapter 4.)

3) Now let's take advantage of the increased bandwidth immediately available to us simply, using our knowledge that nothing magical or miraculous happens in "bringing the SWR down" to 1.0. When we use coaxial line to center-feed a dipole, the operation is usually for one amateur band only. One exception is that satisfactory operation may be obtained on 15 meters with a 40-meter dipole. However, we now have the freedom *to operate anywhere within the entire band*, letting the SWR climb to whatever value it should as the antenna terminal impedance changes with frequency (but still staying within limits that are defined presently). To minimize the increase in mismatch and SWR resulting from the frequency excursion to either end of the band, the dipole should be cut to resonate near the center of the band. On the 75-80 meter band, where the percentage of frequency excursion is the greatest, the mismatch at the ends of the band will be somewhat less severe with a 75-Ω feed line than with a 50-Ω line. Of the smaller sized coax, RG-59 is

preferred over RG-58, because the combination of the somewhat lower maximum SWR and lower matched-line attenuation with the RG-59 permits either a greater frequency excursion away from the self-resonant frequency of the dipole, or a longer line for the same amount of loss. Of the larger size cable, either RG-8 or RG-11 gives nearly equal results, because the matched-line attenuation of the RG-11 is a little greater than with RG-8, thus offsetting the gain resulting from its lower maximum SWR. However, the lower attenuation of the larger cables permits either a greater frequency excursion, or a longer line for the same loss than with the smaller cables, irrespective of their relative power-handling capabilities.

4) The smallest reduction in power that can just barely be detected as a change in level at the receiving station with the AGC disabled is 1 dB. So, to find the SWR which reduces the radiated power by 1 dB, we first use Fig 6-2 to find the attenuation per hundred feet of the correct feed-line type at the desired operating frequency. Then apply the correction for the length of the actual feed line to be used to determine the attenuation, α, of the line. Now go to Fig 6-1 and find the α-loss curve which most nearly corresponds to the value of the feed-line attenuation. Starting where that loss curve crosses the SWR = 1.0 line, follow the curve to the right until 1.0 dB additional loss is indicated on the dB scale on the right side of the graph. Read the SWR at this point from the scale on the bottom. This is the SWR which will reduce the radiated power by the "just barely noticeable" amount at the receiving station, compared with the signal that would be received if the line had been perfectly matched at the load. More exact data is presented later, but Table 6-1 shows SWR values to be expected for a dipole at the ends of the bands when the dipole is cut for resonance at the center of the band. Applying these data to Fig 6-1 readily shows that it requires feed lines of lengths substantially longer than the average to lose enough additional power from SWR ever to be noticed at the receiving station. In other words, a full 1 dB of additional loss will seldom be encountered, and therefore, no "pileup punch" will be sacrificed in obtaining the increased operating-bandwidth flexibility.

5) At an SWR of around 4:1, the additional loss because of SWR just equals the perfectly matched line attenuation for any line. Thus, in effect, a 4:1 SWR multiplies the matched line loss by a factor of two. As an example, this statement means that the power lost in 350 feet of RG-8, or in 174 feet of RG-59, at 4.0 MHz with an SWR of 4:1, will

Table 6-1

Approximate SWR Values at Bands Edges for Wire Dipoles Cut to Resonate at Band Center

Frequency, MHz	*Max. SWR Value*
3.5 to 4.0	5:1 or 6:1 (50-Ω line)
3.5 to 4.0	4:1 or 4.5:1 (75-Ω line)
7.0 to 7.3	2.5:1
14.0 to 14.35	2.0:1
21.0 to 21.45	2.0:1
28.0 to 30.0	3.0:1

have a "just barely noticeable" difference compared to a line having no attenuation loss whatever! This is because these lines each have a matched-line attenuation of 1.0 dB. However, also note that this "just barely noticeable" difference could *not* be noticed when the signal is well up on the S meter, because on many receivers it takes a *6 dB change in signal level for a change of one S unit.*

6) The SWR on the feed line may be monitored to determine that the SWR is within the limit based on the line attenuation, by placing the SWR indicator between the line-matching network and the feed-line input terminals. But remember, *the SWR remains on the line even after the matching network has been properly adjusted.* The match between the transmitter and the matching network may be monitored with the SWR indicator placed between the transmitter and the network. The network is properly adjusted when the forward power is maximum and the power reflected from the network is zero. If the forward power readings are the same as those obtained with a dummy load, and the reflected power reading is zero in both cases, the input impedance of the network is the same as the impedance of the dummy load. If the SWR indicator shows some power being reflected from the matching network and the transmitter still loads properly, obtaining further reduction of the reflected power is probably unimportant. This indication of reflected power is not showing an "SWR," but only the degree of impedance mismatch at the input of the network. If insufficient TVI rejection is obtained with the line-matching network alone, a conventional TVI filter may be used between the transmitter and the matching

network with the same degree of effectiveness as when used in a line that is matched at the load.

As I have now shown, any required matching can be performed at the input to the feed line, instead of at the terminating load such as an antenna. Therefore, no SWR bandwidth limit for amateur use (such as the commonly used, low arbitrary value of 2:1) is realistic unless it is based on the attenuation of the specified feed-line installation and the amount of total attenuation allowed. The arbitrary 2:1 SWR limit came into existence because the matching range of most amateur transmitters was thought to be limited to 2:1 by design, with economics being considered more important than operational flexibility. But simple line-matching networks as described in the bibliography references can extend the inherent matching range of the transmitters to accommodate values of load impedance far beyond the limits defined by a 2:1 SWR.

If it were not for the cost and space factors, line-matching networks could be built into modern transmitters, giving us amateurs back the matching range we were accustomed to having with the old swinging-link method of coupling. We weren't as conscious of SWR before the pi-net coupling replaced the swinging link. Matching at the line input in those days involved basically a simple adjustment of the adjustable link position to achieve the proper degree of coupling to match the resistive component of the load, and retuning the plate-tank capacitor to cancel the reflected reactance. Using this technique, we often loaded our transmitters into lines with high SWR values without even knowing about the SWR. But with the appearance of the SWR indicators after the departure of the link, we "discovered" SWR and then learned how to misinterpret the meaning of the SWR readings.

In conclusion, if the feed-line loss is within your acceptable limits at a given SWR level, determined from consulting Figs 6-1 and 6-2, and if the transmitter can be adjusted to load and tune properly (either with or without an additional line-matching network), operate, and don't worry about the SWR, because you are now using realistic SWR for the *right reasons!*

When the series of articles, "Another Look at Reflections," originally appeared in *QST*, reader response was excellent. But some readers told the author, "Your story is interesting, but you'll never convince me that I won't get out better with a perfect 1.0 SWR." Now I remind any reader who still entertains any skepticism of these entire proceedings concerning SWR that the information presented here is not simply a recitation of my own ideas or opinions, but has been taken directly from the professional scientific and engineering literature (note the extensive bibliography in Chapter 24), and paraphrased specifically for the radio amateur with great care not to change the meaning. Moreover, in striking contrast to the many differing opinions heard on the subject during amateur discussions, there *are no* such differing opinions among the professional sources, because among the professionals (including textbook authors) the principles involved are completely understood and are based on true, proven scientific facts which are not subject to divergent opinions such as we find in politics or religion.

Apparently many have forgotten that this story was told for the amateur in *QST* no less than twice prior to the initial appearance of this series, by two well-known experts in this subject area. They are Mr. George Grammer, W1DF, former engineer and Technical Editor of *QST*, and Dr. Yardley Beers, WØJF, formerly a professor of physics, Chief of the Radio Standards Physics Division, National Bureau of Standards, and Senior Scientist, Quantum Electronics Division of the National Bureau of Standards. Their illuminating contributions, listed as *Refs 3 through 6, 16*, and *22* of the bibliography should be reviewed, even if it means a trip to the library—the trip will be very rewarding.

Chapter 7

My Transmatch Really Does Tune My Antenna

(Adapted from *QST*, August 1976)

Sec 7.1 Introduction

Chapter 4 discusses conjugate matching and introduces the concept of wave interference as the basis for the mechanics of the matching. I used the transmission-line stub form of matching for the example because of the ease in describing the generation of the wave and the wave action involved in the matching process. Chapters 5 and 6 discuss some of the reasons why many have been plagued with myths concerning mismatch and SWR, myths that have prevented an enlightened use of impedance matching in our daily operating procedures concerning antennas and transmission lines. In those chapters I also explained power loss caused by line attenuation, and presented both graphic and mathematical means for determining the additional loss resulting from SWR on the feed line. An additional detailed discussion of the conjugate match may be found in Chapter 19. That chapter explains why a perfect conjugate match can be achieved using a classical generator, but that a true conjugate match seldom exists with the RF transmitter, even though the principles of conjugate matching apply.

This chapter focuses on the use of line-matching networks formed with adjustable capacitors and inductors instead of stubs made from sections of transmission line. Here I show that the wave actions in both of these forms of matching are identical. I will also explain in detail how transmitters are matched to mismatched feed lines with an external matching network as shown in Fig 7-1, or by using the pi-network tank in the transmitter shown in Fig 7-2. But first let's examine one last

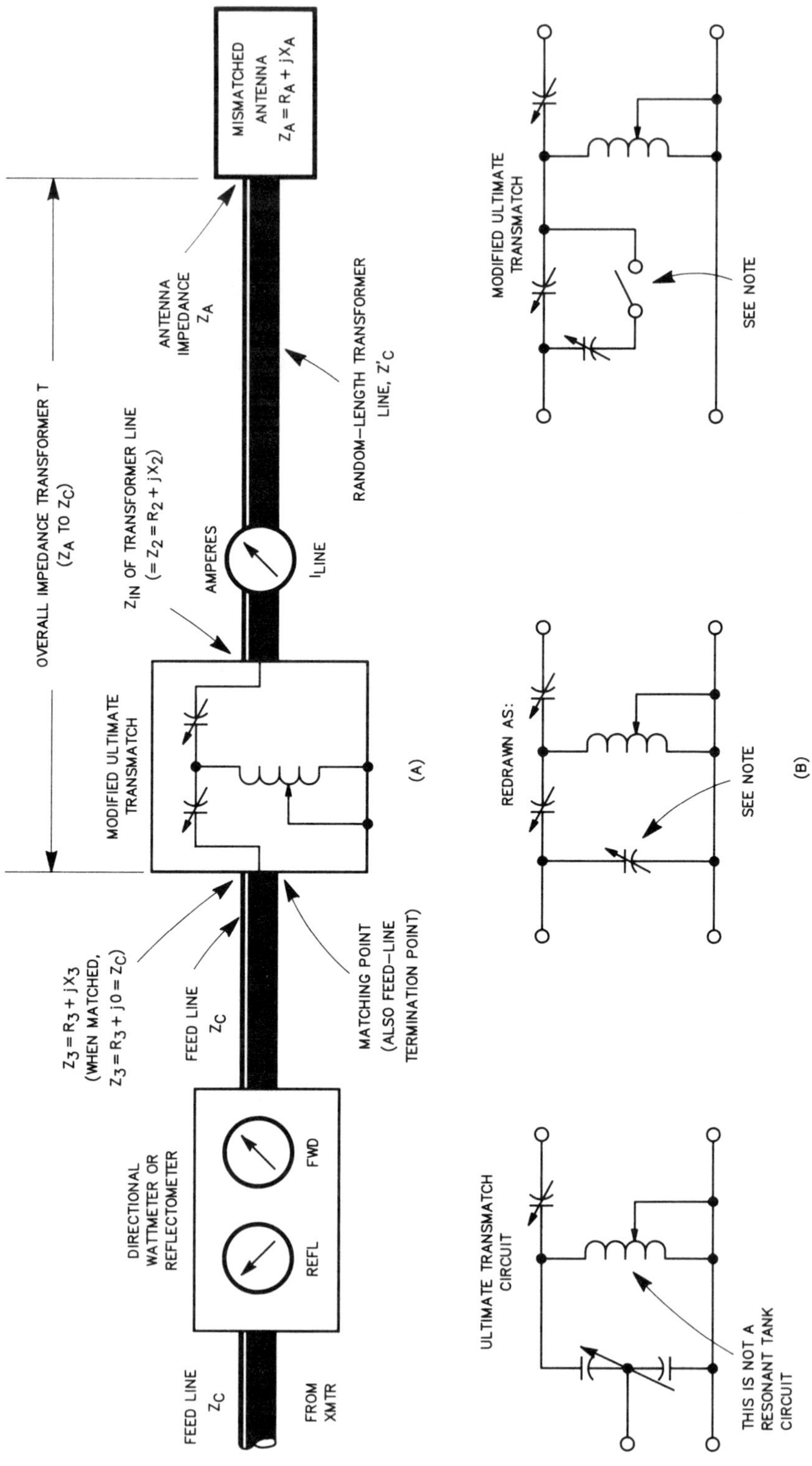
MISMATCHED ANTENNA
$Z_A = R_A + jX_A$
OVERALL IMPEDANCE TRANSFORMER T
(Z_A TO Z_C)
ANTENNA IMPEDANCE
Z_A
RANDOM-LENGTH TRANSFORMER LINE, Z'_C
Z_{IN} OF TRANSFORMER LINE
$(= Z_2 = R_2 + jX_2)$
AMPERES
I_{LINE}
MODIFIED ULTIMATE TRANSMATCH
(A)
$Z_3 = R_3 + jX_3$
(WHEN MATCHED,
$Z_3 = R_3 + j0 = Z_C$)
FEED LINE
Z_C
MATCHING POINT
(ALSO FEED-LINE
TERMINATION POINT)
DIRECTIONAL WATTMETER OR REFLECTOMETER
REFL
FWD
FEED LINE
Z_C
FROM XMTR
ULTIMATE TRANSMATCH CIRCUIT
THIS IS NOT A RESONANT TANK CIRCUIT
REDRAWN AS:
SEE NOTE
(B)
MODIFIED ULTIMATE TRANSMATCH
SEE NOTE

←Fig 7-1—Impedance matching with a matching network external to the transmitter. The impedance of the feed line, Z_C, is usually 50 or 75 Ω in most amateur applications, but the impedance of the transformer line, Z'_C, may be any value, and may be of open-wire or coaxial type. The matching network at A is a modified Ultimate Transmatch (*Ref 41*), as shown at B. NOTE: In the redrawn circuit at B (center) the shunt input capacitor is not needed to obtain a match, and the capacitor range may be increased by reconnecting as shown at the right. The story of how this change came about is told in Chapter 14, Sec 14.4.

reason why so many have been reluctant to operate with an SWR greater than 1:1.

The principle that maximum efficiency is obtained when a feed line is perfectly matched to an antenna—no reflections on the feed line, and a 1:1 SWR—is so well known it hardly needs repeating. Even so, it is recited in practically every good textbook on the subject. So it is important to appreciate that this book makes no statement which violates that principle, nor suggests any disagreement with it. The misconceptions to be clarified concerning lost reflected power stem simply from overuse, or misuse, of the perfect-match principle in its practical applications.

Ironically, textbooks may be a bit responsible for this misuse. While extolling the virtues of the perfect match, many authors fail to explain how much (or how little) one loses when the load is mismatched to the line, if compensated with a conjugate match at the line input. Those authors generally present the case for the ideal load match in terms of single-frequency operation, and practically ignore the unique, multifrequency operations of the radio amateur. We have frequency *bands*, not single, specific frequencies—and we want to operate anywhere within those bands.

Because the antenna impedance changes as we change frequency, relaxation of any feed-line-to-antenna matching restriction is necessary if we are to enjoy such operating freedom. Although many engineering textbooks discuss loss versus load mismatch, few texts discuss multifrequency antenna-matching situations where line-input matching usually exists. Consequently, an overly rigorous application of the perfect-load matching principle has unwittingly been thrust upon us by dozens of misleading statements appearing in various amateur journals.

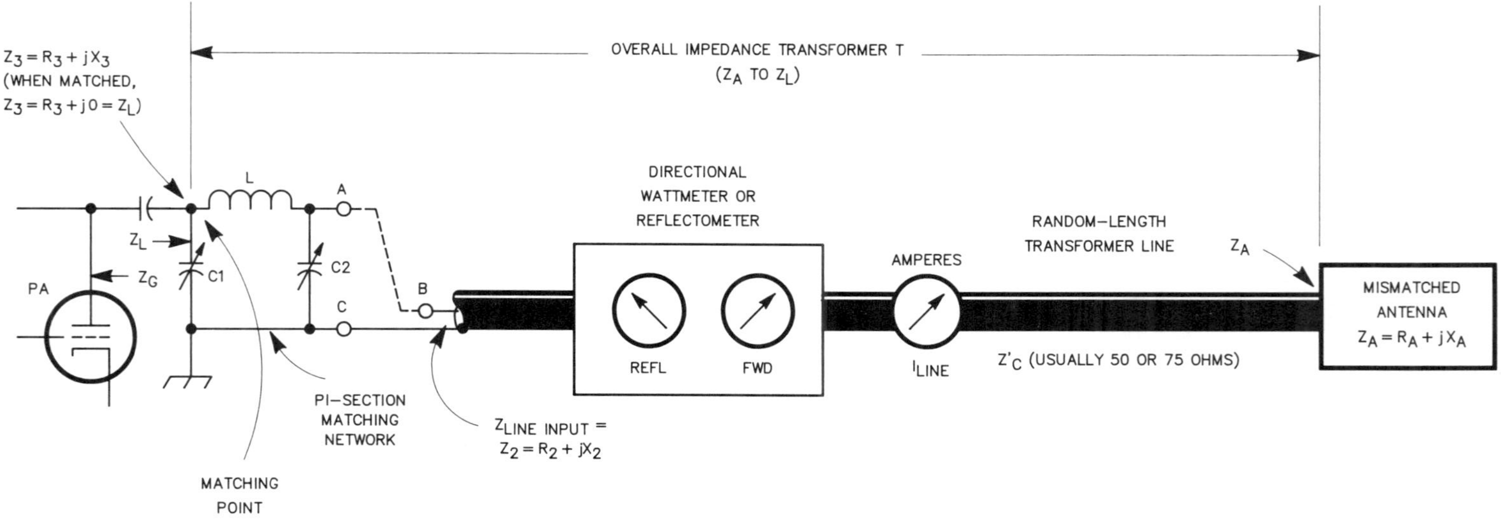

Fig 7-2— Impedance matching with the final-amplifier tank circuit of the transmitter. Z_L is defined as the optimum load impedance of the amplifier—the impedance into which it delivers its maximum available power. This impedance must not be confused with the internal impedance of the amplifier, Z_G. They are not the same.

Add to this a prevalent misconception of pi-network loading principles. Result? The "lost reflected power" syndrome and the mania for low SWR. Thus, providing guidance for the amateur concerning match quality and efficiency in his multifrequency operation is the primary purpose of this writing.

For example, Fig 6-1 graphs transmission loss in dB versus SWR for various values of line attenuation. The graph shows that maximum efficiency is indeed obtained with a perfect match. On the other hand, it also provides dramatic evidence that when using low-loss feed line that is matched at the line input, the difference between having either a perfect load match or a moderately high SWR is insignificant in terms of power transferred to the antenna. In other words, through the use of line-input matching, the antenna accepts the maximum available power from the transmitter, even when the feed line and antenna are mismatched, and with the antenna off resonance. Matching at the input terminals of the feed line allows us to tolerate this load mismatch because the reflection loss caused by the mismatch is compensated by the *reflection gain* provided by the input match. Consequently, the transmitter is properly coupled to its desired load impedance and the reflected power is conserved. It is *system resonance* that underlies these intrinsic characteristics of input matching involving the principles of the conjugate match. And *system resonance* compensates for the effect of the off-resonant condition of the antenna as we move around within a given frequency band.

For the myth believer who is still unconvinced of the wondrous compensating powers of line-input matching, let me quote Everitt on the Maximum Power-Transfer Theorem from classical network theory (*Ref 17, p 49; Ref 69*).[1] The application to feed-line antenna matching is indicated in parentheses: "The maximum power will be absorbed by one network (the antenna) from another (the feed line) joined to it at two terminals, when the impedance of the receiving network (the antenna) is varied, if the impedances of the two networks at the junction are conjugates of each other." Everitt then presents the proof of the theorem.

The expressions in Eqs 6-1 and 6-2, which are plotted in Fig 6-1, illustrate this theorem for the case where a feed line is the sending

1 All references appear in Chapter 24.

network. These expressions show that when the networks are conjugately matched, meaning when the transmitter is tuned to the line, (1) there is no loss at the terminals joining the networks, (2) there is no loss in the sending network, the feed line, when that network attenuation is zero, and (3) if the sending network has attenuation other than zero, the transmission loss results only from the line attenuation. For further discussion on the theorem, I invite you to review Chapters 2, 4, 5, and 19.

The way that conjugate matching and reflection gain are achieved is illustrated in Chapter 4, using the stub form of matching to develop the understanding of the wave action. The stub form was used for the illustration because it is easy to visualize. But since it's not a practical form to use where we make frequent changes in frequency, let's see how conjugate matching is performed using devices which are easily adjusted to perform at any desired frequency. Such devices are the Transmatch, a T network, L, pi, and other types of networks. I will show that these devices can perform the matching function at the *input* of a feed line, and that the feed line can be of *any* length instead of requiring it to be ½ λ. I will also show how, in some instances, matching at the input of the feed line is performed by the transmitter tank circuit itself.

Sec 7.2 Matching at the Line Input

At this point you may ask, "Why match at the *input* of the feed line?" The answer is that without matching at the input, we have very little operating flexibility relative to changing frequency within a band. In the absence of a line-matching network we are restricted to operating in a narrow portion of a band (especially on 80 meters) unless effective measures for broadbanding the antenna itself have been taken. We are restricted because, as we deviate from the resonant frequency of the antenna, the resulting increase in the impedance mismatch between the feed line and the antenna is transferred back to the input of the feed line as an increased mismatch between the transmitter and the feed line. As a result, the load impedance as seen by the transmitter varies beyond acceptable limits; the transmitter fails to load properly, and it can be damaged by overloading or by arc-overs caused by underloading.

These phenomena (plus an unawareness of the remarkable performance capabilities of line-input matching) are largely responsible for the traditional low-SWR mania. On the other hand, simple impedance matching at the input of the feed line provides stupendous improvement in operating flexibility; the matching network compensates for the impedance changes at the line input terminals, and provides the correct load impedance for the transmitter at whatever frequency we select within an entire band. The correct load impedance for the transmitter is obtained by simply adjusting the network, conveniently located at the operating position.

So the next question is, "Why not broadband the antenna and avoid having to retune a matching device when changing frequency?" The answer is that we can, but only to a limited degree. This is because, for example, broadbanding techniques which would permit coupling the average amateur transmitter directly into the feed line over the entire 80-meter band (with no adjustments other than retuning the transmitter) are not practical in the average amateur situation. This includes the coaxial dipole (sometimes called a Double-Bazooka) which, contrary to prevalent opinion, fails to deliver any *significant* bandwidth improvement over a simple dipole when it is fed with the usual 50-Ω feed line.

While the above statement may appear a bit incredible, I wrote a revealing analysis of that antenna and published it in *Ham Radio* (*Ref 62*). (A condensed version of that analysis appears here as Chapter 18.) However, there is a recent development worthy of your attention. In my analysis I presented information concerning the use of a feed line having an impedance Z_C higher than the usual 50 or 75 Ω (such as 144 Ω) for increasing the bandwidth with stubs in the coaxial dipole. Using my analysis, Frank Witt, AI1H, has engineered a novel transmission-line impedance transformer that achieves the required effect of increasing the feed-line impedance. With Witt's transformer, the 2:1 SWR bandwidth of an 80-meter coaxial-stub dipole extends well beyond the range of 3.5 to 4.0 MHz (*Refs 122, 123 and 124*). Witt has presented a practical way of broadbanding the dipole that deserves serious attention.

An explanation of how either a final tank network or an external line-matching network performs conjugate matching from the viewpoint of reflections is really a continuation of Chapter 4. However, to dramatize both the availability and the advantages of matching at the input of the feed line, we digressed in Chapters 5 and 6 to highlight some of the wrong and right reasons for using low SWR. These reasons show that, in the mania for obtaining low SWR on the feed line, we have often put misplaced emphasis on matching at the wrong end of the line. As stated earlier, the principles of wave reflection found in stub matching are the same as in other matching schemes, such as series ¼-λ transformers, L, T, pi networks, etc. As some of the concepts I discuss here were presented in detail in Chapter 4, you may wish to refer to them again.

Sec 7.3 The Intermediate Role of the ¼-λ Transformer

How many amateurs remember the Johnson Q Match? And how many have used a ¼-λ section of 70-Ω transmission line to match a 100-Ω resistive load to a 50-Ω feed line? Both are examples of series ¼-λ line transformers. In addition to requiring an electrical length of 90°, impedance matching is accomplished in these transmission-line transformers because of a specific relationship between their characteristic impedance Z_C, their input impedance Z_I, and their output load impedance Z_L. To perform the matching, the ratio of Z_C to Z_I must be the inverse of the ratio Z_C to Z_L, that is

$$\frac{Z_I}{Z_C} = \frac{Z_C}{Z_L} \qquad \text{(Eq 7-1)}$$

In other words, the impedance required of the transformer is the geometric mean value of the two impedances being matched, stated simply by the well-known expression

$$Z_C = \sqrt{Z_I \times Z_L} \qquad \text{(Eq 7-2)}$$

Both the impedance and the length of the ¼-λ line transformer play important roles in clarifying the principles underlying all forms of line-matching networks. Since these roles are not generally well understood, let's examine them.

Chapter 4 shows that reflections play a necessary role in impedance-matching operations. We saw that impedance matching is obtained by canceling the reflections from a load mismatch by wave interference. The interference is set up by new, separate reflections generated by a separate mismatch introduced at a desired matching point. The mismatch introduced at the matching point is tailored to complement the load mismatch, so that the new reflection has the same magnitude and opposite phase (at the matching point) as the reflection generated by the load mismatch. The reflection generated by this complementary mismatch is called either a complementary, or a canceling reflection. In stub matching, the complementary mismatch is introduced by the stub. While investigating complementary reflections generated by stubs, we observed the wave action through which the ¼-λ transformer performs the matching (see Table 4-1 and Fig 4-2D).

In Chapter 4 we analyzed the effects of using various stub and transformer lengths to match a resistive load to various feed lines having different values of impedance. We found that when the feed-line impedance is equal to Z_I (relative to transformer impedance Z_C and load impedance Z_L in Eqs 7-1 and 7-2), the transformer length became 90°, and the stub length became zero.

In this case the complementary reflection is generated by the mismatch appearing at the junction of the feed line and the input of the transformer. This mismatch results from the abrupt change in impedance (Z_I to Z_C) encountered by the forward wave as it propagates out of the feed line (Z_I) and into the transformer (Z_C). This mismatch is complementary to the resistive load mismatch in magnitude because the ratio between the feed line and transformer impedances (Z_I/Z_C) is the inverse of that appearing between the transformer and the load impedances (Z_C/Z_L), as in Eq 7-1. Thus, the reflections generated by both the load mismatch and the transformer-input mismatch are equal in *magnitude* (as required to obtain perfect cancellation), because both mismatches are equal in magnitude.

Because the input and output mismatches are physically separated by 90° (the length of the transformer), they are also complementary in relation to the *phase* of the reflections appearing at the matching point (also required for cancellation). This is because the wave reflected by the input mismatch has zero distance to travel relative to the matching point, but the 90° length of the transformer results in a travel of 180° for the wave reflected by the load mismatch—90° from the input to the load, plus the 90° return trip. Thus, the load-reflected wave arrives at the matching point with a 180° phase difference relative to the input-reflected wave. We now have two complementary reflected waves—equal in magnitude but opposite in phase at the matching point. Consequently the two waves mutually cancel, resulting in total re-reflection of both waves into the transformer to propagate in the forward direction, as explained in the final paragraphs of Chapter 4. The voltage and current components of both re-reflected waves are in phase with their corresponding components of the source wave. Hence, an impedance match appears at the input terminals of the transformer, all of the power reflected from the load mismatch which reaches the input of the transformer is again on its way to the load, and no reflected wave appears on the feed line.

This ¼-λ line-transformer matching action deserves serious study. Why? Because it provides an intermediate step in understanding how matching can be achieved at the input of a line transformer of *any random length*, and which may have *any impedance* for its terminating load, such as the complex impedance $Z_A = R_A + jX_A$ of a mismatched, off-resonance antenna. And this line transformer is none other than the feed line, so many strive to operate with no reflections by simply restricting its load to a matched, resonant antenna!

Sec 7.4 Input Line-Matching Networks

Now let's examine external line-matching networks from the viewpoint of impedance matching with reflections. Referring to Fig 4-2D, we replace the 90° transformer (T) with the combination of an adjustable matching network and a line of *random* length to connect the mismatched load (antenna) to the network. This arrangement is shown in Fig 7-1A. The line connecting the matching network to the antenna will now be called the "transformer line." The line connecting the transmitter to the matching network is the "feed line," as in Chapter 4. The matching point is defined as the junction of the feed line and the input of the network.

In a manner which is explained later, the transformer line (which can have *any* value of impedance Z'_C) transforms the complex antenna impedance $Z_A = R_A + jX_A$, to a second impedance $Z2 = R2 + jX2$ at the input of the transformer line. The network then transforms Z2 to a third impedance $Z3 = R3 + jX3$ at the matching point (the network input). When the network is correctly adjusted, Z3 is a pure resistance R3, equal to the feed-line impedance Z_C. In mathematical terms, $Z3 = R3 + j0 = Z_C$. Thus, the antenna impedance Z_A is matched to the feed-line impedance Z_C, which is a proper load for the transmitter (*Ref 17, p 243; Ref 69*). Without the matching network, the transmitter load would be impedance Z2, which could deviate far beyond the range of matching capability for most transmitters. However, by adding the network, we obtain the impedance match by simply adjusting the network to transform R2 to equal the feed-line impedance Z_C at the matching point, and to cancel any reactance X2 appearing at the input of the transformer-line to zero at the matching point.

Let's now examine more closely the transformation of impedance

$Z2$, that which appears at the input of the transformer line. The variations of resistance $R2$ and reactance $X2$ of impedance $Z2$ are dependent on three different factors: the antenna impedance Z_A, and both the length and the impedance Z'_C of the transformer line. From Chapter 4 we know that, for given antenna and transformer-line impedances, a length of line can be found that will make $R2 = Z_C$, but will also yield reactance $X2$. The reactance $X2$ requires canceling to obtain a match, for example with a stub, as shown in Fig 4-2A or 4-2B. Another length of line can be found that will make $X2$ become zero, but now $R2$ will not equal Z_C (still no match). *There is no length* that will yield $Z2 = R2 + j0 = Z_C$, unless Z_A equals Z'_C.

This situation illustrates the typical endless cat-and-mouse game we play in trying to load the transmitter by changing line lengths. A fanciful solution to this problem would be a line of *variable* length, plus a device for dumping the unwanted reactance. So how can we possibly solve the problem using a fixed, random-length line? Because, as we now discover, the line-matching network is the star performer! We know from elementary transmission-line theory that a transmission line is made up of an infinite number of tiny, distributed, series inductances and shunt capacitances. So it should not be surprising that we can adjust the *electrical* length of a line of a given *physical* length by adding lumped inductances or capacitances. Indeed, by a selection and adjustment of reactances arranged in an appropriate L, T, or pi configuration, we can simulate a line having *any desired electrical length* without specifying any physical length whatever! (See *Refs 8 through 13; 19, p 115; 21; 22; 24; 30; 31; 41; 61* and *63*.)

When a matching network is adjusted to obtain a match between the antenna and feed-line impedances, it performs the following two feats. First, it creates the effect of stretching the *electrical* length of the transformer line to make it reach the matching point, so that the resistance $R2$ at the input of the *physical* transformer line is transformed to $R3 = Z_C$. And second, it introduces reactance $-X3$ to cancel reactance $+X3$ of the *stretched*-transformer line appearing at the matching point, in the same manner as a stub would perform in stub matching.

The introduction of reactance $-X3$ at the matching point provides the complementary mismatch which generates the canceling reflection having equal magnitude and opposite phase relative to the reflected wave arriving at the matching point from the mismatched antenna. The matching network has thus provided the proper *overall* transformer length to obtain both the required transformation of the resistance component from R_A to R3, and a canceling-phase relation between the load-reflected wave and the canceling reflected wave provided by the complementary mismatch of reactance –X3. Consequently, the network has also transformed a dream into reality by conjuring up the fanciful variable-length line and a means for dumping the unwanted reactance.

An ideal arrangement for observing the action while performing the tuning adjustments of the network includes an RF ammeter in the transformer line to indicate line current (a meter in each conductor if using a balanced, two-wire line), and a dual-meter reflectometer or wattmeter to indicate forward and reflected power simultaneously in the feed line. It is important that the reflectometer be adjusted initially to indicate zero reflected power with the feed line terminated in a pure resistance equal to the feed-line impedance. The tuning adjustments are complete when we obtain maximum current in the transformer line simultaneously with maximum forward and zero reflected power in the feed line. Of course, because of standing waves on the transformer line, we will obtain different values of line current depending on where along the line the ammeter is inserted. However, our only interest is in seeing *changes in relative current* to indicate when maximum network output occurs during the tuning adjustments. Hence, neither the absolute line current, nor where the meter is inserted, is important.

The simultaneous indication of maximum current in the transformer line and zero reflected power in the feed line is meaningful. This condition denotes four significant factors for comparing the wave actions involved in impedance matching with a line-matching network versus matching with the ¼-λ line transformer described earlier. First, proper network adjustment establishes the complementary mismatch between the feed-line termination and the input of the network which produces the canceling reflection at the matching point. Second, the canceling reflection at the matching point is equal in magnitude and opposite in phase relative to the load-mismatch reflection. Third, the canceling reflection and the load-mismatch reflection nullify each other at the matching point. Consequently, a purely resistive impedance equal to the feed-line impedance Z_C appears at the input terminals of the network, while reflections and a standing wave remain on the transformer line. And fourth, observing the transformer-line current rising to maximum while the reflected power in the feed line drops to zero provides visual evidence that the power reflected from the load mismatch is indeed re-reflected by the complementary mismatch at the matching point.

Incidentally, it is a good practice to have an ammeter permanently connected in the transformer line. Here's why. If your matching network effectively comprises more than one L section, you can generally obtain a match (zero reflected power in the feed line) with several different combinations of network L and C tuning. However, minimum network loss (which corresponds to highest maximum transformer-line current) usually occurs while using the maximum C and minimum L at which a match can be obtained. Monitoring the transformer-line current during tune-up lets you select the L-C combination which yields the highest output line current. To ensure quick resetability of the network, and to minimize on-the-air tune-up time, L and C settings for the best combination should be logged whenever the network is tuned to a new frequency. The transmitter should be tuned initially into a dummy load, then the network tuned into the antenna *using the lowest power* at which the reflectometer provides a satisfactory indication.

Sec 7.5 Tank-Circuit Line Matching

Let's now examine the case where the final amplifier tank circuit of Fig 7-2 performs the line-matching function. To differentiate from the external network, I will call this network the "tank network." Comparing Figs 7-1 and 7-2, we see that, in general, the transformations of impedance from Z_A to Z3 are identical whether using the tank network or the external network. The principal difference is the range of the transformation that can be performed by the two networks. With the external network, impedance Z2 is transformed to a value Z3, matching the feed-line impedance Z_C. When the tank network is used alone, as shown in Fig 7-2, impedance Z2 is transformed to directly match the load impedance Z_L of the final amplifier of the transmitter.

Now a requirement for an amplifier (tube or transistor) to deliver its maximum available power into a load (the loaded tank circuit) is for it to see an impedance which we call the *optimum load impedance*, Z_L (not the same as *internal* impedance, Z_G). In practice, impedance Z_L is usually resistive, so that $Z_L = R_L + j0$. Thus, the amplifier is properly loaded when it sees impedance $Z3 = R_L + j0$ (which is $R3 + j0$) looking into the tank network. The amplifier is underloaded with R3 greater than R_L, overloaded with R3 less than R_L, and it will have higher than normal dissipation if Z3 contains any reactance X3.

When using the tank network alone to perform the matching, as in Fig 7-2, the impedance value of Z3 is determined by two factors: the impedance value of Z2 loading the network, and the impedance transformation ratio of the network. The transformation ratio is somewhat variable (using the tuning and loading capacitors, C1 and C2), thus providing a range of impedance-matching capability. The matching range allows impedance Z2 to be any value which the network can transform to the impedance Z3 = $R_L + j0$ by adjusting the tuning and loading controls. Additional details on this point appear in Chapter 13.

Now we can summarize the principal operating conditions involving transformation of antenna impedances to the optimum load impedance for the output amplifier.

Case 1

The antenna impedance $Z_A = R_A + j0$ (resonant) is matched to the transformer line, and the tank network is used alone—no external matcher, Fig 7-2. Here, antenna impedance Z_A equals the impedance of the transformer line, Z'_C, the SWR on the line is 1:1, and impedance Z2 is equal to the line impedance. If the transformer-line impedance Z'_C is the commonly used 50 Ω, then the tank network yields the proper amplifier load $R_L + j0$ with the same tank adjustment settings as obtained when tuning up with a 50-Ω dummy load.

Case 2

The antenna is operated somewhat off resonance. Its impedance, $Z_A = R_A + jX_A$, yields a mismatch to the transformer line so as to transform Z_A to an impedance Z2 that is within the matching range of the tank network (for transformation to Z3 = $R_L + j0$). Again, the tank network is used solo.

Case 3

The antenna is operated off resonance, beyond the matching range of the tank network. Here, the tank network is used in conjunction with an external network, as in Fig 7-1. An impedance Z3 = $R_L + j0$ cannot be obtained with the tank network alone, and the amplifier would be either underloaded, overloaded, or reactively loaded. We remedy this situation by inserting an external line-matching network, which transforms impedance Z2 to an impedance Z3 that is within the range the tank network can transform to $R_L + j0$.

Case 1 needs no comment, so let's refer to Fig 7-2 and examine the action in the tank network as it performs the matching function under the conditions of Case 2. As described in Chapter 4, wave interference establishes an open circuit to reflected waves at the matching point. This interference exists between the wave reflected by the mismatched termination on the line and the wave generated by the complementary mismatch at the matching point in the matching network. Recall that the open circuit so established causes total re-reflection of the reflected wave returning toward the generator.

In the tank network, the open circuit to the reflected wave is established at the tank-network input terminals, fed by the plate of the amplifier tube. Consequently, waves reflected from a mismatched antenna, those causing impedance Z2 to deviate from Z'_C, enter the tank network at its output terminals and become totally re-reflected on arrival at the open-circuited input of the tank. When the network is tuned to resonance, voltage and current components of the reflected wave are re-reflected in phase with the corresponding source-wave components emanating from the amplifier tube. Thus, the amplifier sees a resistive load $Z3 = R3 + j0 = Z_L$ at the input to the tank circuit, and the reflected power reaching the input adds to the power from the amplifier. This is why a directional wattmeter indicates a higher forward power than the amplifier is actually delivering when reflections are present on the transformer line.

Optimum amplifier loading is obtained by adjusting the capacitance of the loading-control capacitor C2 to the value that causes the network to transform R2 to $R3 = R_L$. When R2 changes, following a change in antenna impedance Z_A, C2 can then be varied to modify the network transformation ratio. This transforms the new value of R2 to $R3 = R_L$. The plate tuning capacitor, C1, is then readjusted to return the network to resonance and we again have optimum amplifier loading. The range of R2 values which can be transformed to $R3 = R_L$ by varying C2 is determined by the design parameters of the network (*Refs 4, 63* and *64*).

Now, what happens when the line-input impedance Z2 that loads the network contains reactance X2? This reactance shifts the available range of capacitance provided by the loading-control capacitor, C2. This shift occurs because reactance X2 appears in parallel with the reactance of the loading capacitor. Consequently, for a proper setting of the loading capacitor where X2 = 0, a capacitive X2 increases the effective capacitance C2 (decreasing loading), while an inductive X2 decreases C2 (and increases loading). Thus, to obtain proper loading when X2 is present, the setting of the loading-control capacitor must be shifted from the X2 = 0 position to compensate for the additional line reactance X2. If R2 is within the matching range of the tank network, proper loading will be attainable. This is true as long as X2 doesn't exceed the value which the loading-control capacitor range can accommodate to maintain the value of C2 required to transform R2 to R3 = R_L.

If the line reactance X2 is too large for the loading capacitor to compensate, we have the conditions described in Case 3. However, there are simple remedies for this condition if R2 is within matching range of the tank network, but X2 is not. Here we can simply add a compensating reactance in series with the RF output (between points A and B in Fig 7-2). The added reactance may either cancel the line reactance X2, or merely bring it within the range the tank network can handle. On the other hand, if the resistance component $R2_P$ of the equivalent parallel circuit of impedance Z2 is within the matching range, the compensating reactance should be added in parallel across the RF output between points A and C.

Whether the compensating reactance should be capacitive or inductive can be determined experimentally by trying first one kind or the other to see whether loading is improved or worsened. The correct kind and value of compensating reactance has been found when proper loading can be obtained with a convenient setting of the loading-control capacitor. An excellent discussion of this subject by Grammer appears in *Ref 4, Part 3*. When using this matching technique with the tank network alone, an adjustment of the length of the transformer line can be of great assistance in obtaining values of impedance Z2 which are most favorable to the matching range of the network. This subject of impedance transformation versus line length is discussed in Chapter 11.

When operating under the conditions of Case 2 (Fig 7-2), tuning up into a dummy load requires special care; it can be troublesome since the actual load Z2 at the input of the transformer line often differs widely from that of the dummy load. If the tank network is first tuned into the dummy load and then switched to the transformer line-input impedance Z2, the tank network must be *retuned* to the new impedance Z2! Failure to retune to the actual operating impedance Z2 after tuning up into the dummy load results in an improper load impedance Z3 for the amplifier—it is no longer Z3 = $R_L + j0$ because the loading of the tank network has changed. Without retuning, the amplifier is then either underloaded, overloaded or reactively loaded.

In addition to the possibility of damaging the amplifier if operated in this mistuned condition, you also have less power available! Because of this mistuning, the amplifier *delivers less power* by the amount of the reflection loss resulting from the line mismatch. This power reduction is as shown in Fig 6-1, on the curve labeled "Reflection Loss Without Conjugate Match" and at the appropriate SWR ordinate. For example, if the line SWR is 3:1, the amplifier output will drop by 25% while it sees the improper load. On the other hand, retuning the network to the actual operating impedance Z2 establishes a line-input match that restores the proper amplifier load impedance $R_L + j0$ and the amplifier again delivers its maximum available power. For additional discussion on this point refer to Chapter 6, Sec 6.3.

"But," you ask, "how do we determine when proper loading is obtained, or that impedance Z2 is within the matching range of the network?" The answer is by simply completing a normal tuning and loading operation in which you obtain the same plate current, plate-current dip, and screen current readings that you do with the dummy load. However, tuning- and loading-control settings, and the relative power (output voltage) will generally differ from those obtained with the dummy load, depending on how much Z2 differs from 50 Ω. If normal plate current (and dip) cannot be obtained with any setting of the loading capacitor, Z2 is outside the matching range of the tank network, and we have conditions as defined in Case 3. If proper loading cannot be obtained with the simple series- or parallel-reactance compensating technique described earlier in this section, then an external network is required. However, I want to emphasize that *any* value of $Z2 = R2 + jX2$ can be transformed to a suitable value for loading the tank network by selecting the proper network configuration (*Refs 19, p 115; 30* and *61*).

The range of impedances Z2 that the tank network will transform to the value equal to the load impedance $Z_L = R_L + j0$ of the amplifier raises an interesting point concerning the tuning procedure for the external network. The usual practice is to tune the tank network with the dummy load, and then switch in the external network and antenna and tune for zero reflected power in the feed line (not the transformer line). This procedure should be followed if there is a filter in the feed line. However, in the absence of a filter, it is necessary to adjust the external network only for a line-input impedance Z3 which brings it within the matching range of the tank network. This can be a time-saving feature when changing frequencies during contest operation! If both tank and external networks are adjusted for optimum match at midrange of the intended frequency excursions, in most cases only the tank network needs retuning with changes in frequency. This is true providing the frequency excursions are within the range in which the external network yields a load impedance that the tank network can transform to $R_L + j0$.

Chapter 8

The Reality of Reflected Power

(Adapted from Technical Correspondence, *QST*, February 1983)

Sec 8.1 Introduction

From time to time, articles concerning transmission-line principles and techniques appear in various Amateur Radio journals, promoting the erroneous notion that reflected power is fictitious. (See *Refs 65, 67* and *97.*)[1] Unfortunately, many amateurs are misled by these authors who are misinformed on this subject, while specialists in wave and field propagation are dismayed at seeing this erroneous concept in print. The truth is, reflected power is as real as the power delivered to the line by the transmitter, or that radiated from the antenna.

While there are many analytical proofs that are unfailing, such proofs are often disbelieved by those who do not understand them. However, here's one easy laboratory proof with physical evidence that can't be denied: Energy reflected by a mismatched line termination can be entirely separated from the forward-traveling wave, and can then be dissipated in a temperature-calibrated resistor and accurately measured as I^2R heat.

If the power associated with the reflected energy were fictitious, no heat would be generated! Within the accuracy of the measuring equipment, the power measured as heat in the resistor equals the reflected energy calculated from the known relationship between the impedances of the mismatched load and the line. The power measured in the load equals the source power minus the reflected power, which provides the proof we're looking for. I have performed these measurements many times. (Also see Chapter 3.)

---

1 All references appear in Chapter 24.

A semantic problem with the term "power flow" also fuels the erroneous belief that reflected power is fictitious. This brings us to the question, "Does power flow?" To help us understand the answer, let's examine an analogy that involves current.

When we talk about "current flow," we take the meaning for granted. However, does *current* really flow? The basic electricity sections of engineering textbooks (also *The ARRL Handbook*) say that current does not flow—*charge* flows. Current is defined as the quantity of charge flowing past a point per unit time. However, once we leave basic electricity and move on to circuit analysis, the term "current flow" is used almost exclusively—and yet we know exactly what is meant.

The same problem exists with the term "power flow." Engineering textbooks define power as the "quantity of energy passing a point per unit time." Thus, power does not flow—*energy* flows. However, except when reciting the definition of power, textbooks and journals on wave propagation use the term "power flow" almost exclusively, with only an occasional use of "energy flow." As with "current flow," we know what is meant because of the common usage which generally overshadows the strict definition.

Apparently unaware of the common usage, Woods rejects the use of "power flow," and has distorted its meaning in an attempt to show not only that reflected power is fictitious, but also to challenge proven textbook principles of transmission-line phenomena (*Refs 67, 97*). His definitions of power and energy are correct, but the remaining portions of his articles are not. In addition, Woods' criticism of my statements published in *QST* (repeated here in Chapter 3) concerning directional wattmeters would have us believe that directional wattmeters don't even prove the physical existence of waves on the transmission line (*Ref 67*).

Walter Johnson, the author of the graduate-level text book, *Transmission Lines and Networks* (*Ref 18*), gives us the true facts. He was the chairman of the electrical engineering department at Princeton University when he authored his book. Johnson derived equations involving forward and reflected waves of voltage and current, equations that provide means for calculating both forward and reflected power from values of voltage and current obtained from specific points in the standing waves. These equations are the basis for understanding the operation of all directional wattmeters, and are repeated here as Eqs 3-4 through 3-9. In his lack of understanding, Woods has called Eq 3-9 "fallacious."

Another reason many amateurs (including Woods) believe reflected waves contain no power is because they misinterpret certain characteristics of standing waves (*Ref 97*). First, let me state the facts: Because the voltage and current components of the forward wave are always in phase, and those of the reflected wave are always 180° out of phase, the resulting standing-wave plots of voltage and current are always separated on the transmission line by a *distance, not phase*, of 90°, as shown in Fig 9-3. (Also see Chapter 3.) Line voltage, measured across the line, is the phasor sum of the forward and reflected voltages. Line current, measured in series with the line, is the phasor sum of the forward and reflected currents.

The 90° distance between the voltage and current standing-wave plots conveys no phase information, but at every point along the line where either the voltage or current standing wave is minimum, the phase angle between *line* voltage and *line* current is exactly 0°. At points along the line between the standing wave minima, the phase angle may vary between 0° and plus or minus 90°, depending on the degree of mismatch. The phase angle reaches 90° only when the mismatch results in total reflection of power (an infinite SWR, as shown in Fig 9-3).

Sec 8.2 Standing Waves Versus Reflected Waves

Now for the classic misconception: The 90° distance between the voltage and current standing-wave plots has been misinterpreted as the phase angle between the voltage and current of the reflected wave. From this misinterpretation, at least two authors have drawn the erroneous conclusion that, because the cosine of 90° = 0, reflected waves contain no power! (*See Refs 68* and *97*.) It is the *standing wave* that contains no power, not the reflected wave. The results have been disastrous, because the erroneous conclusions of these authors were published in Amateur Radio journals that enjoy wide circulation.

Further proof that reflected power is real comes from the Poynting Vector Theorem. Consider an imaginary unit-cube with its sides parallel to the X, Y and Z axes, containing a quantity of energy, E. If this cube moves forward with a velocity v, the energy will be carried past a given point along the Z axis at the rate of Ev joules per second, which is equal to the power in watts per unit area XY.

In vector notation $E_X \times H_Y = P_Z$ (watts per unit area). E_X and H_Y are energy field vectors; E and H represent the magnitudes, and X and Y the mutually perpendicular directions of the electric and magnetic fields contained in the cube. The Poynting Vector P_Z represents the power passing the point in the Z direction, at right angles to the XY plane formed by the E and H field vectors. When a flow of energy encounters a change of impedance (a mismatch), the magnitudes of both the E and H fields change, causing a motor-generator effect on the circuit elements. As a result, the mismatch is converted to a virtual source generator (of the reflected wave) in series with a matched load (*Ref 69*).

According to Lenz's law on counter-electromotive force, in any energy not absorbed by (or transmitted past) the mismatch, the vector direction of either the E or the H field must reverse. Which field reverses depends on the character of the mismatch. When either the X or Y direction reverses, the Poynting Vector of the unabsorbed energy must also reverse to the –Z direction—launching the reflected wave. The energy now flowing in the –Z direction is a portion of that energy flowing earlier in the +Z direction, except its magnitude has changed and its direction reversed. An exact analogy is that of light energy reflected from a mirror.

Wave energy reflected from a mismatch can be totally separated from the energy traveling in the forward wave. After separation, the Poynting Vector, P_{-Z}, of the reflected wave is the cross product of the E and H field vectors of the reflected wave: $E_{\pm X} \times H_{\mp Y} = P_{-Z}$ watts per unit area, again proving that reflected power *is* real power.

Chapter 9

Standing Wave Development and Line Impedance

Sec 9.1 Introduction

So far in this book we have discussed the impedance-matching function of the antenna tuner, or Transmatch, from the viewpoint of wave reflections. The wave (or power) reflected by the mismatch at the junction of the feed line and antenna is returned to the antenna because of re-reflection by the antenna tuner. We observed that because of the impedance match achieved by the antenna tuner, power reflected by the mismatch at the feed line and antenna junction adds to the source power in the tuner. This addition enables all the reflected power to be returned to the antenna and radiated, with the exception of that lost because of line attenuation.

The next few chapters explore the background for understanding matching from the viewpoint of impedance. There, we will examine a concept that explains why impedance can exist in the absence of physical resistors, inductors, or capacitors. Once we understand this concept, we can apply it to a practical, mismatched transmission line. In doing so, we will learn why the impedance Z_I appearing at the input of the mismatched line no longer equals the line's characteristic impedance, Z_C, when reflected waves generated by the mismatched load terminating the line arrive back at the input.

However, to fully appreciate the discussions that follow, you must clearly understand that reflected power traveling rearward along a transmission line is *real power*, not reactive power, as some would have us believe. If you are among those who have been misled to believe that reflected power is reactive power, then I suggest you review Chapters

3 and 8. You can learn the truth from those chapters, and thus dispel for yourself the reactive-reflected-power myth. The myth has been perpetrated unwittingly by some writers who drew erroneous conclusions from a basic fact concerning standing waves (*Refs 67, 68, 97*).[1] That fact is this: Standing waves of voltage and current are always separated by 90° in distance along the line. This 90° distance has been misinterpreted to be a 90° phase difference between the reflected voltage and reflected current, which is not true. Those writers then drew the erroneous conclusion that what they thought was a 90° "phase difference" between reflected voltage and current meant that the reflected power is reactive. As explained in detail in Chapter 3, the phase difference between the reflected voltage and current is *always* 180°, *never* 90°, providing evidence that reflected power is real.

Other people have tried to persuade me that reflected power is not real power for another reason. They say that power cannot flow in both directions simultaneously on a transmission line for the same reason that water cannot flow in both directions in a pipe (*Ref 99*). This is a false argument. Using water-flow analogies to explain electrical circuits is fine if the analogies fit, but this one doesn't fit—water flow enclosed in a pipe is not a legitimate analogy for current flow on a transmission line.

The smooth surface of a pond, however, is a legitimate basis for an analogy. Consider this: A pebble dropped at the center of a pond causes ripples, or waves, to travel forward until they reach the edge of the pond. Unless the material of the pond edge totally absorbs the energy in the forward wave, reflected waves containing the unabsorbed energy will originate at the edge and travel back toward the point where the pebble was dropped. The resulting standing wave can then be observed on the surface of the water between the edge of the pond and the point where the pebble was dropped, as the forward and reflected waves pass through each other. Much more is said later concerning the formation of standing waves, and waves traveling in both directions on a transmission line was discussed in detail in Chapter 3.

Before we proceed to the background study of matching from the viewpoint of impedance, let's take a final, summarizing look at matching from the viewpoint of wave action and reflections. Keep in mind

---

1 All references appear in Chapter 24.

that the wave actions described in this summary are responsible for *all impedance matching operations*, regardless of the type of network used to perform it.

Whenever impedance matching is performed at any junction of transmission lines, components, or loads having different impedances, a wave reflection is set up from the mismatch at the junction. With whatever type of matching network is used, the matching is achieved because the network itself presents a complementary mismatch that generates another wave reflection that is complementary to that generated at the junction. These two complementary reflected waves are mirror images of each other, because they have the same magnitude but opposite phase.

The complementary nature of the waves generated by the network is obtained by selecting the appropriate values for the components in the network. When the two reflected waves of opposite phase join at the matching point in the network, they cancel each other. This cancellation creates a virtual open circuit to the rearward traveling waves and causes their total re-reflection—a complete reversal of their rearward direction. The waves then add, producing a resultant wave traveling in the forward direction in phase with the source wave.

Because the network mismatch generated a wave having the same magnitude but opposite phase relative to the wave generated at the junction mismatch, the voltage and current components of the new forward-traveling wave resulting from the re-reflection emerge in phase with the corresponding components of the source wave. Hence, the power in the re-reflected resultant wave adds directly to the source power. Because the power contained in the wave reflected from the junction mismatch is now a portion of the forward-traveling wave, it is evident that the power reflected by the junction mismatch is added to the source power at the matching point, and that no reflected power travels beyond the matching point in the rearward direction. It is for this reason that the matching device transfers the maximum available power from a source to a load of different impedance.

Sec 9.2 Introduction to Line Impedance

Now let's begin our in-depth background study of the impedances encountered on transmission lines. This study will lead to an understanding of the impedance that appears at the input terminals of the line carrying RF power to the antenna. We know that widely differing values of impedance appear at the input terminals with various combinations of line length and antenna impedance. It is these large excursions of line-input impedance relative to the preferred 50 Ω that make loading of the transceiver difficult or impossible, and often require an antenna tuner to accomplish a satisfactory loading. So in this study we'll want to see why these different impedance values appear. We will want to learn how reflected waves of voltage and current change the input impedance from its usual characteristic impedance, Z_C, of 50 or 75 Ω to any one of an infinite number of different possible values. I'll purposely omit any mention of antenna tuners for the present, but after we see why the input impedance changes (which makes the tuner necessary), I'll explain how the tuner transforms the line-input impedance to either 50 Ω or to an impedance that will satisfactorily load the transmitter. Once this is understood, the earlier paragraph on network matching from the wave reflection viewpoint will come into sharper focus.

First we'll consider the conventional arrangement in which an SWR indicator is inserted in the feed line near the transceiver, so that a short section of feed line connects the transceiver to the SWR indicator, and a longer section runs from the indicator to the antenna. See Fig 9-1. We'll assume that the transceiver is tuned to the resonant frequency of the antenna, and that the antenna resistance, R_A (the resonant impedance), is the same value as the characteristic impedance, Z_C, of the feed line. Under these conditions we can say that the antenna is matched to the feed line. We know that when the antenna is matched to the feed line, there are no reflections generated at the antenna end of the line, and the SWR indicator shows a reading of 1:1. But do we really know what this reading of 1:1 means?

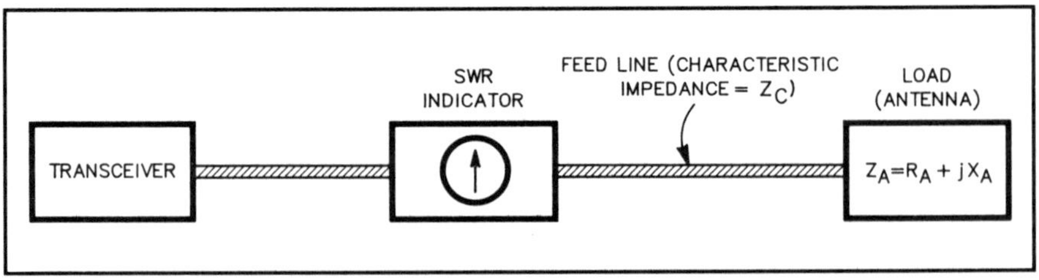

Fig 9-1—The conventional arrangement of connecting a transceiver to an antenna. The SWR indicator is inserted in the feed line near the transceiver. For the text discussion of input impedance, Z_C = 50 Ω.

In answering this question, let's first make sure we really understand what SWR is—what the numbers in the standing-wave *ratio* tell us. The line voltage and line current vary with distance along the line. The numbers in the ratio tell us the relative amount of variation in line voltage (and current) between the minimum and maximum values that appear. But to appreciate the significance of these variations, it is important that we also understand the reason why they occur. So at this point I'll digress to explain this very important concept.

Sec 9.3 Line-Voltage Variation and the Generation of Reflections

For ease in understanding, let's use numbers for the various impedances and voltages—we'll use 50 Ω for the line impedance, and a source-generator voltage of 1 volt. Hence, from Ohm's Law, the line current will be 0.02 ampere when the line is terminated with a matched load. We'll begin by continuing with the condition shown in Fig 9-1, in which the line is terminated with a matched load, except in this case we'll make the load a 50-Ω resistor instead of the antenna. This load gives rise to no reflection, because all the power carried by the line is dissipated as heat in the resistance of the load. And there is also no change in current, as the line current of 0.02 ampere becomes load current. This happens because there was no change in impedance in going from the line to the load. This is an important point to remember. In this case there is no variation in either the voltage or current along the line, so the ratio between the maximum and the minimum of voltage or current is obviously one. We represent this in ratio form as 1:1.

Let's now replace the matched 50-Ω load resistance with a 150-Ω resistance. This new 150-Ω resistance causes a 3:1 mismatch to the line, resulting from the 3:1 ratio between the 150-Ω load impedance and the 50-Ω line impedance. Because of motor-generator action in the load (which was explained in Chapter 3, but which I explain again shortly), the 3:1 mismatch causes the generation of a reflected potential of 0.5 volt. The reflected 0.5 volt adds to the 1 volt of the forward voltage wave at the load point and causes a new potential of 1.5 volts to appear across the load. Keep in mind that the voltage all along the line remains at 1 volt before the first voltage wave encounters the mismatch. As we will see, the line voltage changes after the forward voltage encounters the mismatch.

The constant forward line current of 0.02 ampere encountered an abrupt change of impedance in going from 50 Ω of line impedance to

150 Ω of load resistance. The current suddenly dropped to 0.01 ampere because of the increased resistance. The resulting change in the electromagnetic field caused by the change in current generated the new voltage—the 0.5 volt that is reflected. Thus, motor-generator action generates the reflected voltage.

Another way of visualizing the generation of the reflected voltage is from the viewpoint of a reflection generator connected in series with the load resistance. Because the changing current in the load resistor caused the field surrounding the resistor to change, a new voltage was generated within the resistor. This is because the resistor is a conductor immersed in the lines of force of the electromagnetic field surrounding it, which changed in the same way as if the resistor was a conductor in the rotating armature of a mechanical generator. The only difference is that in the mechanical generator, the field is stationary and the conductor moves, while in the case of the load resistor, the resistor (the conductor) is stationary and the field moves. The result is the same in either case—a voltage is generated.

We may therefore say that a mismatched load is a reflection generator. The voltage it generates is applied between the line terminals at the load end of the line in the same way the source generator applies the original source voltage between the terminals at the input of the line. Once the reflected voltage is applied to the load end of the line, it starts a new wave of voltage and current traveling rearward up the line in the same manner as the forward wave of voltage and current travel toward the load. The two waves slide past each other in opposite directions, as if the other didn't exist, except for one notable result—they produce the standing wave!

Before I go into detail on how the standing wave is formed, this is an opportune time for me to reintroduce rho (ρ), the magnitude of the reflection coefficient. This coefficient has an important relationship in the development of the standing wave. Rho first appeared in Eq 3-1. It is directly related to the SWR, as shown in Eqs 3-2 and 3-3. Reflection coefficient ρ defines the size, or magnitude of the reflected voltage (or current) relative to the forward voltage (or current). However, in practice the reflected voltage is used more often than current when specifying ρ. The assigning of 1 volt for the forward-wave voltage in the earlier paragraph was for the express purpose of simplifying the understanding of the reflection coefficient, because with a forward-wave voltage of 1, the voltage of the reflected wave automatically becomes the reflection coefficient. Thus, the reflected voltage of 0.5 in the example above signifies that the reflection coefficient magnitude, ρ, is also 0.5.

"Reflection coefficient" is a more specific term than SWR, because with the complete complex reflection coefficient we can specify and define impedance, while with SWR alone we cannot. However, we need more than just the *magnitude* of the reflection coefficient to specify impedance. We also need *phase,* or an angle, which SWR does not contain. Sec 1 of Chapter 3 introduced this phase angle as theta (θ). The angle of the reflection coefficient in the example above is $\theta = 0°$ relative to the forward voltage, meaning the reflected wave is exactly in phase with the forward wave. Hence, the 0.5 volt of reflected voltage adds to the 1.0 volt of forward-wave voltage to make the load voltage equal 1.5 volts. We saw in Chapter 3 that the nature of the load determined the magnitude $\rho = 0.5$, but because the resistance of the load was larger than the line impedance Z_C, the angle of the reflection coefficient is $\theta = 0°$.

If the load resistance had been 16.667 Ω (one third the value of the line impedance, instead of three times the value) the magnitude would remain at 0.5, but the angle would be shifted to $\theta = 180°$. In that case the polarity of the reflected voltage is reversed, so it is *subtracted* from the forward voltage, making the load voltage $1 - 0.5 = 0.5$ volt. The reason for this change is explained in Chapter 3, and is also discussed in Chapter 11, which covers the use of the Smith Chart.

We can now make general statements concerning the relationship between the reflection coefficient and the nature of the mismatch which determines it. These statements apply when the load impedance is purely resistive:

1) If the load current becomes less than the transmission-line current (because the load resistance R_L is greater than the line impedance, Z_C), the load voltage rises above the line voltage, and the angles of the voltage and current reflection coefficients are $\theta = 0°$ and 180°, respectively.

2) Conversely, if the load current becomes greater than the line current (because R_L is less than Z_C), the load voltage becomes less than the line voltage, and the angle of the voltage and current reflection coefficients are reversed, becoming 180° and 0°, respectively.

Angles other than 0° or 180° are obtained when the load contains reactance, but that's getting ahead of our story. We'll get into reactance concepts in detail later. Consequently, the concept contained in the above general statements form the basic reason why a standing wave is developed when reflected waves of voltage and current travel in the rearward direction along the line in conjunction with their respective forward-traveling components: While the phase relationship between the forward voltage and current is always 0°, the phase relationship between the reflected voltage and current is always 180°. The 180° phase difference is simply because of the opposite direction of wave travel of the forward and reflected wave. But as we will see, the standing wave is born because of this seemingly small detail!

Sec 9.4 Development of the Standing Wave

We are now ready to examine the development of the standing wave. Our ultimate goal for this section is learning why the input impedance of the line changes in the presence of reflections from a mismatched load termination, but just be patient. We are still laying the groundwork that is essential to a clear understanding of the reason why the impedance changes. As we proceed, it will become evident that the impedance appearing anywhere on the line is directly related to the relative amplitudes of the voltage and current standing waves.

In Chapter 3, I stated that standing waves are formed by the alternate cancellation and reinforcement of the forward and reflected waves as they slide past each other while traveling in opposite directions. As we proceed in this section, I will prove this statement true with an analytic approach. I'll use the vectors appearing on the W2DU Vector Graph of Fig 3-2. For convenience, the graph is reproduced in this chapter as Fig 9-2, and appears at a larger size on a foldout page near the back of the book. These vectors, more properly called "phasors," represent the values of the forward (incident) and reflected voltages and currents appearing on the transmission line, as described in Chapter 3 and as appearing on the graph itself.

As we now look into the development of standing-waves, let's use the same values of impedances, voltages, currents, and reflection coefficients as in the earlier example. Refer again to Fig 9-1, and remember that the load is a 150-Ω resistor. It should be clear that the standing wave originates at the 150-Ω mismatched load terminating the 50-Ω line. To review briefly an earlier discussion, when the forward voltage wave having the amplitude of 1 volt encountered the 150-Ω mismatch, the line current of 0.02 ampere abruptly decreased to 0.01 ampere. This change in current caused a corresponding change in the

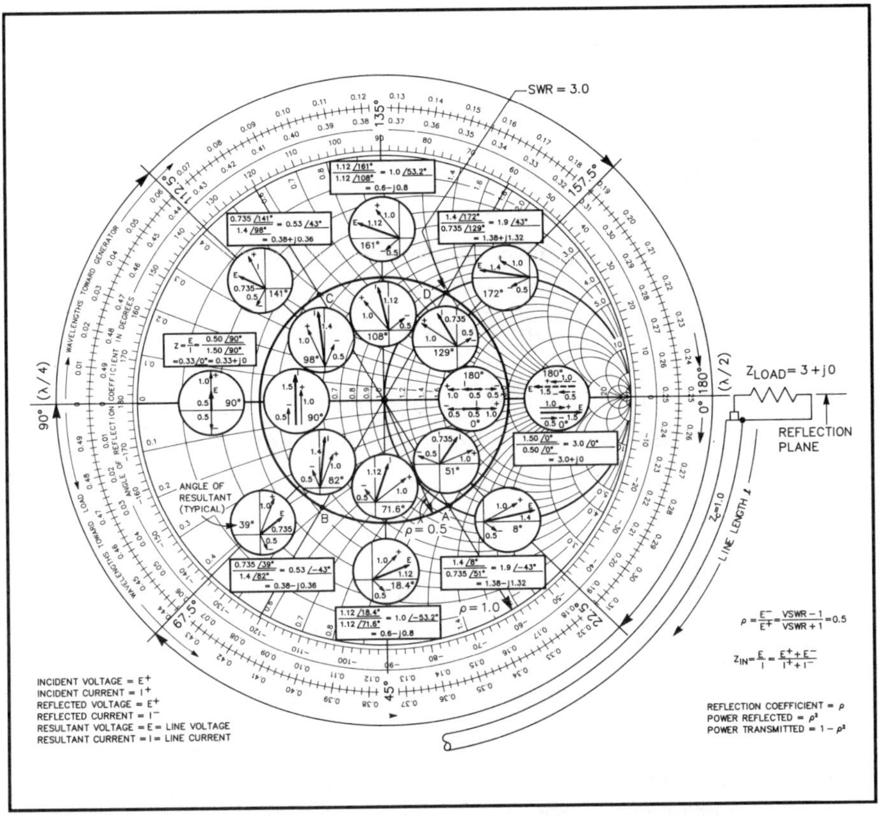

Fig 9-2—The W2DU Reflection Mechanics Vector Graph. This graph is presented as Fig 3-2 with a detailed description of its use in Chapter 3, but is reproduced here for convenient reference. A larger version appears on a foldout page near the back of this book.

electromagnetic field surrounding the load, which induced a new value of 0.5 volt appearing in series with the 150-Ω load resistance. Because the load resistance is greater than the line impedance, the new voltage is in phase with the 1-volt source potential, and adds to the line voltage to produce a total of 1.5 volts across the load resistance. This additional 0.5 volt applied to the terminals at the load end of the line launches a new voltage and current wave into line, which starts traveling in the reverse direction, toward the transceiver (generator).

Because phase plays a prominent role in our discussion of standing-wave development, I'll now digress for a brief discussion of phase, for the sake of clarity. As we all know, the voltage and current waves we are discussing here are sine waves that travel with a finite velocity.

Thus, the waves have a wavelength on the line that depends on the frequency of the wave, or the number of cycles completed in one second. One wavelength on the line is equal to one complete cycle of the sine wave, which is equal to 360° of phase. Therefore, ¼ λ equals 90°, and ½ λ equals 180°, etc. Haven't we all heard how a ½-λ line repeats the load impedance at its input, but reverses the phase, changing it 180°? We will soon understand why. While the magnitude of the reflection coefficient, ρ, is the ratio of the reflected voltage to the forward voltage, as stated earlier the angle θ of the voltage reflection coefficient is equal to the difference in *phase* between the forward and reflected waves.

Both the magnitude and phase of either a voltage or current wave may be represented by a vector line having a length proportional to its magnitude, and a direction representing the angle of the phase. These relationships, and those which follow, are clearly shown in the circular insets of "The W2DU Reflection-Mechanics Vector Graph," Fig 9-2. It may be helpful here to review Chapter 3 for further study of these concepts, which are essential to the understanding of the impedance change at the input terminals of the line when the line is terminated in a mismatch. The vectors of Fig 9-2 (and Fig 3-2) are displayed on a Smith Chart, but don't let that scare you—the explanation of how to interpret it is included. It is also explained in Chapter 10. The Smith Chart presentation actually enhances the appreciation of how the change in impedance occurs. It is important to note that the W2DU Vector Graph was drawn using normalized impedances, where $Z_C = 1$ instead of 50. The values of resistance, reactance and impedance appearing in the graph will conform to the present example simply by multiplying them by 50.

We established earlier for the above example that, at the mismatched termination point (the reflection plane in Fig 9-2), the forward voltage and current and the reflected voltage are all in phase, while the reflected current is 180° out of phase. If we let the reflection plane be a zero-phase reference point, then the forward voltage and current and the reflected voltage appearing there are all at zero phase, and the reflected current is at 180°. Hence, we can assign vector lines to represent the voltages and currents appearing at the zero-phase reference point as follows: Forward voltage and current vectors have length 1 at 0°, reflected voltage is length 0.5 at 0°, and reflected current is length 0.5 at 180°.

If we now move to different points along the line, we see from the vectors shown in the successive insets on the graph that the *magnitudes* of the forward and reflected voltage and current vectors remain constant, but the *relative phases* of all the voltage and current components change. These changes cause their vectors to rotate correspondingly, degree for degree, with changes in position along the line. As we move toward the generator, the phase of the forward wave leads, causing the corresponding vectors to rotate counterclockwise. Conversely, the phase of the reflected wave lags as we move, causing their vectors to rotate clockwise. It is therefore evident that for every degree of motion along the line, the phase relationship between the forward and reflected wave changes two degrees. And therefore, by definition, the angle θ of the reflection coefficient also changes two degrees for every degree of motion on the line.

It is now evident, as shown in the vector plots, that if we move along the line from the 0° reference point to a point 90° toward the generator, the phase of the forward voltage is +90°, while that of the reflected voltage is –90°, for a resulting phase difference of 180°. Likewise the phase of the forward and reflected current wave has also changed by the same amount. However, while the phase of the forward current changed from 0° to +90°, the phase of the reflected current changed from 180° to +90°, which means that the reflected current and the forward current are now in phase at the 90° point on the line. These phase relationships can all be observed in the vector plots at the 0° and 90° points appearing in Fig 9-2.

These new phase relationships at the 90° point on the line are of great significance in the development of the standing wave and ultimately the line-input impedance. Note that the forward and reflected voltage waves are in phase at the zero-phase reference point (where $\rho = 0.5$ at 0°), and the voltages there add vectorially to form a maximum voltage of 1.5. However, at the 90° point (where $\rho = 0.5$ at 180°), the forward and reflected voltage waves are 180° *out of phase*, so by vectorial addition the 0.5 volt of the reflected wave subtracts from the 1 volt of the forward wave, forming a minimum line voltage of 0.5 volt. The maximum line voltage is thus 1.5, and the minimum, at a point 90° away, is 0.5 volt. The voltage standing-wave ratio is therefore 1.5 to 0.5, or 3:1.

Conversely, the current waves behave in the opposite manner. At the zero-phase reference point, the out-of-phase reflected current adds vectorially to the line current to *subtract* from the line current, resulting in a minimum load current of 0.01 ampere (one half of the 0.02 ampere original line current). However, at the 90° point on the line, the reflected current is in phase with the forward current, and adding. This raises the line current 0.01 ampere higher than the original 0.02 ampere, for a maximum line-current value of 0.03 ampere. Thus, we see the current is 0.03 ampere at the maximum, and 0.01 ampere at the minimum. The current standing-wave ratio is 0.03/0.01 = 3, also 3:1.

If we continue on toward the generator another 90° to the 180° point, we will have completed 180° of travel along the line. Then we will have returned to the equivalent of the starting point, where the relative phase relationships are the same as they were at the zero-phase reference point, with the exception that the phase of *all* vectors has changed 180°. This is shown by the positions of the vectors in the insets of Fig 9-2. Hence, we have witnessed the alternate reinforcement and cancellation of the forward and reflected waves through the use of vector representation of the waves.

Several general statements of great importance may now be made, which are verified by observing the relative positions of the vectors in Fig 9-2:

1) At any point on the line where a voltage maximum and a current minimum occur, the angle of the reflection coefficient θ is 0°.

2) Conversely, at any point on the line where a voltage minimum and a current maximum occur, the reflection coefficient angle $\theta = 180°$.

3) A voltage maximum and a current minimum always occur simultaneously at the same point, and vice versa.

4) Points where voltage maxima and current minima occur are always separated by 90° along the line from points where voltage minima and current maxima occur.

5) The SWR is the ratio of the maximum and minimum values of either the voltage or current.

These five statements may be interpreted graphically in Fig 9-2. Graphic illustrations of statements three through five appear in Fig 9-3, which is explained in more detail in Sec 9.6.

So far I have identified and explained the formation of the maximum and minimum points of the standing wave—the successive reinforcement and cancellation mentioned earlier. Now let's examine the smoothly changing voltage and current amplitudes appearing on the line between the maximum and minimum points—amplitudes along the line that form the standing wave.

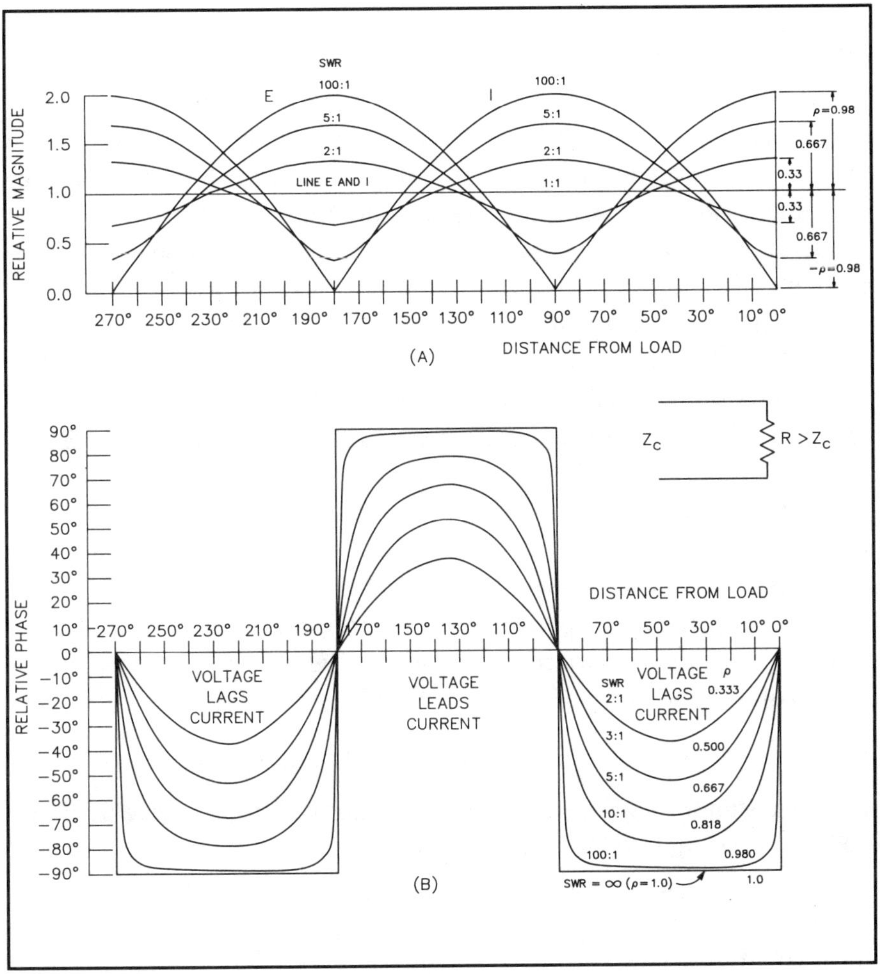

Fig 9-3—Plots of relative magnitude (A) and phase (B) between line voltage, E, and line current, I, versus position along a mismatched transmission line. Curves for different mismatch values are shown for resistive loads (no reactance) greater than Z_C. When the resistive load is less than Z_C, the E and I plots will exchange positions, and the phase will reverse.

The voltage standing wave is a pattern of the successive magnitudes of voltage resulting from the *vectorial* addition of the forward and reflected voltage waves at every point along the line. These two waves slide past each other while traveling in opposite directions. Likewise, the current standing wave is produced by the vectorial addition of the forward and reflected current waves all along the line. Earlier we saw that when the phase was 0° and 180° (at the points of maximum and minimum line voltage and current), we could determine the amplitude of the standing wave by simple addition or subtraction of the forward and reflected waves. However, to find the standing wave amplitude *between* these maximum and minimum points, we cannot simply add or subtract, because the respective reflected voltages and currents are no longer either exactly in or exactly out of phase. The complete vectorial, mathematical treatment is required for all these in-between values of relative phase along the line.

Accordingly, eight vectorial additions were calculated at 22.5-degree intervals along a ½-λ line having a 3:1 SWR. The results of these vectorial additions are shown as insets in Fig 9-2. The voltage E and current I resultants of the vectorial additions are shown by the wider, darker vector lines in the enclosed circles. Both the lengths (magnitudes) and the phase angles of these resultants are indicated in the insets. The magnitudes of the resultants are plotted as the voltage and current amplitudes for the corresponding 22.5-degree intervals along the line to supply the points for forming the resulting smooth, continuous curves of the voltage and current standing wave shown in Fig 3-3.

Sec 9.5 Determining Impedance On The Mismatched Line

We should now be aware that the vector resultants labeled voltage E and current I in Fig 9-2 represent both the magnitude and phase of the actual line voltage, E, and line current, I, appearing at the corresponding points on the line. The length of a vector represents the magnitude, and the angle of direction represents the phase. Voila! The values that these vectors represent are just what we need for determining the impedance appearing at those points on the line. Keep in mind that the line voltages and currents represented by these vectors result from the vectorial addition of the forward (incident) and reflected voltages and currents at each point as the waves slide past each other, traveling in opposite directions.

To determine the impedance at any point on the line, including the input terminals, we simply apply Ohm's Law, $Z = E/I$. For the desired point, we divide the line voltage, E, by the line current, I. Vectorial division is required here to obtain the impedance, because E and I are complex numbers, comprising both magnitude and phase angle, as shown in their vectorial presentation in Fig 9-2.

Of course the resulting impedance is also complex, containing capacitive reactance at points where the angle of reflection coefficient is negative, and inductive reactance where the angle is positive. At points where the angle of reflection coefficient is either 0° or 180°, the resulting impedance is a pure resistance, because, as we found earlier, at these points (and only at these points) the line voltage and current are in phase. This means that the impedance contains reactance everywhere on the line except where the angle of reflection coefficient is 0° and 180°. (The relationship between the angle of the reflection coefficient and reactance is explained in Chapter 10.) To perform the vector division we divide the magnitudes, E/I, to yield the magnitude, Z, of the impedance. We then subtract the phase angle of the current, I, from that of the voltage, E, to yield the phase angle phi (ϕ) of the impedance.

I'll illustrate with an example from Fig 9-2. Using the resultant vectors of voltage, E, and current, I, appearing where the line length is len° = 45°, and where the angle of the reflection coefficient is –90°, we see that the voltage vector has a length of 1.12 normalized units at angle 18.4°, and the current vector is also 1.12 units long, but at angle 71.6°. As stated above, to perform the vector division to obtain the impedance, we simply divide the magnitudes, 1.12/1.12 = 1, and subtract the angles, 18.4° –71.6° = –53.2°. Thus, the normalized line-input impedance in polar form at a point 45° from the 3-Ω load termination is z = 1 Ω at –53.2°. (This division is shown in the adjacent inset box in the graph.)

In a 50-Ω system we multiply the impedance by 50, making the denormalized impedance Z = 50 Ω at –53.2°. Note that these impedances are in polar form. To obtain the resistance and reactance values we need to convert the polar impedances to the rectangular, or complex form, r + *j*x, using the simple trigonometric relationships

$$r = z \cos \phi \qquad \text{(Eq 9-1)}$$

$$x = z \sin \phi \qquad \text{(Eq 9-2)}$$

where

z = the magnitude of the impedance
ϕ = the phase angle of the impedance
r = the resistance
x = the reactance

Solving these relationships for the example yields 1 cos(–53.2°) = 0.6 Ω resistance, and 1 sin(–53.2°) = –0.8 Ω reactance. Denormalizing these values to a 50-Ω system yields R = 30 Ω and X = –40 Ω. In other words, $Z = 30 - j40\ \Omega$, where the minus sign denotes a capacitive reactance. We are following the customary practice here in which symbols for normalized components are shown in lower case, while symbols for actual, nonnormalized components are capitalized.

Summarizing for the 50-Ω system, we see that with a load resistance of 150 Ω terminating a 50-Ω line and producing a 3:1 SWR, the input impedance of the line of length 45° (⅛ λ) is $Z = 30 - j40\ \Omega$. These values can be seen in the W2DU Antenna-Impedance Feed-Line Transformation Graph, Fig 10-1, which is described in Chapters 10 and 12.

Sec 9.6 Development of Reactance on Transmission Lines

From elementary ac theory we know that an impedance contains reactance when the voltage and current are not in phase. When the current lags the voltage, as when flowing through an inductor, the impedance looking into the inductor contains inductive reactance. Conversely, when the current leads the voltage, as when flowing through a capacitor, the impedance looking into the capacitor contains capacitive reactance. So at this point you're probably wondering, "How does a reactance originate when the load terminating a transmission line is a pure resistance?" You'll be happy to know that all the ingredients for obtaining the answer to this question have already been presented.

Recall the electrical configuration of two ac generators—the source generator and the reflection generator, one at each end of the transmission line. The line simply transfers the energy from each generator along the line simultaneously in opposite directions, but with a changing phase relationship with position along the line because of time delay.

Return to the example in Fig 9-2, at the point where the line length len° = 45°. Again observe that the voltage and current arriving from the source generator (the forward, or incident voltage and current—the vectors with the plus signs) are both at +45°, and the reflected voltage and current arriving from the reflection generator (the vectors with the minus signs) are at –45° and +135°, respectively. These are the phase relationships between the voltages and currents from each generator as they arrive at this 45° point on the line. The transmission line has simply carried the energies from each generator to the len° = 45° point on the line while accounting for the change in phase with the time of travel from each generator to this point. Hence, for purposes of analysis we can eliminate the line completely and simply connect the two generators in series, while assigning the same phase relationships between them that previously appeared between the forward and reflected waves at the len° = 45° point on the line.

With this phase relationship, we know the voltages and currents from the two generators connected in series are not in phase. And we know that reactance is developed whenever the voltage and current are not in phase. From the circular insets we see that the resultant phase of the voltages between the two generators is 18.4°, and the resultant phase of the currents is 71.6°, yielding an impedance phase angle of –53.2°, as explained earlier. This means the current leads the voltage by 53.2°, thus producing the 40 Ω of capacitive reactance at the len° = 45° point on the line.

In Fig 9-3 we have a rectangular graph showing the amplitudes of the line voltage and current, E and I (the SWR curves), and the phase angle ϕ between voltage and current. Curves are shown for four different values of SWR versus distance from the load. Hence, the impedance in polar form at any point on the line for these four values of SWR may be obtained by dividing the voltage by the current obtained from the amplitudes in the SWR curves, and adding the phase angle ϕ from the corresponding phase curve. The polar form may then be converted to the rectangular form $R + jX$, as explained in Sec 9.5.

There is one further point concerning impedance on the line. We know that in the presence of reflections, the line impedance is a pure resistance at points on the line where the angle of reflection coefficient $\theta = 0°$ and 180°. The point to note is that, even though the reactance is zero at these two points, the resistance at these points is never equal to the line impedance Z_C in the presence of reflections. The resistance at $\theta = 0°$ is always higher than Z_C, and equals $Z_C \times$ SWR, while the resistance at $\theta = 180°$ is always lower than Z_C, equaling $Z_C \div$ SWR. However, there are points on the line where the resistance, R, does equal Z_C, but reactance is always present at these points, because, as stated previously, reactance is present in the line impedance everywhere on the line except at $\theta = 0°$ and 180°. These points are covered again in Chapter 11 on the use of the Smith Chart.

Chapter 10

An Introduction to the Smith Chart

Sec 10.1 Introduction

In Chapter 9, with a fair degree of difficulty using vector arithmetic, we succeeded in calculating a few line-input impedances. We did this for only one SWR (3:1), and for only eight different line lengths (those appearing in Fig 9-2). However, in amateur operation we know we'll encounter hundreds of different input impedances as we use different antennas, different lengths of transmission line, different bands, etc. Wouldn't it be great if we could determine, at least roughly, what impedances we might encounter with these different configurations when matching the rig to the feed line? Without a computer and a suitable program to run, the impedance-solving problem is tedious at best. There is a practical solution, however—the Smith Chart! In spite of its apparent complexity at first sight, the Smith Chart provides the best solution with the most vivid presentation of changes in line-input impedance as the line length changes.

With the brief explanation of the Smith Chart that follows, you will discover that although it initially looks quite formidable, the chart isn't really all that difficult to use. And once you see how beautifully it demonstrates impedances on transmission lines, you'll find the time spent becoming acquainted with the Smith Chart is well justified.

The W2DU Vector Graph, Fig 9-2, is a basic Smith Chart which I modified by adding vectors representing forward (incident), reflected, and resultant voltages and currents. These added vectors form the basis for developing the standing wave and the impedance functions. (As mentioned in Chapter 9, what I earlier chose to call *vectors* should more correctly be called *phasors*.) The W2DU Antenna-Impedance Feed-Line Transformation Graph, Fig 10-1, is another modification of the Smith Chart designed to enhance the presentation of feed-line imped-

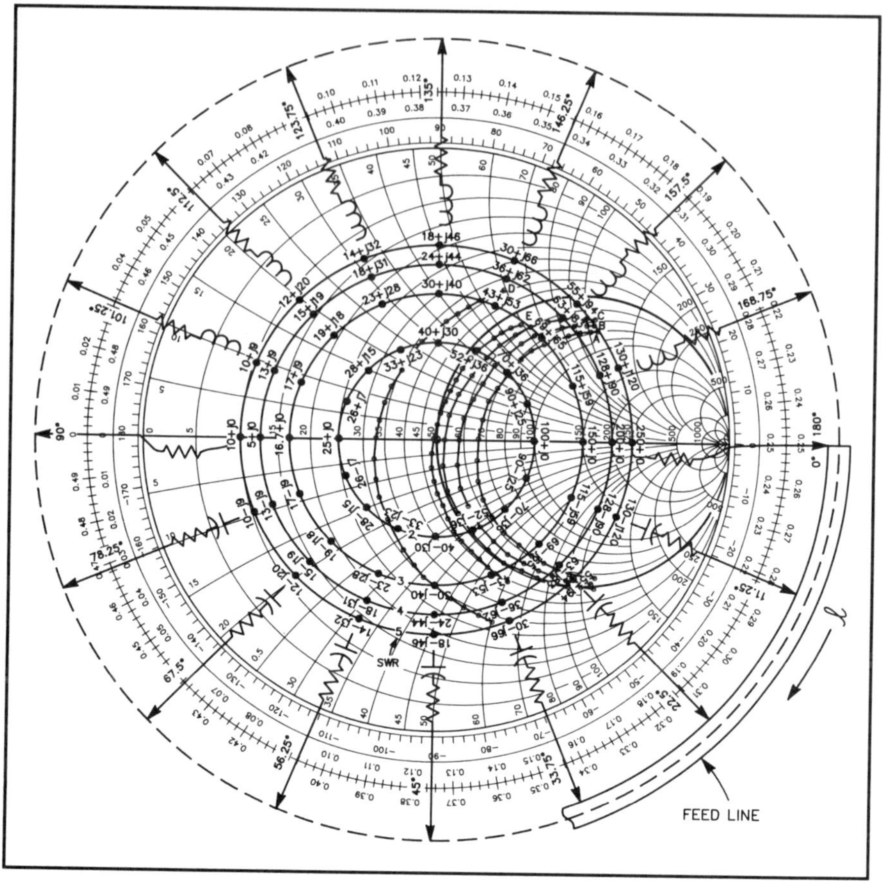

Fig 10-1—The W2DU Antenna-Impedance Feed-Line Transformation Graph. This is a 50-Ω chart. The curves show the impedance of various antennas versus length. Each dot represents a half-percent change from the resonant or center frequency. A larger version of this chart appears on a foldout page near the back of this book.
Curve A—Dipole in free space.
Curve B—Average of curves A and C.
Curve C—W2DU 80-meter dipole, 30 feet above ground, measured with a General Radio 1606-A RF impedance bridge.
Curve D—Vertical radiator over perfect ground (100 radials or more).
Curve E—Radiator of curve D with 16 ohms of ground-loss resistance (typical with 15 radials).

ance transformation. Fig 10-1 will help you understand the matching problem that exists when you attempt to tune and load your rig into a feed line when the SWR is not 1:1. This problem is discussed in detail in Chapter 12.

Fig 10-2 is a basic, normalized Smith Chart with a useful example

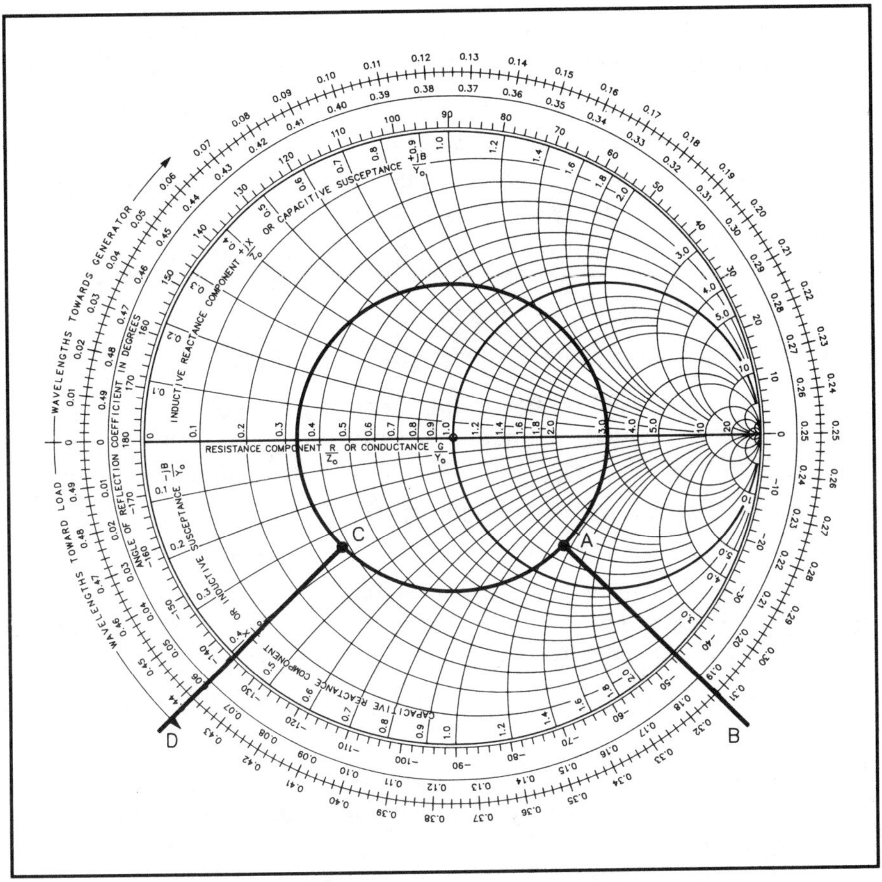

Fig 10-2—A basic, normalized Smith Chart with an example added.

drawn in. Normalizing is explained in Sec 4.2. For convenience, however, I'll repeat portions of the explanation here. Normalized impedances are obtained by dividing the values of all impedances by the system impedance, which is usually taken as the characteristic impedance of the transmission line, Z_C. Normalizing an impedance by dividing it by Z_C amounts to a change of scale such that the unit of impedance is 1 Ω, rather than Z_C Ω. For example, in a 50-Ω system, 150 Ω becomes 3 Ω. And 50 Ω becomes 1 Ω, which appears at the center of the Smith Chart. To convert back to the 50-Ω system, simply reverse the process and multiply the normalized values by 50. For example, the normalized impedance $z = 0.6 + j0.8$ becomes $Z = 30 + j40$.

To normalize a chart to the impedance of any specific line, all impedances on the chart are divided by the line impedance. On a normalized Smith Chart, all values of resistance and reactance are normalized values. Smith Charts are available in both normalized and nonnormalized forms. The normalized form of Fig 10-2 permits use with transmission lines having any and all characteristic impedances, and all impedance values entered or appearing on the chart are normalized to the line impedance. The nonnormalized form is printed with the actual impedances based on the selected line impedance.

Sec 10.2 The Basic Smith Chart

Refer again to Fig 10-2. The basic Smith Chart has two sets of circular scales, one set representing constant resistances, and the other representing constant reactances. The diameter line running horizontally is called the resistance axis, or "real" axis, for reasons that will soon become apparent. The resistance axis represents resistance components, and is graduated in values of normalized resistance. The left of the resistance axis represents zero resistance, and the right represents infinite resistance. The center point of the axis on a normalized chart represents 1 Ω. Figs 9-2 and 10-2 are normalized charts, but the W2DU Antenna-Impedance Feed-Line Transformation Graph, Fig 10-1, is a 50-Ω chart. Thus, the center of the resistance axis on Fig 10-1 represents 50 Ω. The 50-Ω Smith Chart is explained in detail in Chapter 12.

The constant-resistance circles are all centered on the resistance axis, and vary in diameter. The largest circle forms the rim of the graph, representing zero resistance. The smallest circle appears at the right end of the resistance axis, representing z = 50 Ω. The extreme right end of the resistance axis represents an infinite resistance. In addition, all of the resistance circles are tangent to the rim of the chart at the right of the resistance axis. All points along a constant-resistance circle have the same resistive value as the point where it crosses the resistance axis.

The constant-reactance circles are superimposed on the resistance circles, and are all tangent to the resistance axis at the infinite resistance point at the right of the chart. The centers of the reactance circles are off the edge of the chart, but are on a line tangent to the rim at the right of the chart, perpendicular to the resistance axis. The outer rim of the chart is graduated in reactance, and is called the reactance axis. All points along a constant-reactance circle have the same reactance value as the point where it intersects the reactance axis on the rim of the chart. All points on the chart below the resistance axis contain components of capacitive reactance, and those above contain components of inductive reactance. Hence, it is evident that all points on the resistance axis have zero reactance, which is why it is also called the "real" axis.

Because both the resistance and reactance axes go from zero to infinity, any complex impedance can be plotted on the chart. A complex impedance, R + *j*X, may be defined at any intersection of a resistance circle and a reactance circle. The value of the impedance is the complex value of the resistance and reactance circles that intersect. However, remember that the *j* operator must precede the reactance value to denote trigonometric treatment of R and X, rather than simple arithmetical addition, which is a no-no.

There are two peripheral scales on the chart graduated in wavelengths. These indicate position along the transmission line. One scale indicates clockwise direction along the line, or "wavelengths toward the generator," and the other indicates counterclockwise direction, or "wavelengths toward the load." One complete length around the circumference of the chart represents exactly ½ λ of line.

A third peripheral scale indicates angle θ of the reflection coefficient, which is the phase angle between the forward, or incident and reflected voltage wave. Angle θ is directly related to the impedance, as well as to the position of the impedance on the line. Graduations on the bottom half of the chart contain negative values of the angle of the reflection coefficient, from 0° on the right to 180° on the left. Negative values indicate capacitive reactance. The graduations on the top half contain positive values from 0° to 180°, indicating inductive reactance.

If seeing 360° of reflection coefficient angle in traveling a 180° (½-λ) line is confusing, you must recall the discussions in Chapters 3 and 9 in which I explained why the reflection coefficient changes two degrees for every degree of motion along the line. Now for a word of caution: In modifying the Smith Chart in Figs 9-2 and 10-1, I added line-length scales around the chart that don't appear on the basic Smith Chart of Fig 10-2. For example, see the *90°* appearing at the left, and the *0°* and *180°* at the right. Do not confuse these *line-length* degree numbers with those from the ANGLE OF REFLECTION COEFFICIENT scale.

Recall my discussion in Sec 9.4 of reflection-coefficient angles $\theta = 0°$ and 180° relative to the positions of voltage maxima and minima on the line. Now note that $\theta = 0°$ appears on the Smith Chart at the infinite resistance end of the resistance axis (voltage maximum), and that $\theta = 180°$ appears at the zero-resistance end of the resistance axis (voltage minimum). We will see later how the chart shows that the value of SWR determines the actual maximum and minimum voltage values.

I'll now explain the peripheral wavelength scales used to determine either the position of an impedance on a transmission line, or the length of a line. The scales are graduated in decimal values of a wavelength from 0 to 0.5 λ. Both scales start at the left of the chart, and end at the starting point after 0.5 λ of travel around the chart going in opposite directions. I'll explain their use with an example. We'll use Fig 10-2 to determine the electrical length of a transmission line in wavelengths from a measurement of impedance at the input terminals of the line. The electrical length of the line we've chosen is known to be less than ½ λ, so the entire line will fit within one revolution around the chart. How line lengths greater than ½ λ are handled is explained later.

The input impedance of the line was measured with a normalized load impedance $z = 1.38 - j1.32$ terminating the line. This is plotted in Fig 10-2 at point A. This load impedance places us at the angle of reflection coefficient $\theta = -45°$, as the starting point for determining the length of the line. To find the starting point on the wavelengths scale, we extend the radial line through the point where the load impedance appears, on out to the outer scale labeled WAVELENGTHS TOWARD GENERATOR, at point B, where we read 0.313 λ. The measured impedance at the input of the line was found to be $z = 0.38 - j0.36$, normalized. This impedance value is plotted at point C. A radial line through this impedance point on the chart goes through reflection coefficient angle $\theta = -135°$. Extending this radial line to the same outer wavelengths scale we read 0.438 λ, at point D.

Hence, the electrical length of the transmission line is the difference between the wavelength values of 0.438 and 0.313, which is 0.125 λ. This length is ⅛ λ, which is 45° in electrical length and 90° of difference in position along the reflection angle θ scale. These values may be confirmed using Fig 9-2. How to manipulate lengths of feed line, and their impedances, is discussed in Chapter 12.

A wealth of additional information concerning the construction and use of the Smith Chart is available (*Refs 19, 25* through *29, 70 and 71*).[1] Chapter 28 of *Ref 71* by Hall, K1TD, is by far the best I've seen for the amateur.

---

1 All references appear in Chapter 24.

Chapter 11

Using the Smith Chart

Sec 11.1 Introduction

Because impedance along a mismatched transmission line is so directly related to the standing-wave pattern, a logical place to begin discussing the use of the Smith Chart is with entering and plotting SWR data. SWR on a transmission line is represented on the normalized Smith Chart as a circle centered on the resistance axis at r = 1, which is also the center of the chart. The radius of the circle is measured from the center of the chart along the resistance axis to the point where the normalized resistance equals the SWR. For example, as shown in Fig 11-1, a 2:1 SWR circle centered at r = 1 intersects the resistance axis at two points, where r = 2.0, and at 0.5, the reciprocal of the SWR. A 3:1 SWR circle intersects the axis at 3.0 and at 0.33. As stated earlier, the symbols for normalized values of impedance, resistance, and reactance are shown in lower case, while the symbols for the actual, nonnormalized values are capitalized.

There are two very important points to learn about the SWR circle:

1) One trip around the circle represents a movement of 180° along the line, or ½ λ.

2) The constant radius of the SWR circles on the chart indicates that the SWR is *constant* all along the line. If you find this statement confusing because you've witnessed a change in SWR when changing the length of the line, there probably was RF current flowing on the outside surface of the coax, from either an ineffective balun, or no balun at all. This phenomenon is explained in detail in Chapter 21.

Once drawn on the chart, an SWR circle of a given value represents the locus of all complex impedances that appear on a ½-λ section of line having that value of SWR. The impedance values are read from the chart at the points where the SWR circle passes through the intersection of constant resistance and reactance circles. Note, for example,

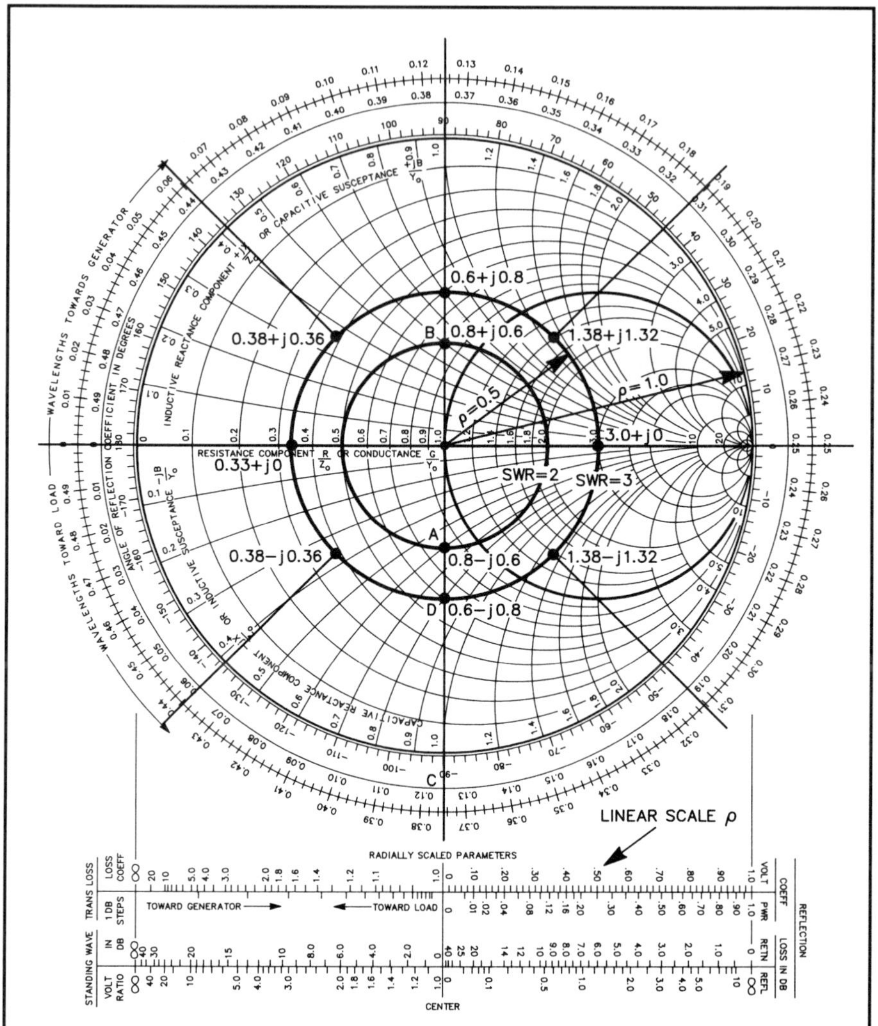

Fig 11-1—A normalized Smith Chart with examples added.

in Fig 11-1 that the 2:1 SWR circle passes through the point where the 0.8 resistance circle intersects the –0.6 reactance circle, point A. The normalized impedance at this point is z = 0.8 – *j*0.6 Ω. As already stated, the 2:1 SWR circle also intersects the resistance axis at two points, where r = 0.5 and r = 2.0. The normalized impedances where it intersects these two points are $z = z_C/SWR = 0.5 + j0$ and $z = z_C \times SWR = 2.0 + j0$ Ω, respectively. These are the *only two* impedances on the entire ½-λ of line that are purely resistive! All impedances everywhere else on the line are reactive—capacitive if below the resistance axis, and inductive if above.

If you haven't already guessed, I'll mention that the SWR-circle and impedance relationship is reciprocal. Not only can all impedances on the line be determined by an SWR circle on the chart, but the SWR of any impedance can also be determined by plotting the impedance on the chart and then measuring the radial distance from the chart center to the impedance point.

For example, assume that the 2:1 SWR circle in Fig 11-1 had not been drawn in. This time let's find the point representing the normalized impedance $z = 0.8 + j0.6$. Being inductive, it lies above the resistance axis, where the 0.8 resistance circle intersects the +0.6 reactance circle. This point is marked in Fig 11-1 as B. Place the centering pin of your drawing compass at the center of the chart, where $r = 1$, and place the pen of the compass on the $0.8 + j0.6$ impedance point. This determines the radius of the SWR circle. After swinging the compass for a complete circle, you'll discover that the SWR is 2:1, because the circle also intersects the resistance axis at 2.0. So you see, any impedance falling on the 2:1 circle can be used to determine the SWR. And furthermore, if any one of those impedances appearing on the 2:1 circle is used to terminate the line, it *will* produce a 2:1 SWR all along the line.

So far I have not mentioned the position of the standing-wave pattern and the distribution of impedance along the line, so now is the time to add position to the puzzle. The first important point to note is that the position of both the standing-wave pattern and the impedances appearing on the line are determined *solely* by the load impedance that terminates the line. *Nothing* can be added at the input terminals that will affect them in any way. Contrary to the belief of many amateurs, an antenna tuner at the line input *does not* reduce the SWR on the line—it only matches the line-input impedance to the source of the RF! The position of impedance on the line is determined by the angle θ of the reflection coefficient. We'll continue by examining the complete complex reflection coefficient.

Every impedance has its own unique complex reflection coefficient, comprising both magnitude ρ and angle θ. The magnitude is directly related to the SWR, as shown in Eqs 3-2 and 3-3. On the Smith Chart, ρ is equal to the radius of the SWR circle measured on a linear scale extending from zero at the center of the chart to 1 at the rim. The linear scale at the bottom of Fig 11-1 is for measuring the radial distances of magnitude ρ on the chart. The decimal value of the magnitude is read directly from this scale. At the center of the chart the radius is zero, indicating zero reflection and a 1:1 SWR. At the rim, or boundary of the chart, where the constant resistance circle is zero, $\rho = 1$, indicating total reflection of all forward traveling waves and an infinite SWR. In other words, the length of the radius extending to the rim of the chart is of magnitude 1.0.

We can also see from Eq 3-3 that when $\rho = 1$, the SWR is infinite, which occurs when the load contains zero resistance. On the other hand, we can find ρ from the SWR by using Eq 3-2. For example, a 3:1 SWR yields $\rho = 0.5$, or half the chart radius. A 3:1 SWR circle (also the $\rho = 0.5$ circle) is shown in Fig 11-1. Note the radial line labeled "$\rho = 0.5$" extending to the SWR circle that is also labeled "SWR = 3." And also observe the radial labeled "$\rho = 1$" extending from the center of the chart to the rim. (All of this information is also shown in Fig 9-2.) In summary, the magnitude ρ of the reflection coefficient for any impedance is equal to the radial length from the center of the chart to the point where the impedance appears on the chart.

So much for magnitude ρ. Now we come to angle θ. To use angle θ of the reflection coefficient for determining the position of an impedance on the line, we draw a radial line from the center of the chart through the point of the impedance, and extend it to the scale on the rim of the chart titled ANGLE OF REFLECTION COEFFICIENT IN DEGREES. The point where the radial line intersects this scale denotes angle θ. For example, Fig 11-1 shows a radial line through point D, the normalized impedance point $z = 0.6 - j0.8$ on the 3:1 SWR circle. This line is extended to reach the θ scale at point C, which shows the radial line at $\theta = -90°$. From Eq 3-2 we find that the 3:1 SWR yields $\rho = 0.5$, and hence the complex reflection coefficient of $z = 0.6 - j0.8$ is $\rho = 0.5$ at $\theta = -90°$. Either set of numbers completely defines the impedance, but angle θ at $-90°$ identifies the relative position of the impedance on the line.

It is interesting to note that this radial line at $\theta = -90°$ also passes exactly through the normalized impedance $z = 0.8 - j0.6$, point A, which is on the 2:1 SWR circle. However, magnitude ρ for an SWR of 2:1 is 0.333 on the magnitude scale. In other words, the impedances $0.6 - j0.8$ and $0.8 - j0.6$ both appear at the same location on the transmission line.

The first occurs with a resistive load that produces a 2:1 SWR, and the second with a resistive load producing a 3:1 SWR.

I stated earlier in this chapter that the load impedance terminating the line determines the position of the impedances all along the line. Let's examine this concept as we direct our attention to Fig 9-2. Here we see line impedances calculated vectorially at eight different points, each point separated by 22.5° in distance along the line. The radial lines extending from the 3:1 SWR circle at each impedance point to the angle of reflection coefficient scale indicate the angle of reflection coefficient, θ, associated with its corresponding impedance. Inspection of the θ angles reveals that the impedance appearing with every 22.5° of increase in distance along the line has a corresponding increase of 45° in angle θ. This reminds us that angle θ changes two degrees for each degree of change in position along the line.

Let's now examine the significance of these θ angles along the line. In determining why the position of impedances along the line is determined by the impedance terminating the line, we must first remember that the impedances anywhere on the line result from the vector division of the voltage, E, and current, I. These vectors result from the vector sum of the forward and reflected waves at each point on the line. In turn, the voltage and current resulting from each vectorial summation is determined by the complex reflection coefficient at each point—in other words, by the relative magnitude, ρ, and phase angle θ of the forward and reflected waves.

Because the reflected waves originate at the impedance terminating the line, we'll let the termination point be a zero-time reference point for the origin of the reflected waves. We know that magnitude ρ and angle θ of the reflection coefficient appearing at the zero-time reference point are determined by the terminating impedance. However, while magnitude ρ remains constant all along the line, angle θ changes two degrees for every degree of distance from the reference point, as stated above. Hence, the phase relationship θ between the forward and reflected waves occurring at the zero-time reference point determines all the other phase relationships occurring at a later time as the waves progress along the line.

For example, refer to Fig 9-2. If we proceed to a point 22.5° along the line from the $3 + j0$ termination where $\theta = 0°$, the forward wave has advanced to +22.5°, and the reflected wave has retarded to –22.5°. This makes the phase of the reflected wave –45° relative to the forward wave, and produces the impedance z = 1.38 – j1.32. In the same manner, at the 45° point on the line, the reflected wave is at –90° relative to the forward wave, producing the impedance z = 0.6 – j0.8. We can see these changes in phase (angle θ) between the forward (incident) and reflected wave in the vector plots appearing in the circular insets of Fig 9-2. It should now be evident that the impedance at any point on the line is determined by the relationship between the forward and reflected voltages and currents. This explains why the terminating impedance, which determines the complex reflection coefficient (ρ, θ) at the zero-time reference point, also determines the reflection coefficient angles and the impedances throughout the entire line.

While the eight line impedances calculated by vector arithmetic are shown in the circular and rectangular insets in Fig 9-2, the impedance points are shown in their correct respective graphic positions on the 3:1 SWR circle. However, because the reading of the values from the chart circles is difficult because of the intrusion of the insets, a 3:1 SWR circle and partial radials at the corresponding 45° angles of reflection coefficient are also shown in Fig 11-1 to allow reading of these normalized impedance values in true Smith Chart fashion.

Let's now see what happens to impedances along the line if we change the position and value of the terminating load impedance in Fig 9-2. Observe that at the len° = 22.5° point, angle θ of the reflection coefficient is –45°, and the normalized impedance z = 1.38 – j1.32. It would be interesting to see what happens to impedances along the line if we were to cut it at this point where len° = 22.5°, eliminate the portion between len° = 0° and 22.5°, and then terminate the new end of the line with an impedance z = 1.38 – j1.32. Because the new load impedance with its own corresponding angle θ = –45° is the same as that which appeared at len° = 22.5° with the 3 + j0 load impedance at len° = 0°, the conditions of reflection at len° = 22.5° are identical to those that existed there before cutting the line. Hence, the values of the reflection coefficient and impedance remain the same as before, *everywhere* along the remainder of the line!

If we now return to the conditions as shown in Fig 9-2, we see that the line can be cut at any point and the impedance appearing at the input terminals of the portion terminated by the 3 + j0 load will be the impedance indicated on the Smith Chart at the point where the cut is made. For example, if the line is cut at len° = 45°, the new normalized input impedance will be z = 0.6 – j0.8, and if cut at len° = 67.5° the impedance would then be z = 0.38 – j0.36, as shown in the rectangular insets of Fig 9-2.

As a result of this discussion, we may now make the following general statement: Any impedance appearing along a terminated line, as indicated on the SWR circle containing that impedance, may be used as a new terminating load at the point on the line where that value of impedance appeared with the original load, and all impedances appearing between the new load and the source will remain unchanged.

Sec 11.2 Four Distinctive Features of the Smith Chart

As we continue examining various uses of the Smith Chart, four interesting features emerge that merit explanation. Two of the four features involve the ½-λ line—first, its capability of repeating the value of its terminating load impedance at its input terminals, and second, its phase-inverting action that reverses the polarities of the voltages and currents appearing at the opposite ends of the line. The third feature is the impedance-inverting action of the ¼-λ line. The fourth feature concerns the appearance of impedances having the same resistance but opposite signs of reactance simply by having opposite signs of their corresponding angles of reflection coefficient θ.

We'll first examine the impedance-repeating action of the ½-λ line, using Fig 10-1.[1] Select one of the impedance points printed out on any of the four SWR circles to represent a starting position on a transmission line. Recall an earlier statement that one complete trip around an SWR circle represents a movement of ½ λ along a line. With your eyes, trace all the way around the SWR circle you selected. No matter where you start, with one complete trip around the circle, the impedance at the starting position on a line (represented by the corresponding point on the SWR circle) is repeated ½ λ away in either direction. This means that if the length of a mismatched line is greater than ½ λ, an impedance appearing at any point farther away than ½ λ from the terminating load will be the same as that appearing ½ λ from that point.

While the impedances repeat themselves *exactly* on lossless lines

[1] In addition to Figs 9-2 and 11-1, the text in this section refers to Fig 10-1, which has not yet been described in detail. To enhance the clarity of the explanations, you may want to skip to Sec 12.1 and read the description of Fig 10-1 before proceeding.

at intervals of ½ λ, they do not repeat exactly on real lines having attenuation. So far, all of our calculations and statements have been made with the assumption that the lines are lossless. However, the lines we use in practice are considered "low loss," so the error in assuming the lines to be lossless is relatively inconsequential. On lines having enough attenuation to make the impedance error significant, the error in the resistance is to bring it closer to the value of the characteristic impedance Z_C of the line, and the error in the reactance is to decrease it.

In addition to repeating the impedance, a ½-λ section of line also reverses the phase, or polarity, of the voltages and currents at opposite ends of the line. To see how the phase inversion is accomplished, it is informative to observe the rotation of the resultant vectors of voltage E (or current I) appearing in the circular insets in Fig 9-2 as we proceed around the chart. Start at the right of the chart where the resultant vectors point to the right, indicating a reference phase of 0° at the 3 + *j*0 load end of the line. Now proceed clockwise around the chart in the 22.5° increments of the insets. Observe the counterclockwise rotation of the resultant vectors with each increment of travel. After completing one trip around the chart, note that the resultants are now pointing to the left, having rotated 180°. This signifies a change in phase of 180° relative to the starting point. Keep in mind that one trip around the chart represents a travel of ½ λ along the line.

Now we come to the third feature, the impedance-inverting action of the ¼-λ line that makes it so useful as an impedance transformer. The wave actions that give this length of line its special impedance-transforming properties are described in detail in the final paragraphs of Chapter 4, and the impedance-inverting action is shown mathematically in Eq 7-1. On the Smith Chart, the impedance inversion is obtained numerically by observing the values of the impedances appearing on an SWR circle at the opposite ends of a diameter line. For example, in Fig 11-1 we find the normalized capacitive impedance z = 0.8 – *j*0.6 at point A, on the radial line at the reflection coefficient angle θ = –90°. However, if we follow the radial line up, crossing the r = 1 point in the center and continue until it reaches the 2:1 SWR circle *above* chart center, we find the normalized *inductive* impedance 0.8 + *j*0.6, at point B. This is the reciprocal of the initial capacitive impedance on the lower half. Hence, by going half way around the SWR circle, corresponding to a change of ¼ λ on a transmission line, we have inverted the impedance. It should be noted that the inverted normalized impedance is also the normalized *admittance* of the initial capacitive impedance.

It is now interesting to note that if we continue around the second half of the SWR circle and arrive back at the point of beginning, we have reinverted the inductive impedance back to the original capacitive impedance we had at the starting point. Hence, two ¼-λ lines connected in series (a ½-λ line) can be likened to a seesaw.

Let's try another example, this time using Fig 10-1. At reflection coefficient angle $\theta = -135°$ on the 5:1 SWR circle, we see the impedance $Z = 12 - j20\ \Omega$. Diametrically opposite, at $\theta = +45°$ on the 5:1 SWR circle, we find the impedance $Z = 55 + j94\ \Omega$. These two impedances are also reciprocals of each other. If we used an impedance $Z = 12 - j20$ to terminate a 50-Ω, ¼-λ line, the impedance at the input terminals would be the reciprocal, $Z = 55 + j94\ \Omega$.

A special case of impedance inversion occurs along the resistance axis, where the reactance is zero. This is the condition that most often prevails when the ¼-λ line is used as an impedance-matching transformer. In most cases the transformer is used to convert one purely resistive impedance to that of another purely resistive impedance. It is evident that both of these resistances will lie on the resistance axis.

Recall from earlier discussions that the right-hand portion of the resistance axis represents maximum impedance points on the line (where $\theta = 0°$), and that the left-hand portion of the axis represents minimum impedance points (where $\theta = 180°$). To illustrate, we can see that, on any of the SWR circles, the resistance appearing on the resistance axis to the right of the center of the chart is $R = 50 \times$ SWR, while on the same circle to the left of chart center, the resistance is $R = 50 \div$ SWR.

Let's now use this information to work out an example of impedance matching with a ¼-λ section of transmission line. Assume that we have two transmission lines of different impedances, and we want to couple them together through matching. One is a 25-Ω line, and the other is 100 Ω. From Eq 7-2, we find the geometric mean value between these two impedances to be 50 Ω, which means that a 50-Ω line of ¼ λ will perform a satisfactory match with a 2:1 SWR on the ¼-λ transformer line. However, we may also observe on Fig 10-1 that, based on a 50-Ω system, both 25 Ω and 100 Ω lie on the 2:1 SWR circle at opposite ends of a diameter that lies on the resistance axis. This is another way of telling us that the ¼-λ, 50-Ω line will provide a match between the 25-Ω and the 100-Ω lines, the impedance match we want.

Now we'll consider the fourth characteristic of transmission lines that appears as an interesting display on the Smith Chart. This concerns the points on the line where the magnitudes and both the resistance and reactance components of two different impedances are equal, respectively, except that the reactances are of the opposite sign. These impedances occur at two different points on the same SWR circle, with each point on opposite sides of the circle equidistant from either end of the resistance axis. In other words, this condition exists for any two points having the same angle θ of the reflection coefficient, but having the opposite sign of θ. For example, on the 3:1 circle of Fig 10-1 at θ = –135°, the impedance is Z = 19 – *j*18. And at θ = +135° the impedance is Z = 19 + *j*18 Ω.

Sec 11.3 A Reader Self-Test and Minimum-SWR Resistance

Everybody knows that when a 50-Ω transmission line is terminated with a pure resistance of 50 Ω, the magnitude of the reflection coefficient, ρ, is 0, and the SWR is 1:1. Right? Of course! With that in mind, here is a little exercise to test your intuitive skill.

If we insert a reactance of 50 Ω in series with the 50-Ω resistance, the load becomes Z = 50 + *j*50 Ω. The SWR will be 2.618:1. Now for the question. With this 50-Ω reactance in the load, is the SWR already at its minimum value with the 50-Ω resistance, or will some other value of resistance in the load reduce the SWR below 2.618:1? You say the SWR is already the lowest with the 50-Ω resistance, because, after all, the line impedance, Z_C, is 50 Ω? Sorry, wrong. With reactance in the load, the minimum SWR always occurs with a load resistance greater than Z_C. In fact, the more reactance, the higher the resistance required for minimum SWR. For any specific value of reactance in the load there is one specific value of resistance that produces the lowest SWR. I call this resistance the "minimum-SWR resistance." Finding the value of this resistance is easy. First you normalize the reactance, X, by dividing it by the line impedance, Z_C: $x = X/Z_C$. Then solve for the normalized value of resistance, r, from Eq 5-1, which is repeated here.

$$r = \sqrt{x^2 + 1} \qquad \text{(Eq 5-1)}$$

Let's try it on the example above. The normalized value of 50 Ω of reactance, X, is x = 1. Substituting in Eq 5-1, $r = \sqrt{2} = 1.414$. So the true value of the minimum-SWR resistance is $1.414 \times 50 = 70.7\ \Omega$. While the 50-Ω resistance yields a 2.618:1 SWR, the 70.7-Ω resistance in series with the 50-Ω reactance yields an SWR of 2.414:1. Not a great deal smaller, but still smaller than with the 50-Ω resistance.

So let's try a more dramatic example, this time with a 100-Ω reactance, which has a normalized value x = 2.0. With a 50-Ω resistance, the SWR is now 5.828:1. However, with the normalized minimum-SWR resistance, $r = \sqrt{2^2 + 1} = \sqrt{5} = 2.236$. Multiplying by 50, we get R = 111.8 Ω. With this larger resistance in series with the 100-Ω reactance, the SWR is reduced from 5.828:1 to 4.236:1. The results of this exercise didn't turn out quite the way you expected, did it?

These results seem incredible, so how do we know they're true? By a simple demonstration using the Smith Chart. The Smith Chart provides an elegant way to demonstrate why the SWR is not at a minimum with 50 Ω of load resistance when reactance is also present in the load. I'll demonstrate using the values in the last example, but first let's normalize them. The 50-Ω resistance, R, becomes r = 1, and the 100-Ω reactance, X, becomes x = 2. This impedance, $1 + j2$, is plotted at point A in Fig 11-2. Next we draw a radial line from the center of the chart to the point on the 2.0 reactance-circle segment where a line perpendicular to the radial becomes tangent to the reactance circle. This point is identified as B in Fig 11-2. The minimum-SWR resistance is then found on the resistance circle to which the radial is tangent, 2.236 in this case. This resistance circle will have a normalized value greater than 1 (meaning greater than 50 Ω).

Here is the proof. We know the length of the radial from the center of the chart to the 2.0 reactance circle is proportional to the SWR. So for minimum SWR, we're looking for the shortest possible radial that will reach the 2.0 reactance circle. The radial is shortest when it is perpendicular to a line tangent to the circle. To prove this, swing the radial in either direction from Point B in Fig 11-2. The radial will no longer reach the 2.0 reactance circle, because the circle curves away from the point of tangency. Hence, the radial would require lengthening to again reach the 2.0 circle, which means an increase in SWR. Using the shortest length that will reach the 2.0 reactance circle, we see the radial is tangent to the r = 2.236 circle, for an SWR of 4.236:1. When the radial is swung and extended to reach the 2.0 reactance circle at r = 1 (point A), the length has increased to indicate an SWR of 5.828:1.

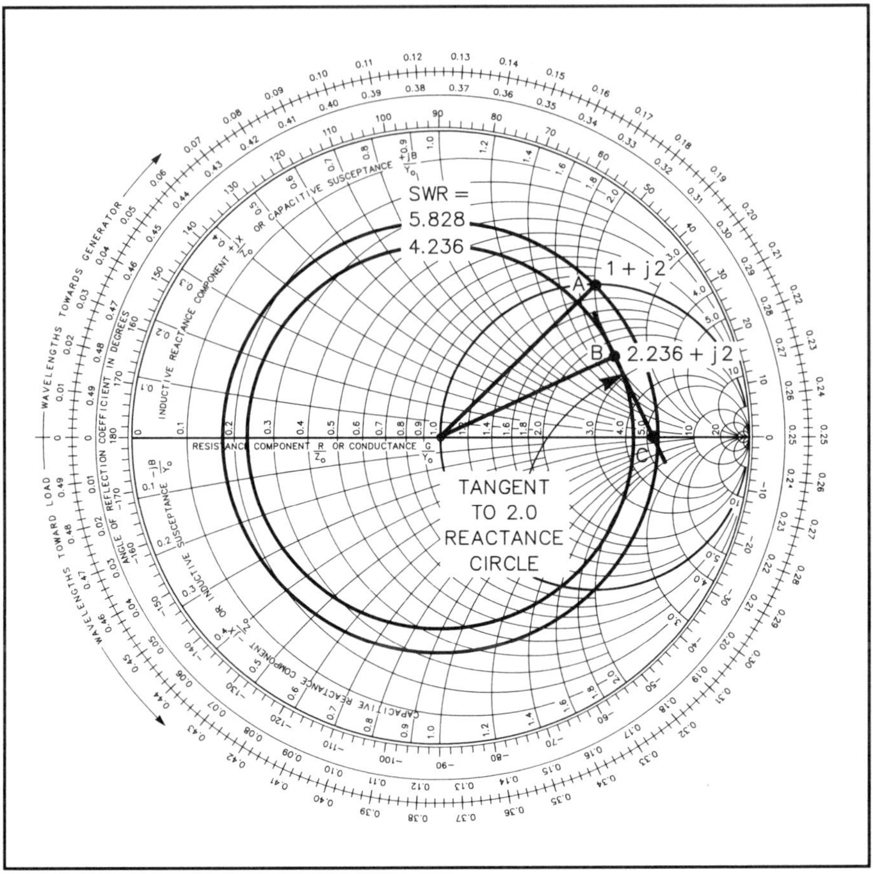

Fig 11-2—For the same reactance component in a load (x = 2 in these plots), the SWR is lowest when the resistance component is greater than the line impedance, Z_C (r > 1). The minimum-SWR resistance is that value producing the lowest possible SWR.

Hence, the resistance yielding the lowest SWR with a reactance of 100 Ω is 111.8 Ω, and the case is proved.

Incidentally, the line tangent to the reactance circle is coincident with the radius of the minimum-SWR resistance circle. In Fig 11-2, that radius is line segment CB, where C is at the center of the minimum-SWR resistance circle. Hence, the radial line and the radius of the r circle are mutually perpendicular.

Using the Smith Chart, we have proved geometrically that the minimum-SWR resistance is greater than Z_C when the load contains reactance, and the Smith Chart is also seen to reveal the value of the resistance. In addition, Eq 5-1 is also seen to yield values of resistance which are in complete agreement with those obtained graphically with the Smith Chart. If you are interested in pursuing the subject further, I have performed a rigorous mathematical proof of the relationship shown in Eq 5-1 in Appendix 1. In addition, Appendices 2 and 3 include equations for determining the exact value of SWR produced on a line when terminated with any complex load R + *j*X. Appendix 4 includes a program listing for hand-held calculators that yields the SWR from R + *j*X. All appendices are contained in Chapter 23.

Another interesting relationship resulting from this exercise is shown in Eq 5-2, repeated here.

$$SWR_M = r + x \qquad \text{(Eq 5-2)}$$

where

SWR_M = SWR obtained with the minimum-SWR resistance
r = the normalized minimum-SWR resistance
x = the normalized reactance

Eq 5-2 shows that to determine the value of the SWR with the minimum-SWR resistance in the load, we simply add the normalized values of the resistance and the reactance in the load. I have never seen a reference to the minimum-SWR resistance in relation to terminating loads in any of the engineering or academic literature. Therefore, I believe I am the originator of the relationships represented by Eqs 5-1 and 5-2 (which I first published in *QST* in April 1974), and both the geometrical and mathematical proofs.

What is the significance of the minimum-SWR resistance? First, it relates directly to the shape of the SWR curve when you plot the SWR measured on your dipole feed line over a range of frequencies that includes the resonant frequency of the dipole. Of significance is the discussion in Sec 5.6 on loss resistance being separable from the radiation resistance in antenna terminal impedance, where Eqs 5-1 and 5-2 were introduced. In addition, the significance of the minimum-SWR resistance phenomenon is evident in the SWR plots of Figs 15-1, 20-2, and 20-3. Those plots of dipole SWR (without traps) show that the SWR values are asymmetrical with respect to the minimum value obtained at the resonant frequency; for the same frequency change from resonance, higher values appear below resonance than those above resonance.

The reason for the asymmetry is that the resistance component of the dipole terminal impedance increases with frequency (as also shown in the figures), and thus the resistance is higher above resonance than it is below resonance. On the other hand, the reactance values are seen to be nearly identical on opposite sides of the resonance point, except for the opposite sign. Hence, for a given frequency excursion off resonance in either direction, the SWR above resonance is less than that for the same excursion below resonance, because the resistance is higher above resonance. This effect follows directly from the minimum-SWR resistance phenomenon described above.

Chapter 12

Antenna Impedance and the 50-Ohm Smith Chart

Sec 12.1 The W2DU Transformation Graph

We are now ready to examine the W2DU Antenna-Impedance Feed-Line Transformation Graph, Fig 10-1, in more detail. This graph has three significant features that will help you use the antenna tuner in matching the rig to the feed line. First, the Smith Chart of Fig 10-1 is a 50-Ω chart, which means that all resistance and reactance components of impedance are shown as actual, non-normalized values. Note that the resistance scale at the center of the chart reads 50 instead of 1.0.

Second, Fig 10-1 has four SWR circles to assist in determining the range of impedances you might expect to encounter at the input terminals of the feed line. As shown in the lower section of the chart, separate circles have been drawn for SWR values of 2:1, 3:1, 4:1, and 5:1. The special feature of these SWR circles is that the actual values of impedance appearing on each circle are printed out at intervals of 22.5° in angle of reflection coefficient θ all around each circle. These are the impedances that will appear at every 11.25-degree interval along each ½-λ section of the feed line for each of the four values of SWR shown. For all practical purposes, these impedance values cover the complete range of impedances one is likely to encounter at the input terminals of a 50-Ω feed line of any length, and for all SWR values up to 5:1. These impedance values, along with their corresponding values of θ, also appear in Table 12-1.

And third, Fig 10-1 shows curves labeled A through E, which plot impedance versus frequency for both dipole and vertical antennas. From

Table 12-1

Impedances Along a 50-Ω ½–λ Line for Various SWR Values

Angle of Reflection Coefficient θ°	*SWR* *2:1*	*SWR* *3:1*	*SWR* *4:1*	*SWR* *5:1*
0	100 + *j*0	150 + *j*0	200 + *j*0	250 + *j*0
−22.5	90 − *j*25	115 − *j*59	128 − *j*90	130 − *j*120
−45	70 − *j*36	69 − *j*65	63 − *j*83	55 − *j*94
−67.5	52 − *j*36	43 − *j*53	36 − *j*62	30 − *j*66
−90	40 − *j*30	30 − *j*40	24 − *j*44	18 − *j*46
−112.5	33 − *j*23	23 − *j*28	18 − *j*31	14 − *j*32
−135	28 − *j*15	19 − *j*18	15 − *j*19	11.5 − *j*20
−157.5	26 − *j*7	17 − *j*9	13 − *j*9	10.3 − *j*9
180	25 + *j*0	16.7 + *j*0	12.5 + *j*0	10 + *j*0
157.5	26 + *j*7	17 + *j*9	13 + *j*9	10.3 + *j*9
135	28 + *j*15	19 + *j*18	15 + *j*19	11.5 + *j*20
112.5	33 + *j*23	23 + *j*28	18 + *j*31	14 + *j*32
90	40 + *j*30	30 + *j*40	24 + *j*44	18 + *j*46
67.5	52 + *j*36	43 + *j*53	36 + *j*62	30 + *j*66
45	70 + *j*36	69 + *j*65	63 + *j*83	55 + *j*94
22.5	90 + *j*25	115 + *j*59	128 + *j*90	130 + *j*120
0	100 + *j*0	150 + *j*0	200 + *j*0	250 + *j*0

Going from θ = 0° at the first line of the body of the table to 0° at the bottom represents 180° (½ λ) of travel along the line. Impedance values are obtained from the SWR circles of Fig 10-1.

these impedance plots you will be able to determine the approximate impedance at the input terminals of the feed line with various feeder lengths, and with various degrees of off-resonance operation of the antenna.

Sec 12.2 Antenna Impedance versus Frequency

We'll now examine the plots of antenna impedance versus frequency shown as curves A through E in Fig 10-1. The types of antennas represented in the curves are described in the figure caption. The reason for including these plots is to demonstrate how the impedance appearing at the input terminals of a feed line can be predicted with a fair degree of accuracy with the Smith Chart if the antenna impedance and feed-line lengths are known.

The dots in all five curves have a special significance because they represent both impedance and frequency. The impedance indicated at each dot is read from the Smith Chart in the conventional manner. However, the frequency corresponding to the antenna impedance for a given length of antenna is indicated by the spacing of the dots. The dots appearing on the zero-reactance axis (the resistance axis) represent the impedance of an antenna cut to its resonant length. In other words, the dots appearing on the resistance axis indicate the antenna impedance while operating at its resonant frequency, whatever that frequency might be. The spacing between each dot represents a 0.5% change from the resonant, or center frequency.

Here's an example of how it works. Curve C represents data measured on an 80-meter dipole, using a General Radio 1606-A RF bridge. The dipole was cut to resonate at 3.75 MHz, the center of the 80-meter band. The length is 125 feet, from the formula length = 468/f(MHz). Thus, the dot in curve C which lies on the resistance axis indicates an antenna-terminal impedance $Z_A = 52 + j0\ \Omega$ at the resonant frequency. Let's now examine the frequency and impedance at the point on curve C ten dots below or in the downward direction from the resonant-frequency dot. First, ten dots at 0.5% per dot means the frequency is 5% below the resonant frequency of 3.75 MHz. The new frequency is thus f = 3.75 – 0.05(3.75) = 3.5625 MHz. The tenth dot was chosen to make the example easy, because it lies on the 5:1 SWR circle. This dot tells us that a typical 80-meter antenna cut to resonate at 3.75 MHz has close to a 5:1 SWR at 3.5625 MHz. The position of this tenth dot also indicates the impedance of this antenna to be approximately $Z_A = 50 - j90\ \Omega$ at 3.5625 MHz. As this is the impedance appearing at the terminals of the antenna, it is also the impedance that our feed line sees as a terminating load impedance.

Sec 12.3 Determining Line-Input Impedance

To determine the impedance an antenna tuner must match at the *input* terminals of the feed line, we need to know only the electrical length and characteristic impedance of the feed line. Because we are working with a 50-Ω Smith Chart, it goes without saying that we're using a 50-Ω feed line. Also, for the purpose of making the example easy to solve, I'll select a line length that results in a line-input impedance that is easy to read from the chart. Therefore let us select a line having a physical length of 113.0 feet. We'll use RG-8 coax, which has a velocity factor of 0.659, so the effective *electrical* length of the line is 113/0.659 = 171.5 feet.

We now need to find the decimal value in wavelengths of the electrical length of 171.5 feet in order to enter the wavelength scale on the chart. One wavelength at 1.0 MHz is 983.57 feet, so one wavelength at 3.5625 MHz = 983.57/3.5625 = 276.1 feet. To find the wavelength, λ, of the line, we divide its electrical length in feet by the number of feet in one wavelength at 3.5625 MHz. Therefore the wavelength of the line is 171.5/276.1 = 0.621 λ. Since 0.621 λ is greater than ½ λ (0.5 λ), we subtract 0.5 λ from 0.621 λ: 0.621 – 0.5 = 0.121 λ. We do this because 0.5 λ takes us completely around the chart with 0.121 λ left over.

We now draw a radial line from the chart center through the antenna impedance Z_A = 50 – *j*90, extending to the outer wavelengths scale, where we read 0.317 λ on the scale labeled WAVELENGTHS TOWARD GENERATOR. This is the starting point for determining the distance along the wavelengths scale on the chart between the point representing the antenna impedance Z_A and the impedance Z_{IN} at the input terminals of the feed line. So next we add to 0.317 λ the effect of the electrical length of the feed line. Because the feed line is 0.121 λ longer than ½ λ, we simply add 0.121 λ to 0.317 λ, which brings us to 0.438 λ on the wavelengths scale. To find the impedance at the input terminals of the feed line, we now draw a new radial line from the center of the chart to the 0.438 λ point on the wavelengths scale. The impedance we're looking for is at the point where the new radial crosses the 5:1 SWR circle, which we see is approximately Z_{IN} = 12 – *j*20 Ω. The 0.621-λ feed line has thus transformed the antenna impedance Z_A = 50 – *j*90 Ω to Z_{IN} = 12 – *j*20 Ω. This is the impedance the antenna tuner will then transform to 50 Ω for the line running to the transmitter.

With the procedure just described, the dipole-impedance data appearing in Fig 10-1 can be applied for general use to obtain the approximate input-terminal impedance of a feed line of any length used to feed any thin-wire dipole, within 6% of antenna resonance. Furthermore, this dipole impedance data can also be used with feed lines having any characteristic impedance simply by normalizing the antenna-impedance data obtained from the 50-Ω W2DU chart to the impedance of the line to be used, and then processing the normalized data on a normalized Smith Chart by following the procedure outlined above.

Turn your attention now to impedance curves D and E for the vertical radiators, and observe that the impedance $Z = 48 + j0$ appearing on curve E (where the curve crosses the resistance axis) comes very close to a perfect resistive match to 50 Ω. Note that this curve represents a radiator having only a few radials, resulting in a ground-loss resistance of around 16 Ω. Many amateurs using ground systems such as this one are apparently happy with the near-perfect match they get at resonance. However, they are blissfully unaware that the true radiating efficiency of this system is only about 67%, while believing the efficiency is close to 100% because of the perfect match. Unfortunately this is not true. The radiation resistance of the resonant ¼-λ vertical radiator is approximately $Z = 32 + j0$ Ω, as shown in curve D, which results in about a 1.5:1 mismatch. Consequently, the total load resistance of 48 Ω in curve E results from adding the 16 Ω of ground-loss resistance to the 32-Ω radiation resistance of the radiator. As a result, the power from the transmitter divides in a 2 to 1 ratio, with only 67% radiating from the antenna, and 33% lost in heating the ground, amounting to a 1.76 dB loss.

On the other hand, when the radiator with the good ground is properly matched at the input of the feed line, the power lost to any unwanted resistance from the 1.5:1 mismatch is negligible. It is interesting to note that even if one didn't match the 1.5:1 mismatch, the maximum loss would be only 0.177 dB more than the matched attenuation of the feed line. A more detailed discussion of this subject of ground loss with vertical radiators appears in Sec 5.4. I think you'll find it quite illuminating.

Before finishing this subject, let's now be completely candid. The ground-loss exercise just completed was straightforward engineering, pure and simple. No mention was made concerning the practical effect of the loss of radiated power from the ground resistance. From the point of view of the FCC in approving an AM broadcast radiator, the strict engineering approach would prevail. In fact, the FCC requires 90 radials for such systems, and most AM stations have installed 120 to optimize the efficiency. However, when we amateurs think of received signal levels in terms of 6 dB per S unit on the S meter, we see that a difference of 1.76 dB is insignificant when the signal level is already S8 or S9. This is, of course, the reason why it makes sense to operate with a vertical antenna even when it isn't practical economically to install more than a few radials. This is why there are so many amateurs who are completely happy with their verticals. On the other hand, the 1.76-dB difference does become significant when using the vertical for low-signal-level DX. Some avid DXers have the fortunate real estate situation where they can squeeze every bit of radiation out of their verticals by having the radials in salty sea water. What a dream! So now it's up to you to make your own informed decision on the number of radials you'll put down.

Chapter 13

The Line Matching Problem

Sec 13.1 Introduction

In Chapter 12 we identified the range of impedances likely to be encountered at the input terminals of a 50-Ω feed line for any length of line, and with any antenna impedance that produces mismatches up to a 5:1 SWR. The next step is to provide the means for transforming the line-input impedances that are outside the matching range of the transmitter to an impedance that will be a satisfactory load for the transmitter. We don't need to look very far for such a device, because impedance-matching networks have been performing such transformations for as long as there have been radio transmitters.

We must keep in mind that the main goal is to permit the output stage of the transmitter to deliver all of its available RF power to the load, which in our case is the antenna. All that is required for the output stage to do this is for it to see its optimum load impedance. To avoid any misunderstanding, the optimum load impedance is, by definition, that impedance into which an amplifier will deliver its maximum available power, while *staying within the limits* of maximum rated plate current. As we will discover very soon, the matching device we are looking for is simply an extension of the tank circuit in the transmitter, if it has one. (For those that don't, there is further explanation in Sec 13.4.) So to achieve a better understanding of how the matching device performs, we must first digress briefly to discuss the role of the pi-network tank circuit used in transmitters with tubes in the final output stage.

Sec 13.2 Matching with a Pi Network

In transmitters using tubes in the final output stage, the conventional pi-network tank circuit shown in Fig 7-2 performs two functions:

It filters out harmonic energy delivered by the output stage, and it matches its output impedance to the tube load impedance. When the output of the network is terminated with an impedance within its matching range, and is properly tuned or adjusted, its input terminals present the optimum load impedance for the tubes to work into. Let's illustrate with an example.

The optimum load impedance of the commonly used pair of 6146 tubes in parallel is around 2000 Ω resistive. Let's initially place a 50-Ω resistive dummy load on the output of the pi network. We use the relative power meter to indicate proper adjustment of the network while obtaining the optimum load impedance for the tubes. The relative power indicator on most commercial rigs is simply an RF voltmeter connected across the output terminals of the pi-network tank circuit. Using the conventional tuning procedure, we slowly advance the output loading control to increase the output power, while simultaneously rocking the plate-tuning control gently back and forth to maintain the dip in plate current and thus maintain resonance in the tank circuit. (Resonance occurs at the bottom of the dip.)

As we continue to advance the output loading control, we pass through a peak in the output power, after which the power *decreases* with any further advancing of the loading control. This means we have gone past the optimum load impedance. The plate current dip will also have become very broad and shallow. Before reaching the peak output, the tubes were underloaded because the load impedance presented by the input of the network was greater than 2000 Ω, and the plate current was also less than normal for the maximum availability of power. After we passed the peak output, the tubes were overloaded, because their load impedance was then lower than 2000 Ω, and the plate current was greater than normal, most likely exceeding the rated or normal value.

The optimum load impedance of 2000 Ω is obtained when the output loading control is adjusted precisely at the peak of maximum power output, and the plate current at the dip is the normal loaded value. Because our adjustment took us past the point of peak output, we'll now readjust the controls for maximum output power (or for the maximum power available at the maximum rated plate current) with the network still terminated with the 50-Ω dummy load. With this condition, we note the positions of the plate tuning and output loading controls.

Next, let's replace the dummy load with a 50-Ω resistor in series with a 50-Ω capacitor, an impedance $Z = 50 - j50\ \Omega$, which produces a

2.618:1 mismatch. The impedance of this new load is higher than the 50-Ω dummy load, so for the same output power, the relative power indicator shows a higher level than with the dummy load. This is why it's called a *relative* power indicator. Remember, it's a voltmeter indicating voltage across an impedance, and not a true power meter.

The same maximum output power as with the dummy load will be indicated by obtaining the same loaded plate current as with the dummy load. On applying power with the new load impedance, the plate current is no longer at a dip minimum. Why? Because the capacitance in the new load reflects capacitance into the tank circuit so that it is no longer in resonance—the tank is now detuned. Resonance is restored by retuning the plate-tuning capacitor for the dip. The output loading-control capacitor (paralleling the output terminals) also requires readjustment to decrease its capacitance. This readjustment allows a higher RF voltage to appear across the increased load impedance, and again transfers the maximum available power into the 50-Ω resistance in the new capacitive load. When the maximum power output has again been obtained by following the same procedure as before, the tubes are again seeing their optimum load impedance of 2000 Ω. Further, the plate current at the dip is the same value as with the 50-Ω dummy load, and the actual power absorbed by the 50-Ω resistance in the capacitive load is the same as that with the dummy load. No power is absorbed by the 50-Ω capacitor.

The important message of this exercise is that the change in positions of the pi-network tank-circuit controls compensates for the change in the load impedance at the output terminals of the tank circuit. The output load impedance was changed from 50 Ω resistive to a capacitive $50 - j50$ Ω, while the impedance compensation obtained by changing the plate tuning and loading of the tank circuit maintained the 2000-Ω input impedance required as the load for the tubes. This example shows that the pi network can match impedances of widely differing values, and that by adjusting the variable components of the network, different values of impedance terminating the network can be transformed to a desired impedance at the input terminals.

Now to another important point that is not generally well understood, because the manufacturers of amateur transmitters imply in their instruction books that the input impedance of the feed line connected to the RF output terminals should be 50 Ω. (Some manufacturers allow for a range of impedances that would produce an SWR of 2:1, referenced to 50 Ω.) The important point is that most transmitters using tubes in the output stage have pi-network tank circuits with impedance-range adjustment capabilities. Most often those capabilities can accommodate mismatches greater than 2:1. Many rigs will accommodate any impedance found on the 3:1 SWR circle, and even some impedances on the 4:1 circle.

So whatever the load on the tank circuit might be, with even a 4:1 SWR, the tuning and loading control range may allow you to obtain a dip in plate current at the same value as with a 50-Ω dummy load. If so, the tubes are seeing their desired load impedance and are operating at the same temperature as with the 50-Ω dummy load; no overheating will result from the SWR on the feed line! The tubes are completely unaware of what impedance is loading the output of the pi-network tank circuit. Sec 7.5 contains additional information concerning matching with the tank circuit.

Sec 13.3 The Antenna Tuner

When the external load impedances are outside the range of those that can be matched by the transmitter's own pi network, an additional matching network is all that is required to match the optimum load impedance of the tubes. Such a network simply extends the impedance range of the pi-network tank circuit. In ARRL publications, this additional matching network is called a *Transmatch*, meaning to match the transmitter to the transmission line. Of course the other popular name for the line-matching network is *antenna tuner*. I rather like antenna tuner because this name is also descriptive of its performance. (You'll often see *ATU* in foreign publications, for *antenna tuning unit.*)

The antenna tuner really does tune the antenna to resonance, in spite of opinions to the contrary of those who are unaware of the principles of conjugate matching. The tuner obtains a match, by which all reactances throughout the entire antenna system are canceled, including that of the off-resonant antenna, thereby tuning it to resonance. It should also be noted here that when the pi network in the transmitter alone matches the feed-line input impedance to the output-tube impedance, the pi network performs the match in an identical manner, and again, the off-resonant antenna is tuned to resonance. If you are interested in the definition and other features of conjugate matching, I refer you to Chapter 19, based on information that I wrote earlier for the ARRL. That information appears on page 16-10 in all issues of *The*

ARRL Handbook (*Ref 1*) starting with the 1986 edition, and on page 25-1 in the 15th edition of *The ARRL Antenna Book* (*Ref 71*).[1] Also see Chapter 17. A review of Chapter 7 will also be informative.

From discussions both on the air and at hamfests, I know there are many amateurs who still refuse to use an antenna tuner because they are adamant in their belief that the tuner doesn't tune the antenna. So before leaving the subject of conjugate matching, I'd like to make two more attempts to persuade nonbelievers into accepting the truth concerning tuning the antenna. First, they should also read "My Feedline Tunes My Antenna" by Byron Goodman, W1DX (*Ref 7*). And second, there are thousands of AM broadcast stations throughout the world using vertical tower radiators for their antenna. The height of the majority of these towers is determined not by the height at which they are resonant, but by the height which obtains the desired vertical radiation pattern. Therefore, nearly all AM station towers are not resonant at their operating frequency, but have a reactance component in their input-terminal impedance. How are these antennas made resonant so they will radiate the total available power from the transmitter? You guessed it—by matching with an antenna tuner! Think about it.

Sec 13.4 Fixed-Tuned Rigs

Here's another item that should surely thrill the nonbeliever. So far we have discussed only transceivers using tubes and pi-network tank circuits. Keep in mind that with these transceivers, all reflected power reaching the tank circuit is conserved—it is totally re-reflected to the antenna, and the only power lost is in feed-line attenuation. Sadly, we now come to the newer solid-state rigs that have no adjustable tuning and matching components. The tuning is fixed. The manufacturers have eliminated adjustable components to reduce costs, and to make room for more sophisticated circuits. The circuits containing the output transistors are designed to work only into a 50-Ω load, period!

Does the circuit have a feature which re-reflects the reflected power to conserve it, as does the pi-network? Hardly! So what happens to the reflected power reaching the output circuit if the feed line is mismatched? All reflected power is lost! The new design routes all reflected power into a circuit which reduces the RF output power in

---

1 All references appear in Chapter 24.

proportion to the amount of reflected power, to protect the output transistors from overloading because of detuning. With too much reflected power (too high an SWR), the output stage shuts down almost completely! I'm sure you all understand what this means—if the amateur with a solid-state rig wants to retain the across-the-band operating flexibility he had with his conventional tube rig, he's forced to install an external matching unit—the antenna tuner. Didn't I tell you this would thrill the nonbeliever?

Chapter 14

Matching Networks

Sec 14.1 Development Of The L Section

The antenna tuner, or Transmatch, in amateur use is fundamentally the same type of matching network as the pi-network tank circuit of the transmitter. It uses the same circuit components—inductors and capacitors, and the circuit action is identical. The only difference is in the circuitry arrangement of the components, although some tuners also use the pi configuration. The circuitry arrangement and component values in the tuner are determined primarily by the range of impedances the tuner is required to accommodate, and the frequency response. All such matching networks have a common element—the simple L section, which is the building block for all more complicated networks. For example, the pi network comprises two L sections with their series reactances connected together, and the T network comprises two L sections with their parallel reactances connected together. But what is an L section? And what does it do? To answer these questions we begin with an examination of some relationships involving resistance and reactance that are pertinent to the circuit action in the L section.

The basis for matching impedances using L sections as building blocks lies in an equivalence between series and parallel circuits. For any circuit containing resistance and reactance in series, there is an equivalent circuit containing resistance and reactance in parallel that has the same impedance and phase angle, when the circuits have an applied voltage of the same frequency. The basic series and parallel circuits are shown in Fig 14-1. Because the same principles apply whether the reactances are inductive or capacitive, the reactances are shown as blocks in the figure. However, the reactances in both circuits are the same kind, that is, if the series reactance, X_S, is inductive in A, the parallel reactance, X_P, in the equivalent parallel circuit in B is also inductive, and vice versa.

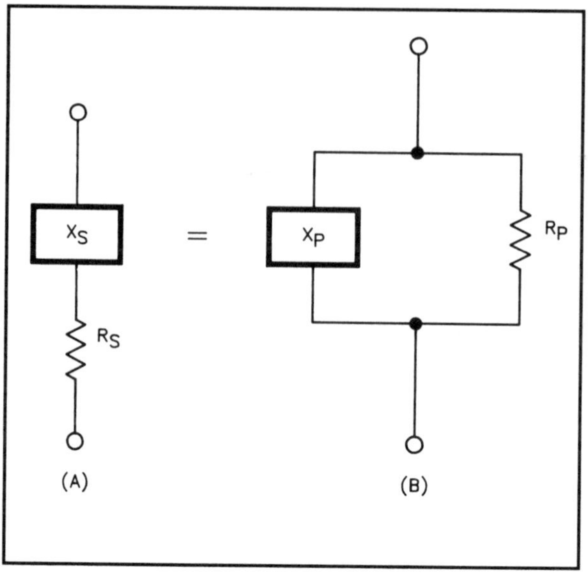

Fig 14-1—At A, the basic series circuit, and at B, the equivalent parallel circuit. X_S and X_P may be either inductive or capacitive, but they must be the same kind of reactance for circuit equivalence.

The value of R_P is always larger than R_S, and X_P is always larger than X_S. And herein lies the basis for impedance transformation: For a given R_S, the higher the value of X_S, the higher will be the values of R_P and X_P. Hence, R_P can be transformed to any value higher than R_S by selecting the proper value of X_S. Because the primary function in matching impedances is the transformation from one resistance level to another (and then canceling the reactance), we find that we can transform levels of resistance as desired by selecting the appropriate value of X_S in series with R_S.

So how do we determine the component values in the two equivalent circuits? Evidently the actual values in the equivalent circuits are related by the quantity Q. When the series and parallel circuits are equivalent, the following equations involving Q provide the necessary relationships between the two circuits to determine all of the component values.

$$Q = \frac{X_S}{R_S} = \frac{R_P}{X_P} \qquad \text{(Eq 14-1)}$$

$$R_P = R_S(Q^2 + 1) \qquad \text{(Eq 14-2)}$$

$$\frac{R_P}{R_S} = Q^2 + 1 \qquad \text{(Eq 14-3)}$$

$$Q = \sqrt{\frac{R_P}{R_S} - 1} \qquad \text{(Eq 14-4)}$$

$$X_S = QR_S \qquad \text{(Eq 14-5)}$$

$$R_S = \frac{R_P}{Q^2 + 1} \qquad \text{(Eq 14-6)}$$

$$X_P = \frac{R_P}{Q} \qquad \text{(Eq 14-7)}$$

Let's now calculate some component values with an example. This example is illustrated in Fig 14-2. We'll use the same values as used with the pi network in matching a 50-Ω resistive load to the 2000-Ω load impedance of parallel 6146 tubes in Chapter 13. We want to transform the 50-Ω resistance to 2000 Ω, so we let R_S = 50 Ω, and R_P = 2000 Ω. From Eq 14-4, we see that R_S and R_P determine the value of Q, so $Q = \sqrt{2000/50 - 1} = \sqrt{39} = 6.245$. Next, the series reactance, X_S, required to transform 50 Ω to 2000 Ω is found from Eq 14-5; $X_S = 6.245 \times 50 = 312.25$ Ω. To complete the calculations, we determine X_P from Eq 14-7; $X_P = 2000/6.245 = 320.26$ Ω. Thus, we can transform the 50-Ω resistance, R_S, to 2000 Ω by inserting a reactance, X_S, of 312.25 Ω in series with R_S. Are we finished? Not quite. Recall that we had to introduce a reactance X_S to obtain the transformation of resis-

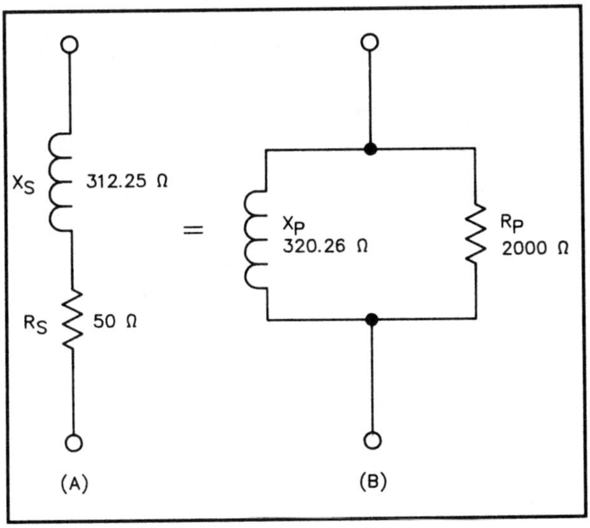

Fig 14-2—Equivalent series and parallel circuits. Component values are related by equations given in the text.

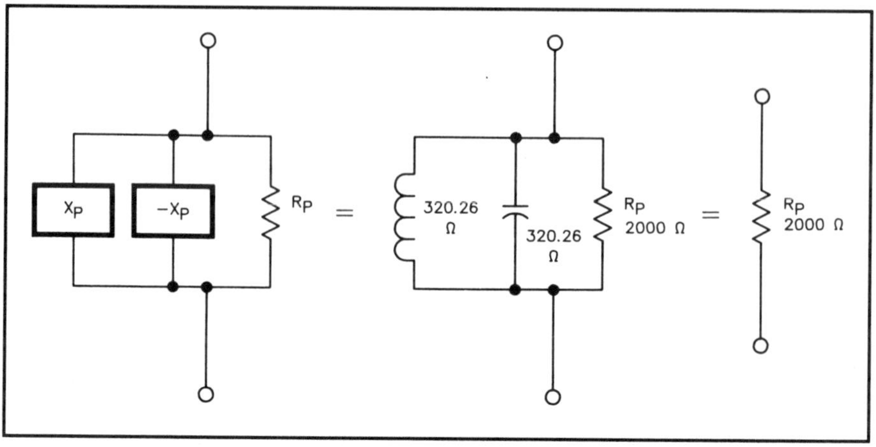

Fig 14-3—Reactance cancellation by adding a parallel reactance of the opposite kind.

tance, and as we have done nothing so far to eliminate it, the resulting parallel reactance, X_P, is still present. However, it is a simple matter to eliminate X_P. By placing a reactance of the same value as X_P, but of the opposite kind ($-X_P$) in parallel with X_P, all of the reactance is eliminated from the equivalent parallel circuit, and only the resistance, R_P, remains. This effect is illustrated in the diagrams of Fig 14-3.

Keep in mind that the series circuit in Fig 14-2A is the only physical circuit that exists. The parallel equivalent circuit of Fig 14-2B does not exist, although if it did it would exhibit the same impedance and phase angle as the series circuit. With the $-X_P$ added to cancel the reactance, X_P, the resulting physical circuit is that shown in Fig 14-4.

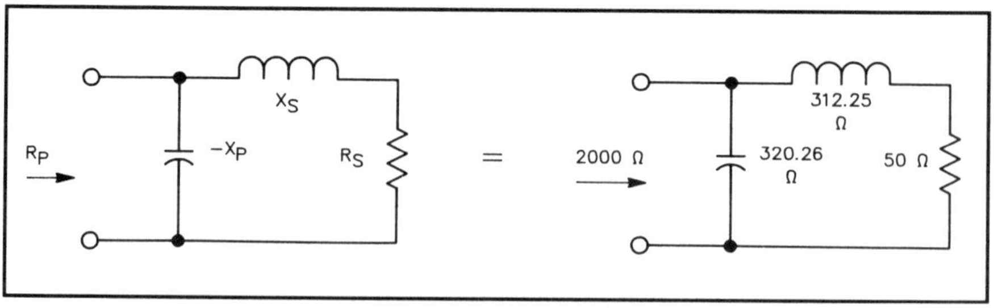

Fig 14-4—The resulting L-section matching circuit for the example given in the text. This circuit is developed from those of Figs 14-2 and 14-3.

R_P is now shown with an arrow to indicate the resistance that a source of power would see when looking into the terminals. The circuit of Fig 14-4 is an L section which evolves from the equivalence of the series and parallel circuits, but appears physically as the original series circuit of Fig 14-2 with the $-X_P$ canceling reactance added. As we can see, the L section of Fig 14-4 has transformed a pure resistance of 50 Ω to another pure resistance of 2000 Ω by means of a series reactance and a parallel canceling reactance. While X_S and X_P are shown as inductive, either inductive or capacitive reactances would perform equally from the viewpoint of impedance matching. The choice of which kind to use depends on the particular application of the circuit.

The L section can be used to match any two values of resistance. However, there is a limitation that sometimes restricts the practical usage of the single L section alone. The limitation is that the Q of the single L section is determined solely by the values of the load and input resistances, R_S and R_P, and in many instances the resulting Q may not be a satisfactory value. In such instances, more complicated circuits comprising two or more L sections may be used to remove the Q limitation, and also to permit other factors to be considered in the design of a matching circuit.

Sec 14.2 Design of the T Network

Now that we have the procedure and equations required to design L sections, we are ready to design T and pi networks. The step-by-step procedure is the same for any network calculation. We'll start with the T network because we're interested both in extending the range of the pi-network tank circuit of the tube rig and in providing a matching unit for the solid-state rig that requires a 50-Ω resistive load, as initially discussed in Sec 13.4. The T network is especially convenient for matching relatively low values of impedance in our range of interest, such as those appearing on the SWR circles of Fig 10-1. The T network is also the choice of most tuner manufacturers for reasons that are discussed later.

As stated earlier, the T network consists of two L sections with their parallel reactances connected together. This is shown in Fig 14-5. The impedance transformation in the T network is performed in two steps. As drawn in Fig 14-5, the right-hand L section (section 1) transforms the resistance in the load up to a value higher than the load and input resistances. The left-hand section (section 2) then transforms

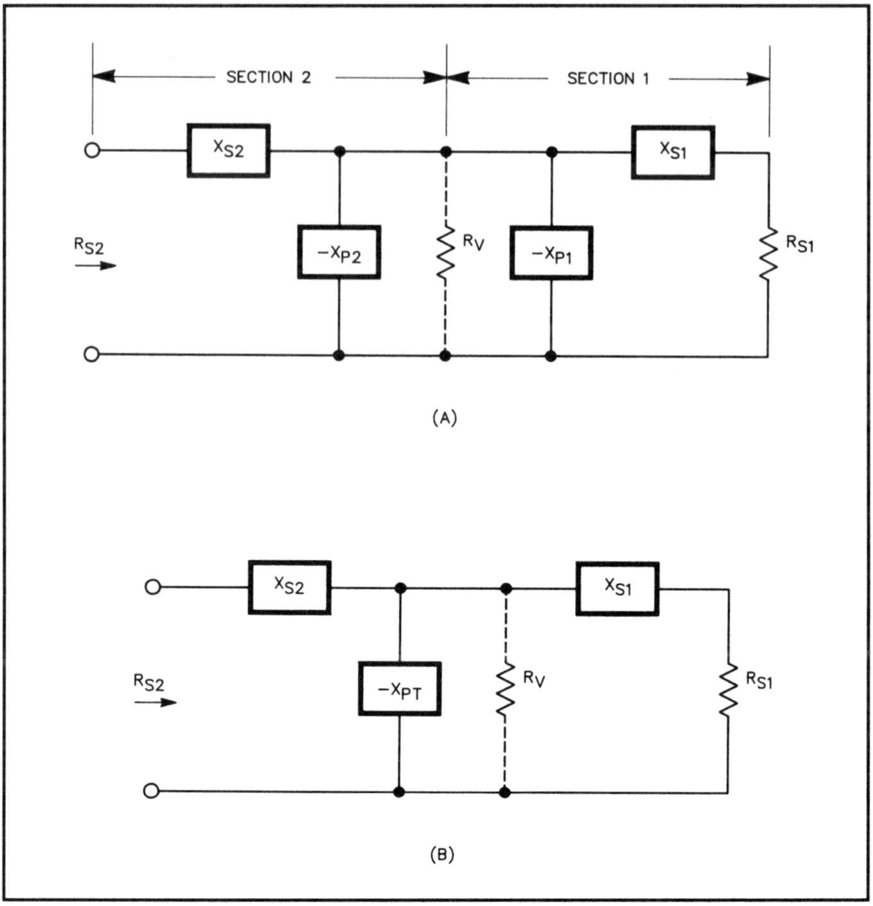

Fig 14-5—The T network, formed from two L sections. The circuits of A and B are equivalent.

that value down to the desired input resistance. The series arm of section 1 contains the load impedance that is to be transformed to the desired input resistance. Any reactance in the load becomes a part of series reactance X_{S1}. The resulting series network reactance is calculated later. The resistance in the load, R_{S1}, is transformed in section 1 to R_P, which is required to be higher than both the load resistance, R_{S1}, and the desired input resistance, R_{S2}.

R_P is a virtual resistance, not a physical resistor, so it is shown as R_V in Fig 14-5. R_V is treated as R_P in the equations, and is determined by R_{S1} and Q1 from Eq 14-2. Q1 can be any value that makes R_V higher than both R_{S1} and R_{S2}. Unless a high Q is desired for selectivity and harmonic rejection, Q is usually kept low to minimize losses from circulating currents. X_{P1} is determined by R_V and Q1 with Eq 14-7. The virtual resistance, R_V, becomes the load for section 2, where it is transformed down to the desired value of input resistance R_{S2}. X_{S1} is determined by R_{S1} and Q1 from Eq 14-5. If the reactance in the load is the same kind (same sign) as the series network reactance, the load reactance is subtracted from X_{S1} to obtain the network reactance; if of the opposite kind, the load reactance is added to X_{S1}. Because R_V and R_{S2} have already been determined, the Q of section 2 is also determined, obtained from Eq 14-4. X_{S2} is then obtained from R_{S2} and Q2 by using Eq 14-5. X_{P2} is determined by R_{P2} and Q2 from Eq 14-7.

Because X_{P1} and X_{P2} are in parallel (Fig 14-5A), a single reactance equal to the total value of the combination will suffice. This reactance is shown as X_{PT}, Fig 14-5B. If the two parallel reactances of X_{PT} are of the same sign, the parallel value is equal to the product of the two reactances divided by their sum; if they have opposite signs, divide the product by their difference. If the signs of the two reactances X_{P1} and X_{P2} are opposite, the net reactance is of the same sign as the smaller reactance.

The completed network is shown in Fig 14-5B. It is evident that there are several possible combinations of inductive and capacitive arrangements that may be used in a T network. If the objectives of the T network are to reject harmonics as well as to match impedances, the choice of reactances would be series inductors and parallel capacitors. Unfortunately, the opposite arrangement is most prevalent in amateur practice, even with the smaller amount of harmonic rejection available. However, one rule must be strictly followed for any combination of the network to perform correctly. This rule is that the two reactances in a single L section must be of the opposite sign.

Now let's work out an example using data from Sec 12.3. In the example there, the antenna impedance $Z = 50 - j90$ was transformed to $Z = 12 - j20$ through a 113-foot length of RG-8 coax. (We obtained the line-input impedance $Z = 12 - j20$ from the 5:1 SWR circle in Fig 10-1.) The line-input impedance, $12 - j20$, is the load to be matched to 50 Ω resistive. This load establishes $R_{S1} = 12\ \Omega$, and $R_{S2} = 50\ \Omega$. The results of this example are shown in Fig 14-6.

First we need to select a value for Q1 that makes R_V greater than R_{S2}, so we'll let Q1 = 5. From Eq 14-2, $R_V = 12 \times 26 = 312\ \Omega$. The

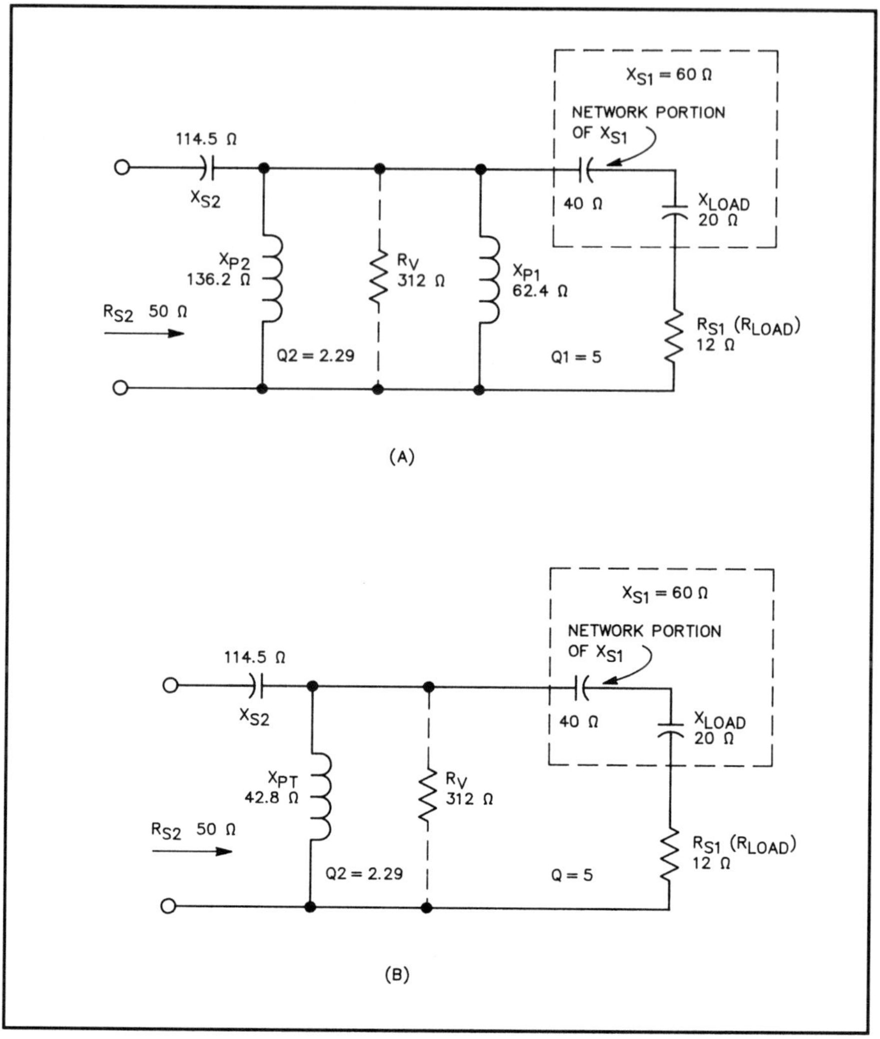

Fig 14-6—Two L sections combined to form a T network in a practical example. The circuits of A and B are equivalent, where the inductance X_{PT} is the resultant reactance of the two inductances X_{P1} and X_{P2} connected in parallel.

reactance X_{S1} required to transform 12 Ω to 312 Ω is found from Eq 14-5; $X_{S1} = 5 \times 12 = 60\ \Omega$. We must now account for the 20-Ω capacitive line-input reactance appearing in the load. Because this reactance appears in the series arm of L-section 1, it is a part of X_{S1}. We may choose capacitive reactance for the series element X_{S1} in Fig

Table 14-1

Component Values for the Example of Fig 14-6 at 3.5 MHz

Circuit Designation	*Reactance*	*Value*
$X_{S1} - X_{LOAD}$	−40 Ω	1136.8 pF
X_{S2}	−114.5 Ω	397.1 pF
X_{PT}	+42.8 Ω	1.95 µH

14-6. Then the capacitive reactance in the load must be subtracted from X_{S1} as calculated above to obtain the series reactance in the physical network. Therefore, the reactance of the capacitive series network element is $X_{S1} - X_{LOAD} = 60 - 20 = 40\ \Omega$. If the reactance in the load had been inductive (the opposite of the series network reactance we chose), it would have been added to the network reactance. Next, from Eq 14-7, we obtain $X_{P1} = R_V/Q1 = 312/5 = 62.4\ \Omega$. This completes the calculations for L-section 1.

The Q for section 2, Q2, is predetermined because R_V and R_{S2} are already determined. So from Eq 14-4, $Q2 = \sqrt{312/50 - 1} = \sqrt{5.24} = 2.29$. The series X_{S2} required to transform R_V, 312 Ω, down to 50 Ω is found from Eq 14-5; $X_{S2} = 2.29 \times 50 = 114.5\ \Omega$. From Eq 14-7 we obtain $X_{P2} = R_V/Q2 = 312/2.29 = 136.2\ \Omega$. For the total parallel reactance,

$$X_{PT} = \frac{X_{P1} \times X_{P2}}{X_{P1} + X_{P2}} = \frac{62.4 \times 136.2}{62.4 + 136.2} = 42.8\,\Omega$$

Finally, one more step is needed to convert the reactances to the circuit constants of the components in a matching unit. The constants for the operating frequency may be calculated from Eqs 14-8 and 14-9. The results using 3.5 MHz are given in Table 14-1.

$$C = \frac{159{,}150}{fX} \qquad \text{(Eq 14-8)}$$

$$L = 0.15915 \times \frac{X}{f} \qquad \text{(Eq 14-9)}$$

where

C = capacitance, pF
L = inductance, μH
X = reactance, Ω
f = frequency, MHz

The circuit constants appearing in Table 14-1 are realistic values, and should be within the range consistent with most commercially made tuners. It is important to realize that the antenna impedance with the 5:1 SWR used in the example is a realistic impedance for operation around 3.55 MHz with an 80-meter dipole cut for resonance at 3.75 MHz. The example also demonstrates that matching the 5:1 SWR with a tuner is practical, and should be routine.

For the example, a lower value of Q1 could have been chosen to satisfy the requirement that R_V be higher than R_{S1} and R_{S2}. However, some of the resulting reactances would have been lower than those found in practical tuners, so the more realistic Q was chosen. Because Q1 in the tuner is determined by R_{S1} and the setting of the output capacitor providing X_{S1}, here is a tip that will lead to the capacitor setting for the most efficient operation of the tuner. The minimum Q1 of the tuner (which is what we want) is obtained with the output loading capacitor set for its maximum capacitance. Try first to obtain a match with the maximum setting. If no match can be obtained, decrease the output capacitor setting in small increments until a match can be obtained. The match that can be obtained with the maximum possible capacitance in the output loading capacitor results in the lowest Q and maximum efficiency of the tuner. This subject is also touched on in the final paragraphs of Chapter 16.

Sec 14.3 Roller Inductor or Tapped Inductor

There seems to be the notion that the antenna tuner must be adjusted for a perfect 1:1, 50-Ω input impedance. Of course if the transmitter has been tuned into a 50-Ω dummy load, it's nice to be able to switch directly into a 50-Ω tuner input impedance without having to retune the transmitter. However, the tuner does not need to be adjusted for a perfect 50-Ω input impedance to obtain the optimum performance. Any tuner input impedance that the transmitter can load into with the same plate current at the dip as obtained with a 50-Ω dummy load is satisfactory, and will result in the same output power as when the tuner has been adjusted for a 50-Ω input impedance. The point of this last statement is to focus on the issue of which is better, a roller inductor or a tapped inductor.

It is true that, with the continuously variable roller inductor for X_{PT}, the parallel reactance in the tuner can be reduced exactly to zero for a perfect match. However, if a tapped inductor has taps at least at every other turn, obtaining a minimum reactance with a tapped inductor rather than zero reactance with the roller inductor will result in no difference in efficiency in the tuner. This is because any small residual reactance in the tuner is canceled by tuning for the resonant plate-current dip in the transmitter, and the indication of a small SWR at the input of the tuner is insignificant.

Let's return briefly to the subject of choosing which arrangement of reactances to use in the T network. There is a dramatic difference in harmonic rejection between networks of series-L parallel-C circuits versus series-C parallel-L equivalents. This difference is shown graphically in Figs 16-2 and 16-3. Those figures show the frequency response of both arrangements of the C and L reactances, and clearly indicate the superiority of the series-L parallel-C arrangement for harmonic rejection. My personal thoughts concerning the manufacturers' use of the less desirable of the two arrangements is that they initially used McCoy's Ultimate Transmatch circuit (*Ref 41*) because it was popular.[1] They probably didn't explore the use of T networks. In 1976 I showed in *QST* that the Ultimate is simply a T network with an unneeded capacitor—the shunt section of the split stator. (See Fig 7-1.) After that, the manufacturers dropped the split stator in favor of the single series input capacitor, probably without any engineering thought concerning which L-C arrangement would be best suited for amateur use.

BASIC computer programs were used to generate the frequency response data and input impedances shown in Figs 16-2 and 16-3, as well as Tables 16-1 and 16-2. The programs are presented in Chapter 15, and are available on diskette. These programs are for general use, enabling you to design or analyze T networks with various values of Q and reactive elements.[2]

---

[1] All references appear in Chapter 24.

[2] Programs in BASIC are available on ARRL diskette for the IBM PC and compatible computers. See information on an early page of this book.

Sec 14.4 The Ultimate Transmatch and the T Network

Servicing or repairing one's transceiver these days is practically impossible for the average amateur because of its sophisticated high-tech design. Building one's own equipment anymore is almost out of the question. However, because of its relative simplicity, the antenna tuner remains attractive to those who still like to build their own. There are also many fine factory-built tuners on the market, so there is a choice of whether to build or buy.

There are a few different circuit arrangements available among manufactured tuners, but the majority now use the T network. The T network has been around in the commercial field for a long time, but only recently has it been the preferred form for amateur use in antenna tuners. From the amateur viewpoint, the T network is an outgrowth of the well known Ultimate Transmatch circuit introduced by Lew McCoy, W1ICP, in *QST* (*Ref 41*), and shown here in Fig 7-1B. McCoy's circuit included a split stator capacitor for the input element, which has since been replaced with the single-element capacitor, as I have shown in Fig 7-1 for a reason that will soon become apparent. The Ultimate circuit was the favored arrangement among tuner manufacturers until I came along and upset the apple cart, with the result that most present-day tuners embrace the more simple T network with the single-element input capacitor. How this change came about involves a story I believe you'll find interesting.

While I was serving as editor of antenna and transmission-line articles for *Ham Radio* during the 1970s, it became necessary to explain the operation of the Ultimate Transmatch to a would-be *Ham Radio* author who submitted a circuit he claimed to be superior to the Ultimate. On analyzing the Ultimate circuit, I observed that the lower portion of the split stator capacitor shunting the output of the transmitter contributes nothing to the matching function. Without the lower portion, the circuit becomes a simple T network, as shown in Fig 7-1B. An experiment using a General Radio 1606-A RF impedance bridge verified the analysis. I then notified Dentron, for whom I was a consultant. They were ecstatic because they were about to reorder a large batch of the more expensive split-stator capacitors. Shortly thereafter, during a trip to Newington, I demonstrated my experiment in the ARRL Lab for some of the League staff, including Doug DeMaw, W1FB. Jim Fisk, then editor of *Ham Radio*, was also present. All were surprised to learn that the shunt section of the split-stator capacitor was unnecessary. I published my finding in *QST* in 1976 (Fig 7-1B here). Doug reported on my ARRL Lab demonstration in his Product Review of the Murch UT 2000-B antenna tuner in *QST* (*Ref 72*), and McCoy reported on the modification in *CQ* (*Ref 73*).

Dentron was the first to go to the simple T network, while others followed. It is also interesting to note that there is a similar situation in the Johnson Match Box tuners, which use a differential capacitor in the output circuit. The instruction book for the tuner says the capacitor is a balanced voltage divider, but I don't believe this is true, because the two capacitor portions shunting the balanced output terminals to ground can be disconnected without affecting the match. This is the same as in eliminating the shunt capacitor in McCoy's Ultimate. Evidently the capacitor portions coupling the terminals of the tank inductance to the output terminals are simply the series-arm reactances of balanced L sections in what I believe to be a modified unbalanced-to-balanced T network.

Sec 14.5 The Pi Network

The component values in the pi network are calculated using the same procedure as with the T network. The basic difference between the two networks is that, while the parallel reactance arms of the L sections are connected together in the T network, the series reactance arms are connected together in the pi network. And because the virtual resistance is in the series arms of both L sections in the pi arrangement, R_V must be lower than both the input resistance and the load resistance.

The pi network arrangement may be used in the Transmatch, or for other general uses, but its most prevalent use is in the output tank circuit in rigs having tubes in the final amplifier. In output tank service, the pi network must filter out the harmonic energy generated in the amplifier, in addition to providing a match between the output tube and the load. The requirement for harmonic filtering places two restrictions on the design: (1) the network must be a parallel-C series-L, low-pass configuration, and (2) the Q must be selected to achieve a predetermined value of harmonic rejection. In output tank service, a Q of 12 is generally considered appropriate.

We'll now calculate the values needed for a pi network to match a 2000-Ω tube load resistance R_{P2} to an output load resistance R_{P1} of 50 Ω with an input Q (Q2) of 12. The results of this example are shown

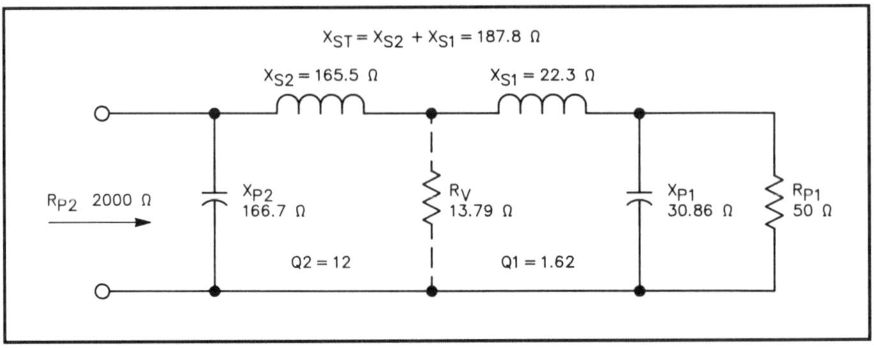

Fig 14-7—Two L sections combined to form a pi network. The component values are calculated as described in the text.

in Fig 14-7. With a Q2 value of 12, we find X_{P2} from Eq 14-7; X_{P2} = 2000/12 = 166.7 Ω. Because the virtual resistance R_V is in the series arms of the L sections, we treat it as R_S in the equations. Therefore we find R_V from Eq 14-6; R_V = 2000/145 = 13.79 Ω. X_{S2} is then found using Eq 14-5; X_{S2} = 12 × 13.79 = 165.5 Ω. This completes the calculations for the input L section. Because Q2 is greater than 10, the two reactances in the input section are nearly equal, but of opposite sign. Hence, the input section is equivalent to the standard parallel resonant tank circuit having the same Q.

We begin the calculations for the output section by finding the output Q (Q1) from R_V and R_{P1} using Eq 14-4; $Q1 = \sqrt{50/13.79 - 1}$ = 1.62. We now find X_{S1} using Eq 14-5; X_{S1} = 1.62 × 13.79 = 22.3 Ω. To complete the network calculations, we find X_{P1} using Eq 14-7; X_{P1} = 50/1.62 = 30.86 Ω. Because R_V is a virtual resistance and does not appear as a physical resistor, the values of X_{S1} and X_{S2} can be added to determine the total series reactance X_{ST}, permitting use of a single reactor for the series element. Thus, X_{ST} = 165.5 + 22.3 = 187.8 Ω.

As stated earlier, if the network is designed for general application, any combination of capacitive and inductive elements may be used for the reactances as long as the one cardinal rule is followed: The two reactances appearing in a single L section must be of the opposite sign to obtain resonance. However, if the design is for output tank service, the parallel elements must be capacitive and the series element inductive to obtain the required low-pass circuit action.

Elmer Wingfield, W5FD, has focused on an interesting aspect of pi network design in an article he published in *QST* in 1983 (*Ref 74*). In the T- and pi-network calculations we made in the examples above, the total operating Q (Q_O) of the networks is the sum of Q1 and Q2. The equations we used to determine the component values have been in general use for years, and they are correct. However, as Wingfield points out, in some instances the equations have been applied incorrectly, notably in the use of the input Q (Q2) as the operating Q (Q_O). When the input Q is 10 or more, the error in misusing Q2 as Q_O is negligible. An example is in our design of the pi network where we chose a Q of 12, but wound up with a total Q_O of 12 + 1.62 = 13.62. The error in this case is negligible, but when the ratio of the input to output resistance is less than 10, the error in using Q2 as Q_O becomes significant. For example, when the input and output resistances are equal, Q1 = Q2, so when taking Q2 as Q_O the error is 100%, because Q_O is twice the value of Q1 and Q2.

Wingfield has derived equations for obtaining the component values of the pi network based on the actual value of Q_O. You will find them useful when designing a pi network where the ratio of the input to output resistances is less than 10. This is an excellent reference. He has included in his *QST* article a computer program in BASIC for performing the calculations. If your copy of Wingfield's article is no longer available, you can obtain a photo copy from ARRL Headquarters.[3] However, I have included a copy of Wingfield's BASIC program in Chapter 15 and on disk, along with my own network programs in BASIC, using the old standard equations presented here.

Other references to the use of L sections in matching networks are found in *Refs 4, 61, 63, 64,* and *74*. Particular note should be taken of the reference by Grammer, W1DF (*Ref 4*), because it is by far the most clear, concise, and illuminating treatise on network design that I've seen published. In three installments, Grammer presents not only the "whats," but also the important "whys" of the circuit functions in the various network combinations. For the reader who wants to learn the "whys" of what you're reading here, the Grammer articles are a must. If you can't find a library copy, photo copies are available.[4]

---

3 ARRL HQ, 225 Main St, Newington, CT 06111. A nominal fee is charged for preparing and mailing photo copies.

4 See footnote 3.

Sec 14.6 Matching at the Load

One of the goals of this book is to provide enlightenment concerning the matching of impedances at the input of a feed line, in contrast to matching at the load, or antenna. However, in the discussion of inductively loaded HF mobile antennas in Sec 6.4, I stated that the matching could be performed either at the line input or at the load, depending on which was most convenient for the operator. So before leaving the subject of matching networks, I'll include a novel, yet simple method of providing L matching at the input terminals of inductively loaded HF mobile antennas. This method is not new. What follows is based on data that I contributed to *The ARRL Handbook*. It appears in Chapter 33 of all editions since 1986 (*Ref 1*).

With this method, which is especially adaptable to loading coils that have no means for either adjusting or tapping into the inductance, only one circuit element for the L network is required, the shunt element. Let's assume we'll use a capacitance for the shunt element. All that is required to determine the value of the capacitance is a reasonably accurate measurement of the resonant SWR. By extending the whip slightly beyond its natural resonant length, the series inductance component of the L network appears inherently in the feed-point impedance. At the correct length, the resulting parallel reactance component, X_L, causes the parallel resistance component, R_A, to equal the feed line characteristic impedance, Z_C Ω. The match is accomplished by canceling the parallel inductive reactance component, X_L, with the shunt capacitance of equal but opposite reactance.

To perform the matching operation, first resonate the antenna at the desired frequency by adjusting the whip length for minimum SWR without the capacitor. The approximate value of X_L with the whip lengthened to make R_A equal to Z_C may now be found from

$$X_L = Z_C \frac{\sqrt{SWR}}{SWR - 1} \qquad \text{(Eq 14-10)}$$

where SWR is that obtained at resonance. The reactance, X_C, of the shunt matching capacitor is the negative of X_L. The capacitance may be determined for the desired frequency, f, from

$$C(pF) = \frac{10^6}{2\pi f X_C} \qquad \text{(Eq 14-11)}$$

Form a capacitor from a combination of fixed mica capacitors in parallel, and connect the combination across the antenna input terminals. Finally, increase the whip length in small increments until the minimum SWR is reached—it should be very low.

If a lower SWR is desired, a trimmer capacitor may be added to the fixed capacitor bank, and by alternate adjustment of the trimmer and whip length, a perfect 1:1 match can be obtained. Once the size of the capacitor has been established for a given band, the antenna can then be matched at other frequencies in the band by simply adjusting the whip length for minimum SWR. If an inductive rather than a capacitive shunt element is preferred, replace the capacitor with an inductor having the same absolute value of reactance, and shorten the whip instead of lengthening it from the resonant length.

Chapter 15

Computer Programs for Impedance Matching

Sec 15.1 Introduction

This chapter was written to provide computer assistance in solving impedance-matching problems and for transforming transmission-line terminal impedances from one end of the line to the other. The chapter contains listings of nine programs written in the BASIC language. The programs are available on ARRL diskette for the IBM PC and compatible computers. (See information on an early page of this book.) These programs will enhance your enjoyment of experimenting with impedance-matching circuitry.

Five program listings relate to the design of T and pi networks, two of which include network input impedance versus frequency, and the resulting frequency response of the network. One additional program calculates transmission-line constants using data obtained from measurements made on the line. The three remaining programs transfer a known complex impedance from one end of a line to the other. The impedance-transformation calculations include the effect of line attenuation. With the detailed instructions provided here, these program listings should be helpful and easy to use. Two program listings for hand-held calculators are also included.

Sec 15.2 Designing T and Pi Matching Networks

T.BAS is listed as Program 1. This program is for designing or analyzing T networks, using the conventional equations appearing in Chapter 14. The prompts ask for the input and output resistances to be matched, and the Q of the output L section. The screen output yields the

Q of the input L section, and the reactance values of the components. The configuration can be either C-L-C or L-C-L. If it is desired to calculate using input Q instead of output Q, simply reverse the input and output resistances at the prompts, and then reverse the input and output series reactances when reading the screen output. The value of Q shown in the input Q column is the output Q.

PI.BAS, Program 2, is for designing pi networks, also using the same conventional equations from Chapter 14. As in the T network, the configuration can be either L-C-L or C-L-C. However, if the network is for a final-amplifier tank circuit, the configuration must be the low-pass C-L-C configuration to obtain the required harmonic rejection. In this program, the input Q is called for in the prompts. However, if you desire to calculate using the output Q, the same reversal technique described for the T network will yield the desired results.

The third and fourth network programs are listed as Program 3, T-CLC.BAS, and Program 4, T-LCL.BAS. Both programs are for designing and analyzing T networks using the conventional equations, but with two added bonuses: calculation of the input impedance of the network versus frequency, and frequency response of the network in decibels. T-CLC.BAS is for the C-L-C configuration, and T-LCL.BAS is for the L-C-L configuration. Both configurations yield the same component values of reactance. But because the C-L-C circuit is a high-pass filter and the L-C-L is low pass, the network input impedances are different, as are the frequency responses. These two programs were used to calculate the input impedances and frequency responses of the two T networks shown in Figs 4 and 5 in my Technical Correspondence item, "Wave Reflections in Attenuators, Filters, and Matching Networks," *QST*, November 1981 (included here as Figs 16-2 and 16-3). However, I want to point out a printing error in the *QST* article—the frequency response data in Table 1 goes with Fig 5, and that of Table 2 goes with Fig 4. This correction appears in "Feedback," *QST*, February 1982.

The capability of a computer program to produce frequency response data is significant; this enables us to determine the amount of harmonic rejection available with various values of Q in a network designed to provide a perfect match between the source and load at the fundamental frequency. So now you might ask, "What is the importance of knowing the input impedance of the network?" It is significant, because the key to harmonic rejection is mismatch. The network is designed to provide a match for delivery of the maximum available power, but *only at the fundamental frequency*. This is shown in Tables 16-1 and 16-2 as a perfect match only at the normalized frequency, f = 1. At all other frequencies, the input impedance is mismatched to the source, and the greater the difference in frequency relative to the fundamental f = 1, the greater the mismatch. Hence, the greater is the rejection of off-frequency power. Thus, the attenuation curves in Figs 16-2 and 16-3 are the direct result of the degree of mismatch occurring at each off-resonant frequency listed in the input-impedance tables. Note the dramatic difference in the attenuation of the high frequencies between the C-L-C and L-C-L configurations, as pointed out in Chapter 14. Further discussion appears in Chapter 16.

The last network program listing, Program 5, is W5FD.BAS. This program is for pi networks in which the total operating Q (Q_O) is used in the calculations. This program uses equations derived by Wingfield, W5FD (*Ref 74*),[1] to yield exact values of reactance using Q_O. It is especially important to use this program when the ratio of input to output resistances is less than 10, as explained in detail in Chapter 14.

Sec 15.3 Antenna Impedances From Measured Line-Input Impedances

Now we come to what for me is the most exciting and challenging aspect of impedance-matching exercises—the rapid transforming of complex impedances from one end of a transmission line to the other. It is exciting because the almost instantaneous calculation, either by computer or hand-held calculator, makes it practical to obtain antenna-terminal impedances covering a wide range of frequencies that would be impractical and too time consuming to perform manually on the Smith Chart. I'm referring to the determination of the resistance and reactance components of antenna terminal impedances at many frequencies across a band. These impedances are obtained by measuring impedances at the input of the feed line, and then transferring the measured impedances through the line to the antenna. It is challenging to obtain antenna-terminal impedance data that is sufficiently close to reality to be satisfying and rewarding. To do so requires an understanding of transmission-line principles, some math, the ability to take measure-

1 All references appear in Chapter 24.

ments with a good degree of precision, and the availability of impedance-measuring equipment that obtains acceptable precision in the measurement data.

With respect to the measuring equipment, I'm thinking in terms of RF bridges in the same league as the General Radio 1606-A impedance bridge. Perhaps there are noise bridges that have the precision required to obtain sufficient measurement accuracy, but at present I don't know of any that are qualified. So if you have access to the proper equipment and have the desire to learn the impedance at the antenna terminals instead of just the line SWR, the computer listings that follow should give you a lot of pleasure. Tables 15-1 through 15-5, along with Fig 15-1, are examples of information you can obtain with these programs. So are Tables 20-1 through 20-5 and Figs 20-1 through 20-3.

Sec 15.3.1 Calibration of the Feed Line

To obtain the terminal impedance of an antenna with acceptable accuracy requires that the measuring equipment be located at a sufficient distance from the antenna so as not to distort the RF field surrounding it. Hence, a feed line is required. To obtain a mathematical solution to the problem of transferring an impedance measured at the input terminals of the feed line to the impedance at its load terminals first requires calibration of the feed line. For the calibration, we need to know the true, measured characteristic impedance Z_C of the line—not just the nominal value of 50 or 75 Ω as listed in the manufacturer's catalog. We also need to know the electrical length and attenuation of the line. These calibration data are obtained by first making short-circuit and open-circuit impedance measurements on the feed line alone, that is, measuring the impedance at the line input with the far end terminated first in a short circuit, and then in an open circuit. Then the measured data is processed with TRANSCON.BAS, Program 6. This program yields Z_C, the electrical line length, and the line attenuation. An equivalent program listing for Hewlett-Packard hand-held calculators is given near the end of this chapter as Program HP1. When you run the computer program, supply the measured short- and open-circuit resistances and reactances at the prompts.

The short-circuit impedance measurement (Z_{SC}) and the open-circuit measurement (Z_{OC}) should be made on frequencies at which the electrical length of the feed line is fairly close to odd multiples of ⅛ λ. Frequencies that make the line length close to ⅛ λ and ⅜ λ are preferable for obtaining the best accuracy. For a detailed discussion on this subject, refer to *Ref 70, p 134.*

For verification that the calibration data is accurate at the antenna operating frequencies, measurements should also be made at frequencies of two successive odd multiples of ⅛ λ bracketing the antenna frequencies. The reason for using ⅛ λ multiples is that at these lengths the Z_{SC} and Z_{OC} measurements yield reactances X_{SC} and X_{OC} that are of similar magnitude, and are nearly equal to Z_C. Acceptable accuracy will be obtained if the ratio between X_{SC} and X_{OC} does not exceed 3:1. However, the resistance components R_{SC} and R_{OC} will be very low, with typical values as shown in Table 15-1, a feed-line calibration data printout from TRANSCON, Program 6. The lower these resistance values, the lower the line attenuation. The resistances would be zero on lossless line.

Mathematically, line reactance X_{SC} = Z_C tan len°, and X_{OC} = Z_C cot len° on lossless line, where len° is the electrical line length. However, the measured values will differ slightly on real lines with attenuation. Note that X_{SC} equals Z_C when len° is ⅛ λ (45°), because tan 45° = 1. An approximate value of Z_C may be found from the expression $Z_C = \sqrt{X_{SC} \times X_{OC}}$. TRANSCON (Program 6) yields the exact value of Z_C because the program includes the effect of resistances R_{SC} and R_{OC} resulting from line attenuation.

A short explanation at this point concerning the sign of the reactances X_{SC} and X_{OC} may be helpful. A short-circuited line shorter than ¼ λ is inductive, a positive reactance. An open-circuited line shorter than ¼ λ is capacitive, a negative reactance. Adding a ¼-λ section to lines shorter than ¼ λ reverses the sign; inductive becomes capacitive, and vice versa. Therefore, short-circuited lines of lengths greater than ¼ λ but less than ½ λ are capacitive, and open-circuited lines greater than ¼ λ but less than ½ λ are inductive. Adding ½ λ to any length of line will not change the sign of the reactance, because, as we learned earlier, a ½-λ line simply repeats the impedance seen at its load. Knowledge of these facts will assist you in understanding the results of your measurements. Note that the values and signs of the data entered for the X_{SC} and X_{OC} reactances appearing in Table 15-1 conform with the facts just outlined. Additional discussion on the circuital effects of shorted and open-circuited lines may be found in *The ARRL Antenna Book* (*Ref 71, p 24-11*).

Table 15-1

Printout of Feed Line Calibration Data from TRANSCON Program

This printout shows the results from three successive sets of data input. Measurement results are entered from the keyboard at the prompts (=?). See text for details.

```
FREQUENCY =? 1.9
Rsc =? 2.Ø
Xsc =? 5Ø.Ø
Roc =? Ø.6
Xoc =? -48.95
          CHARACTERISTIC    ELEC.LINE     LINE     LENGTH      ATTENUATION
            IMPEDANCE        LENGTH      ATTEN.    PER MHZ     PER SQR FREQ.
  FREQ.      ZC      ANGLE    DEGREES       DB        PHI          DB   A
----------------------------------------------------------------------------
  1.9ØØ    49.494   -Ø.794   45.31448    Ø.11343   23.84973     Ø.Ø8229

FREQUENCY =? 5.65
Rsc =? 2.1
Xsc =? -5Ø.177
Roc =? 2.5
Xoc =? 47.778
          CHARACTERISTIC    ELEC.LINE     LINE     LENGTH      ATTENUATION
            IMPEDANCE        LENGTH      ATTEN.    PER MHZ     PER SQR FREQ.
  FREQ.      ZC      ANGLE    DEGREES       DB        PHI          DB   A
----------------------------------------------------------------------------
  5.65Ø    49.Ø18   -Ø.299  134.3Ø46Ø    Ø.2Ø436   23.77Ø73     Ø.Ø8598

FREQUENCY =? 5.7
Rsc =? 1.9
Xsc =? -48.42
Roc =? 2.58
Xoc =? 49.47
          CHARACTERISTIC    ELEC.LINE     LINE     LENGTH      ATTENUATION
            IMPEDANCE        LENGTH      ATTEN.    PER MHZ     PER SQR FREQ.
  FREQ.      ZC      ANGLE    DEGREES       DB        PHI          DB   A
----------------------------------------------------------------------------
  5.7ØØ    48.994   -Ø.369  135.3161Ø    Ø.19837   23.73966     Ø.Ø83Ø9
```

We'll now examine the various data appearing in Table 15-1. The values shown there are typical. They result from open- and short-circuit impedance measurements on 41 feet of RG-214 coax plus a commercially made W2DU HF balun.[2] Real transmission lines have attenuation loss, so the characteristic impedance Z_C of real lines has a reactance component (negative), as well as the resistance component. This information, shown in the second and third columms in the table (Z_C), is in polar form. Moving to the right, we see the electrical line length in degrees, followed by the line attenuation in dB at the measurement frequency. However, for use with the impedance transformation programs to be described shortly, we need the data converted to line length per MHz (PHI) and attenuation per square root of MHz (A). That way, during the impedance transformation, the program multiplies PHI by the frequency of each antenna measurement, and multiplies A by the square root of each measurement frequency, so we have the resulting TRANSCON line-length and attenuation values corrected for each antenna-measurement frequency.

When collecting data, it is customary to take several measurements, rather than a single one, so each measurement provides an accuracy check on the others. In this way, we can spot a wild error that might otherwise go unknown and cause much distress later. TRANSCON has an elegant way of spotting measurement errors. If our measurements were perfect, with zero error, all values in the PHI column in Table 15-1 would be identical for all frequencies of measurement. And so would the values of A. Hence, a large deviation in either of these two columns indicates either a measurement error or an equipment malfunction. However, note in the table that the difference in the values in those two columns is very small, indicating good agreement between the successive measurements of Z_{SC} and Z_{OC}, even with large differences in measurement frequency.

Values of length per MHz, PHI = 23.7°, attenuation per square root of MHz, A = 0.082 dB, and the characteristic impedance, Z_C = 49,

---

[2] The W2DU balun is a choke balun made of a 10.5-inch length of 50-Ω RG-141 coaxial line with 50 no. 73-mix ferrite beads surrounding the coax. This is a current balun, and contains no ferrite-core transformer which would introduce serious errors in the impedance measurements. For details on the development of the W2DU balun, see Chapter 21. That chapter also explains why it is absolutely necessary to use a choke balun of the W2DU type if meaningful antenna impedance data is to be obtained. It tells why measurements on an antenna made either with no balun, or with a ferrite-core balun, yield erroneous data. For additional information concerning errors produced by ferrite-core balun transformers, see *Ref 118*.

obtained using program TRANSCON (as shown in Table 15-1), are the feed-line calibration data used in transforming impedances from one end of the transmission line to the other. Programs for this purpose are described in the next section.

Sec 15.3.2 Transmission-Line Impedance-Transformation Programs

LINTRNSM.BAS, LINTRNSA.BAS, and WRITE.BAS (Programs 7, 8 and 9), compute the unknown impedance appearing at one end of a transmission line when the impedance at the opposite end is known. The programs also yield the reflection coefficient and SWR at both ends of the line. The programs have built-in compensation to correct for the changes in electrical length and attenuation with changes in the frequency of the impedance measurements. The programs compute using the basic hyperbolic transmission-line equation shown in the heading box in the program listings. With LINTRNSM (the final letter M for manual), the impedance at a given frequency appearing at one end of the line is entered one line at a time from the keyboard, and the output is immediately sent to the screen and printer. With LINTRNSA (the final letter A for automatic), multiple impedance and frequency data are taken from a data file, and the calculated data outputs are automatically sent to the screen and printer, line by line in columnar form. Both programs operate from prompts.

Before using the impedance-transformation programs, you will need to determine the electrical length, attenuation, and characteristic impedance Z_C of the feed line at some frequency f, using the TRANSCON program to calibrate the feed line, as discussed in Sec 15.3.1. The first prompt appearing in the LINTRNSM program asks if the transformation is from the input to the load. Answer **y** if yes; **n** means the transformation is from the load to the input. If the answer is **n**, the program reacts to transform the impedance in the opposite direction. The next prompt asks for PHI, the electrical line length per MHz (from TRANSCON). The next prompt asks for A, the attenuation per square root of MHz. A is the dB value appearing in the right-hand column of the TRANSCON calibration printout (see Table 15-1). The program corrects for the electrical length and attenuation at the frequency of each antenna measurement by multiplying PHI times the measurement frequency, and A times the square root of the measurement frequency.

The next prompt asks for Z_C, which is printed out in the TRANS-

CON calibration printout. The next prompt asks for the frequency f of the first antenna measurement. The next two prompts ask for the measured impedance data, R and X, in that order, after which the program runs. After the calculated data of the first line is printed out, the next prompt asks for the frequency f and the data of the next antenna measurement, and so on. To enter calibration data for a new transmission line, type Control-C and RUN. The prompts will then ask for the new calibration data.

LINTRNSA runs from a data file stored on a disk, which includes the feed-line calibration data and a set of related impedance data at selected frequency intervals. The data file must be configured in the format shown in Table 15-2. The file begins with a text description of the test setup. The text is limited to 255 characters. No carriage returns are allowed until the end of this text description. A carriage return within the text will halt the program. The data begins on the line following the text description, and contains three items of data per line. The first data line contains PHI (the electrical length per MHz), the attenuation factor A (the attenuation per square root of MHz), and Z_C (the characteristic impedance of the line), in that order, obtained from the calibration data printout using TRANSCON. Each item of data must be separated by a comma, with a carriage return at the end of the line after the third data item. The second line, and all subsequent lines, each contain the measurement frequency, f, the line-input resistance, R_I, and the input reactance, X_I, in that order, with the same comma and carriage return sequence as in line 1. Spaces may follow the commas for easier readability of the data entries. The last data line must contain zeros—0,0,0.

Before you run LINTRNSA, select a file name and configure the test description and data in a disk file as instructed above. The first LINTRNSA prompt asks for the file name of the data file. It then asks if the transformation is from the input to the load. Answer **y** if yes; **n** means the transformation is to be from the load to the input. The program then takes the PHI, A, and Z_C calibration data from the disk file, followed by the impedance data obtained at all the frequency points measured. The program continues to run until it reaches the last data line, containing the zeros. The format of the data printout is shown in Table 15-3. The values in this table result from using the input data file illustrated in Table 15-2. Input data can be seen in the data printout of Table 15-3: descriptive text at the beginning, data line 1 in the feed line

Table 15-2

Data File Format for Use with Program LINTRNSA.BAS

No carriage returns are permitted inside the text description. Otherwise, a carriage return exists at the end of every line. This input data is contained in a disk file, available from the ARRL. See information on an early page of this book.

```
125' W2DU 80-m Dipole, Height 40', Calibrated RG-214 + W2DU HF Balun,
RF Bridge General Radio GR-1606-A, Receiver Kenwood TS-530, Signal
Generator Boonton 250A RX Meter. 80 METERS
23.7, 0.082, 49
3.45, 5.9, 6.9565
3.475, 6.7, 8.2014
3.5, 7.7, 9.4286
3.525, 9.0, 10.9220
3.55, 10.5, 12.3944
3.575, 12.25, 13.9021
3.6, 14.8, 15.1389
3.625, 17.9, 16.0
3.65, 22, 16.5753
3.675, 26, 15.7823
3.7, 30.9, 13.5676
3.725, 35, 9.7987
3.75, 37.2, 4.8533
3.76, 37.5, 1.8617
3.77, 37.8, -0.0398
3.775, 37.6, -1.457
3.78, 37.2, -2.5661
3.785, 37.2, -3.6988
3.79, 36.9, -4.7493
3.8, 35.6, -6.7105
3.825, 32, -10.8497
3.85, 28, -13.5065
3.875, 23.95, -14.4516
3.9, 20.6, -14.359
3.925, 17.5, -13.8854
3.95, 15.1, -12.9114
3.975, 12.95, -11.9497
4.0, 11.4, -10.375
4.025, 10.15, -9.2422
4.05, 9.1, -8.2716
4.075, 8.16, -6.9939
0,0,0
```

calibration section, and data in the remaining lines in the first three columns.

The printout format of LINTRNSA is the same for LINTRNSM, except that LINTRNSM prints out one line at a time as data is entered

Table 15-3

Printout of LINTRNSA Program for an 80-Meter Dipole

Input data is that shown in Table 15-2.

```
FILE --> a:dipole.dat <--
125' W2DU 80-m Dipole, Height 40', Calibrated RG-214 + W2DU HF Balun,
RF Bridge General Radio GR-1606-A, Receiver Kenwood TS-530S, Signal
Generator Boonton 250A RX Meter. 80 METERS

TRANSFORMING Zi = Ri+jXi MEASURED AT LINE INPUT TO ZL = RL+jXL at LOAD

FEEDLINE CALIBRATION:
LINE LENGTH IN DEGREES/MHz, PHI = 23.7
LINE ATTENUATION IN DB/SQR(MHz), A = .082
LINE IMPEDANCE, Zc = 49
```

F	Ri	Xi	RHOi	SWRi	RHO L	SWR L	RL	XL
3.450	5.90	6.957	.7889	8.47	.8171	9.933	55.22	-147.31
3.475	6.70	8.201	.7653	7.52	.7927	8.649	56.21	-136.31
3.500	7.70	9.429	.7370	6.60	.7635	7.457	57.75	-125.48
3.525	9.00	10.922	.7026	5.72	.7279	6.350	58.16	-112.97
3.550	10.50	12.394	.6655	4.98	.6896	5.443	58.40	-101.44
3.575	12.25	13.902	.6256	4.34	.6483	4.687	58.15	-90.45
3.600	14.80	15.139	.5704	3.66	.5912	3.892	59.37	-78.39
3.625	17.90	16.000	.5084	3.07	.5271	3.229	60.58	-66.58
3.650	22.00	16.575	.4345	2.54	.4505	2.640	60.62	-53.76
3.675	26.00	15.782	.3640	2.14	.3774	2.212	61.86	-42.99
3.700	30.90	13.568	.2791	1.77	.2894	1.815	62.32	-30.65
3.725	35.00	9.799	.2021	1.51	.2096	1.530	62.64	-19.44
3.750	37.20	4.853	.1478	1.35	.1533	1.362	63.60	-9.32
3.760	37.50	1.862	.1346	1.31	.1397	1.325	64.36	-3.88
3.770	37.80	-0.040	.1290	1.30	.1339	1.309	64.14	-0.33
3.775	37.60	-1.457	.1327	1.31	.1376	1.319	64.46	2.26
3.780	37.20	-2.566	.1400	1.33	.1453	1.340	64.98	4.39
3.785	37.20	-3.699	.1433	1.33	.1487	1.349	64.63	6.49
3.790	36.90	-4.749	.1511	1.36	.1567	1.372	64.70	8.54
3.800	35.60	-6.711	.1766	1.43	.1832	1.449	65.74	12.93
3.825	32.00	-10.850	.2468	1.66	.2561	1.688	67.15	24.37
3.850	28.00	-13.507	.3194	1.94	.3314	1.992	67.87	35.86
3.875	23.95	-14.452	.3889	2.27	.4036	2.353	69.44	47.23
3.900	20.60	-14.359	.4478	2.62	.4648	2.737	71.15	57.90
3.925	17.50	-13.885	.5067	3.05	.5260	3.220	72.28	69.85
3.950	15.10	-12.911	.5548	3.49	.5760	3.717	74.00	81.09
3.975	12.95	-11.950	.6020	4.02	.6251	4.334	74.42	93.31
4.000	11.40	-10.375	.6365	4.50	.6610	4.899	78.61	105.24
4.025	10.15	-9.242	.6670	5.01	.6928	5.510	80.00	116.23
4.050	9.10	-8.272	.6943	5.54	.7212	6.175	79.96	126.62
4.075	8.16	-6.994	.7195	6.13	.7475	6.920	82.50	139.11

from the keyboard. The data in Table 15-3 results from line-input impedance measurements made at frequencies listed in the column headed F. Measured input impedance data are listed in Columns Ri and Xi. RHOi, RHO L, SWRi, and SWR L are the magnitude of reflection coefficient and the SWR values at the line-input terminals and at the load (antenna), respectively. RL and XL are the resistance and reactance components of the dipole antenna impedance obtained by transforming the measured line-input impedance along the line to the antenna.

As mentioned earlier, the data shown in Tables 15-1 through 15-5, and 20-1 through 20-5, and in Figs 15-1 and 20-1 through 20-3 were obtained using the computer programs TRANSCON and LINTRNSA. The input data was obtained by measuring 40- and 80-meter antennas at W2DU through a feed line calibrated in characteristic impedance, electrical length, and attenuation, using the procedure described in Sec 15.3.1. The curves in the figures represent raw, unsmoothed data obtained directly from measurements using a General Radio 1606-A RF bridge. The smoothness of these curves attest to the quality of the results that can be obtained when using precision measuring equipment.

While LINTRNSA prints out the calculated data to hard copy, WRITE.BAS (Program 9) sends the output data to the screen and to a disk file. When you run WRITE, the program prompts you for the name of the data input file, and then for the name of the file in which you wish to save the calculated data. From there on, use the same procedure as for LINTRNSA.

Sec 15.4 Programs for Hewlett-Packard Hand-Held Calculators

If you prefer to use a hand-held calculator instead of a computer for solving the transmission-line problems, two calculator program listings are included that can substitute for the TRANSCON and LINTRNSM computer programs. They are written in reverse Polish notation for Hewlett-Packard calculators. I use HP-15C and HP-34C calculators. Other models may have different register designations than those shown in the programs.

The first is Program HP1, called TRANSMISSION-LINE CONSTANTS. It yields the same line-calibration data as TRANSCON (except for PHI and A—instead, the total line length and attenuation are yielded). The second, called TRANSMISSION-LINE IMPEDANCE TRANSFORMATION and listed as Program HP2, yields the same data provided by LINTRNSM and LINTRNSA, plus the angles of the reflection coefficient at both the input and output of the line, and the electrical length of the line at the frequency of each impedance measurement. The listings are written for calculators with twenty registers, ten numbered 0 through 9 and ten more that are accessed by first pressing a decimal point and then numbers 0 through 9. In the text that follows they are designated as R_0 through R_9 and $R_{.0}$ through $R_{.9}$.

Sec 15.4.1 Program TRANSMISSION LINE CONSTANTS

First, review Sec 15.3.1, "Calibration of the Feed Line." Second, load the calculator memory from the program listing. Next, load storage registers R_1 through R_4 with R_{SC}, X_{SC}, R_{OC}, and X_{OC} in that order, using data from TRANSCON, Program 6 (see Table 15-1), and then run the program. Remember that one of the reactance values must be negative. At the first pause (at program line 19), the line impedance, Z_C appears. To check the accuracy of the program so far, compare the reading with the Z_C value printed in the TRANSCON calibration printout. Also compare the next two readings with the line attenuation and line length values appearing on the same calibration printout. If there are no program errors, the results will be identical to the values in the printout to at least seven significant figures. It should be noted that the computer and the calculator each run on two completely different algorithms. Because the computer and calculator programs yield identical answers, each answer verifies the other, and the accuracy of both programs is proved.

At the second pause (at program line 75), the line attenuation in dB appears. At the third pause (at program line 82), the line length len° in degrees appears. If the sign of the number is positive and less than 90°, it indicates the correct length of the line if the actual line length lies between 0° and 90°. If the sign of the number is negative, press R/S one more time to add 180° to the reading of len°. The resulting number will be positive, and indicates the correct length if the actual line length lies between 90° and 180°. Note that the vital line constants that appeared at the pauses remain available because they are stored in the memory registers: Z_C in R_5, line attenuation in $R_{.2}$, line length len° in $R_{.3}$, and len° + 180° in $R_{.4}$.

If the actual line length is greater than 180° (½ λ), there is more work to be done. As we know, a ½-λ line repeats its load impedance, so the line lengths indicated by the calculator are ambiguous if the line is longer than ½ λ. Therefore, to obtain the correct reading we must first determine the approximate length in degrees from a physical measurement of the line. To do this, first measure the length of the line in feet, and then divide by the decimal value of the velocity factor—0.659 for polyethylene dielectric, 0.695 for Teflon, or 0.775 for foamed polyethylene.[3] This yields the electrical length of the line in feet, len_{FEET}. At 1 MHz, 360° = 1 λ = 983.57 feet. Now find len_F, the length in feet of 360° at the frequency f of the line measurement; $len_F = 983.57/f$. To find the electrical length of the line in degrees, len°, multiply the electrical length of the line in feet by 360 and divide by len_F; len° = $(360 \times L_{FEET})/len_F$. These calculations yield a length close enough to the true length to learn in which quadrant the real line length lies.

If the real length lies between 0° and 90°, the correct length is indicated by len° at program line 82. If the real line length lies between 90° and 180°, the length indicated at program line 82 will be negative, and len° + 180° (at line 88) will yield the correct answer. If the actual length lies between 180° and 270°, subtract len° from 180°, observing the sign of len°. If the length exceeds 270°, set up a four-quadrant diagram and draw in a radius line at len° from the reading at program line 82. Next determine which quadrant the real line length falls into from the calculation based on the physical measurement, noting the total number of degrees required to go around the four quadrants. Once you have found the correct quadrant, add 180° as needed to the len° to place the corrected len° value in the same quadrant as that obtained from the physical length, and then add 360° to len° for every complete revolution the line length made around the four quadrants.

After the true electrical length len° has been determined, divide this length by the line-measurement frequency, f, to obtain PHI, the length in degrees per MHz. Next divide the line attenuation α from register $R_{.2}$ by $\sqrt{f}$ to obtain the attenuation factor A, which is the attenuation per square root of MHz. We now have the required line calibration data—PHI, A, and Z_C—to be used with the program TRANSMISSION-LINE IMPEDANCE TRANSFORMATION, in the same manner as the TRANSCON calibration data is used with LINTRNSM and LINTRNSA.

---

3 The factor for foamed polyethylene is between 0.75 and 0.81, depending on the amount of air in the foamed material.

Sec 15.4.2 Program TRANSMISSION-LINE IMPEDANCE TRANSFORMATION

Before using this calculator program, it will be helpful to become familiar with Sec 15.3.2, the part explaining the computer program LINTRNSM. Although LINTRNSM computes from a different algorithm, it has the same purpose, achieves the same results, and operates from the same line calibration and input data. The algorithm used in this calculator program first converts the known impedance at one end of the line into its complex reflection coefficient, magnitude ρ at angle θ. It then modifies ρ in accordance with line attenuation factor A, the line attenuation per square root of MHz at the impedance measurement frequency f, and it modifies θ in accordance with PHI, the line length in degrees per MHz. It then converts the modified values of ρ and θ to the unknown impedance Z, appearing at the opposite end of the line. Consequently, as in LINTRNSM, the line length and attenuation are corrected for changes with frequency.

On lossless lines, magnitude ρ is constant all along the line. However, because of line attenuation, ρ is smaller at the input end of the line than at the load. Therefore, when transforming a known input impedance to determine the load impedance, input ρ is divided by the decimal attenuation factor A to enlarge it. Conversely, when transforming a known load impedance to determine the input impedance, the load ρ is multiplied by A to reduce it. Hence, the selection of multiplication or division by A is determined by which direction the impedance transformation is to be made. In LINTRNSM the selection is made at the prompt. In this calculator program the selection is made in the program at line 38. When going from the input to the load, key in division at line 38; when going from the load to the input, key in multiplication. Further, the value of the reflection coefficient angle θ increases when going from the input to the load, and vice versa. So for PHI to apply the proper correction to θ for the change in line length with frequency, load +PHI into the storage register $R_{.1}$ when transforming from the input to the load, and –PHI when going from the load to the input.

After loading the program into the memory, also load the remaining line-calibration data into storage registers—the attenuation factor A into register $R_{.0}$, and line impedance Z_C into register R_0. Next, load f_{MHz}, the frequency of each impedance to be transformed, into register R_1.

Don't forget to reload the new value of f with each new impedance. To run the program after the value of f is loaded, key in the resistance component R of the impedance to be transformed and press ENTER. Then key in the reactance component X, and execute the program. When execution is completed, the output data will appear at the various storage registers as indicated in the program listing, and as follows: unknown R at register R_8, unknown X at R_9, input SWR at $R_{.3}$, load SWR at $R_{.4}$, input ρ at R_4, input θ at R_5, load ρ at R_6, load θ at R_7, and line length in degrees at frequency f at $R_{.2}$.

To determine if the program is entered correctly, try it out using data from Table 15-3. Set program line 38 to multiply, and use +PHI. Load the PHI, A, and Z_C feed-line calibration data, and the f, R_I and X_I data from any of the lines in the table. After you run the program, the calculator output data should agree exactly with that of the printed computer output data in Table 15-3.

Sec 15.5 Determining Antenna Measurement Accuracy

In most of the measurements we make on antennas, the accuracy of the results is generally taken for granted within the known accuracy of the measuring device. In some cases, the accuracy of the measuring device is overlooked, or even ignored. However, in the area of quantitative measurement there are often procedures for checking the results to determine whether they are correct, or whether unknown errors in either the measuring device or the measurement procedure are producing answers that are incorrect, or even completely unrealistic. Without some sort of check, one may never know that the results are not what they should be. Fortunately, there is an elegant procedure for determining the accuracy of the measurements when transferring an impedance measured at the input terminals of a feed line to the terminals of the antenna.

I stated earlier, in the description of the TRANSCON program, that in measuring antenna impedance through a feed line, the electrical length, characteristic impedance, and attenuation of the feed line must be determined before measuring the antenna impedance. Accuracy of the values of these feed-line parameters *is of utmost importance*. The elegant feature for determining the accuracy of the measurements lies in the fact that any error from incorrect feed-line calibration will become evident if the antenna is measured at least two times, using a different length of feed line during each measurement. If errors exist in the calibration of the feed lines, the maximum divergence of the errors in the resulting antenna impedances will be obtained if the lengths of the two feed lines differ by ¼ λ. In fact, the best spread in the error determination is obtained if one line is ¼ λ and the other ½ λ. In other words, the maximum difference in the indicated antenna impedances because of feed-line errors will appear when you measure with these two specific lengths of feed line. The correct value of antenna impedance will then lie somewhere between the two spread values of the impedances obtained with the two different feed lines.

On the other hand, remember that the true antenna impedance cannot be found with an accuracy any better than that of the impedance-measuring device. So far, I have not become aware of any noise bridges that have the accuracy required to make satisfactory measurements of the type we've been discussing. I hope that RF bridges having the required accuracy, and within an acceptable price range for the amateur, will be available in the future.

I have included an example of this procedure, shown in the data of Tables 15-4 and 15-5, and in Fig 15-1, which plots the data. The details of the equipment used during the measurements are given below. The data in Table 15-4 was obtained using a feed line that is ¼ λ at 3.797 MHz; the feed line in Table 15-5 is ½ λ at 3.782 MHz, close enough in frequency to obtain the desired results.

For Table 15-4

125-foot W2DU 80-meter dipole, height 40 feet. Feed line: ¼ λ at 3.797 MHz, calibrated RG-214 plus W2DU HF balun.[4] RF bridge: General Radio 1606-A. Receiver: Kenwood TS-530S. Signal generator: Boonton 250A RX Meter.

For Table 15-5

125-foot W2DU 80-meter dipole, height 40 feet. Feed Line: ½ λ at 3.782 MHz, calibrated SF-214 plus W2DU HF Balun. RF bridge: General Radio 1606-A. Receiver: Kenwood TS-530S. Signal generator: Boonton 250A RX Meter.

Compare the values of antenna terminal resistance R_L and reactance X_L of Table 15-4 with those appearing in Table 15-5. The same

---

[4] See footnote 2 on page 15-7.

Table 15-4

Measured Antenna Impedance Data Transformed Through a Quarter-Wave Calibrated Feed Line

Feed line calibration:

Line length in degrees/MHz, PHI = 23.7 (¼ λ at 3.797 MHz)
Line attenuation in dB / $\sqrt{MHz}$, A = 0.082 (0.160 dB at 3.797 MHz)
Line impedance, Z_C = 49 Ω

Freq	R_I	X_I	RHO_I	SWR_I	RHO_L	SWR_L	R_L	X_L
3.450	5.90	6.957	0.7889	8.47	0.8171	9.933	55.22	−147.31
3.475	6.70	8.201	0.7653	7.52	0.7927	8.649	56.21	−136.31
3.500	7.70	9.429	0.7370	6.60	0.7635	7.457	57.75	−125.48
3.525	9.00	10.922	0.7026	5.72	0.7279	6.350	58.16	−112.97
3.550	10.50	12.394	0.6655	4.98	0.6896	5.443	58.40	−101.44
3.575	12.25	13.902	0.6256	4.34	0.6483	4.687	58.15	−90.45
3.600	14.80	15.139	0.5704	3.66	0.5912	3.892	59.37	−78.39
3.625	17.90	16.000	0.5084	3.07	0.5271	3.229	60.58	−66.58
3.650	22.00	16.575	0.4345	2.54	0.4505	2.640	60.62	−53.76
3.675	26.00	15.782	0.3640	2.14	0.3774	2.212	61.86	−42.99
3.700	30.90	13.568	0.2791	1.77	0.2894	1.815	62.32	−30.65
3.725	35.00	9.799	0.2021	1.51	0.2096	1.530	62.64	−19.44
3.750	37.20	4.853	0.1478	1.35	0.1533	1.362	63.60	−9.32
3.760	37.50	1.862	0.1346	1.31	0.1397	1.325	64.36	−3.88
3.770	37.80	−0.040	0.1290	1.30	0.1339	1.309	64.14	−0.33
3.775	37.60	−1.457	0.1327	1.31	0.1376	1.319	64.46	2.26
3.780	37.20	−2.566	0.1400	1.33	0.1453	1.340	64.98	4.39
3.785	37.20	−3.699	0.1433	1.33	0.1487	1.349	64.63	6.49
3.790	36.90	−4.749	0.1511	1.36	0.1567	1.372	64.70	8.54
3.800	35.60	−6.711	0.1766	1.43	0.1832	1.449	65.74	12.93
3.825	32.00	−10.850	0.2468	1.66	0.2561	1.688	67.15	24.37
3.850	28.00	−13.507	0.3194	1.94	0.3314	1.992	67.87	35.86
3.875	23.95	−14.452	0.3889	2.27	0.4036	2.353	69.44	47.23
3.900	20.60	−14.359	0.4478	2.62	0.4648	2.737	71.15	57.90
3.925	17.50	−13.885	0.5067	3.05	0.5260	3.220	72.28	69.85
3.950	15.10	−12.911	0.5548	3.49	0.5760	3.717	74.00	81.09
3.975	12.95	−11.950	0.6020	4.02	0.6251	4.334	74.42	93.31
4.000	11.40	−10.375	0.6365	4.50	0.6610	4.899	78.61	105.24
4.025	10.15	−9.242	0.6670	5.01	0.6928	5.510	80.00	116.23
4.050	9.10	−8.272	0.6943	5.54	0.7212	6.175	79.96	126.62
4.075	8.16	−6.994	0.7195	6.13	0.7475	6.920	82.50	139.11

Table 15-5

Measured Antenna Impedance Data Transformed Through a Half-Wave Calibrated Feed Line

Feed line calibration:

Line length in degrees/MHz, PHI = 47.599 (½ λ at 3.782 MHz)

Line attenuation in dB / $\sqrt{MHz}$, A = 0.1334 (0.259 dB at 3.782 MHz)

Line impedance, Z_C = 50.7 Ω

Freq	R_I	X_I	RHO_I	SWR_I	RHO_L	SWR_L	R_L	X_L
3.450	368.00	–76.232	0.7668	7.58	0.8118	9.627	53.17	–144.35
3.475	287.00	–131.799	0.7464	6.89	0.7904	8.541	52.49	–133.09
3.500	219.50	–140.286	0.7209	6.17	0.7636	7.459	53.13	–122.72
3.525	167.50	–132.340	0.6917	5.49	0.7327	6.483	52.91	–111.52
3.550	133.00	–117.042	0.6569	4.83	0.6960	5.579	53.30	–100.75
3.575	110.00	–100.000	0.6142	4.18	0.6510	4.730	54.34	–89.95
3.600	93.50	–84.028	0.5650	3.60	0.5989	3.987	55.13	–78.96
3.625	81.20	–69.241	0.5079	3.06	0.5385	3.334	55.52	–67.64
3.650	74.00	–55.890	0.4431	2.59	0.4699	2.773	57.09	–56.92
3.675	68.50	–43.265	0.3689	2.17	0.3913	2.286	58.30	–45.61
3.700	65.90	–31.622	0.2904	1.82	0.3081	1.891	60.49	–34.51
3.725	64.00	–20.993	0.2131	1.54	0.2261	1.584	61.93	–23.47
3.750	63.00	–10.533	0.1418	1.33	0.1505	1.354	63.01	–12.03
3.770	63.50	–2.782	0.1147	1.26	0.1217	1.277	64.31	–3.30
3.775	64.00	–0.927	0.1162	1.26	0.1234	1.281	64.92	–1.18
3.778	63.90	0.265	0.1152	1.26	0.1223	1.279	64.83	0.18
3.780	64.40	1.111	0.1194	1.27	0.1268	1.290	65.37	1.15
3.785	64.30	3.170	0.1214	1.28	0.1289	1.296	65.23	3.52
3.790	65.00	5.013	0.1308	1.30	0.1389	1.323	65.90	5.66
3.795	65.50	6.983	0.1406	1.33	0.1492	1.351	66.30	7.94
3.800	66.00	8.684	0.1503	1.35	0.1596	1.380	66.66	9.94
3.825	69.00	18.562	0.2152	1.55	0.2285	1.592	68.08	21.42
3.850	73.90	28.442	0.2872	1.81	0.3050	1.878	69.93	32.94
3.875	80.00	38.452	0.3548	2.10	0.3770	2.210	71.21	44.40
3.900	90.00	47.949	0.4171	2.43	0.4432	2.592	74.16	55.90
3.925	101.00	52.994	0.4547	2.67	0.4832	2.870	77.27	63.78
3.950	117.00	64.557	0.5150	3.12	0.5474	3.419	78.66	77.74
3.975	135.50	70.692	0.5543	3.49	0.5893	3.870	80.43	88.29
4.000	162.50	73.000	0.5925	3.91	0.6300	4.406	83.73	100.43
4.025	195.00	67.826	0.6255	4.34	0.6653	4.976	86.79	112.58
4.050	230.00	50.864	0.6533	4.77	0.6950	5.557	89.44	124.28
4.075	261.00	15.828	0.6757	5.17	0.7190	6.116	93.03	135.62

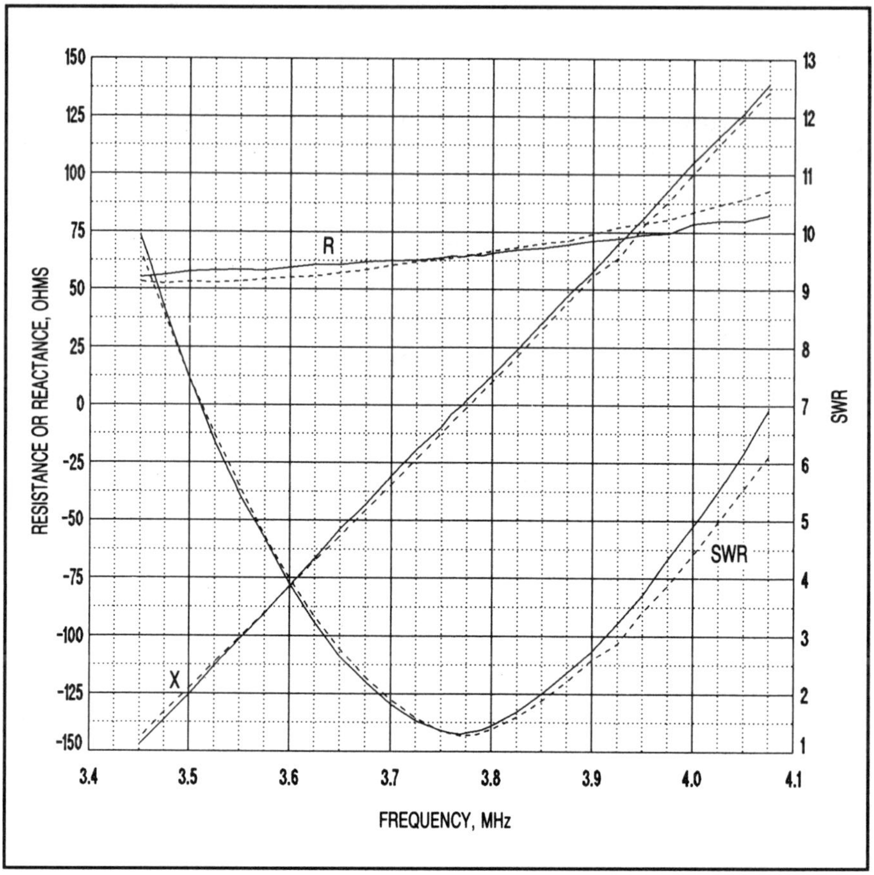

Fig 15-1—A comparison of resistance R, reactance X, and SWR of a 125-foot 80-meter dipole determined while measuring the impedances through different lengths of calibrated feed line. The solid lines indicate values obtained with a ¼-λ line (data from Table 15-4), and the broken lines indicate values obtained with a ½-λ line (data from Table 15-5). The calibrated feed lines include the 10.5-inch length of RG-141 coax inside a commercially made W2DU balun. See footnote 2 on page 15-7 for balun description. *(Computer-generated plot courtesy of Rick Maxwell, WB4GNR.)*

antenna was measured with two different line lengths, ¼ λ (results shown in Table 15-4) and ½ λ (results shown in Table 15-5). You'll note differences in the antenna impedances obtained with the different lines at each frequency. These differences result from the total error present in the measurements of the feed lines during their calibration, plus any error in the line-input measurements and in the mathematical transfer of the measured data to the antenna terminals through the LINTRNSA computer program. After examining Tables 15-4 and 15-5, and Fig 15-1, I believe you will agree that the differences between the measured impedances of my 80-meter dipole using the two different lengths of feed line are indeed small, which proves that a high degree of measurement accuracy can be obtained using the computer programs and the procedures I've presented.

Sec 15.5.1 Proof of Balun Effectiveness

The close agreement in the data obtained from the two different lengths of feed line shown in Fig 15-1 proves the effectiveness of the balun in eliminating antenna current from the outer surface of the coax shield. If substantial current was flowing on the coax shield because of an ineffective balun (or with no balun at all), the outer surface would be a part of the antenna radiating system. In that case, both the impedance and the SWR of the "antenna" would be affected by the length of the feed line. And if that were the case, there would be no agreement in the impedance or SWR data appearing in Fig 15-1, because the impedances and SWR values measured with the two different lengths of feed line would be drastically different.

This is the reason why many see a change in SWR when they change the length of the feed line! When a balun in the system is performing effectively, the *SWR will not change* when length of the feed line is changed. This is explained in more detail in Chapter 21.

Sec 15.5.2 Comparison of Data From Different Feed Line Lengths

There are two more points that are of interest. It is often said that a ½-λ feed line is necessary to measure the impedance or SWR of an antenna. This is not true. I have proved this to be incorrect with the evidence shown in Fig 15-1. However, the first point of interest here appears in the resistance R_I and reactance X_I columns in Table 15-5. These are the values of resistance and reactance measured at the input terminals of the feed line with the antenna connected at the other end. The values of resistance R_L and reactance X_L are those appearing at the load end of the line, which are at the input terminals of the antenna. These values indicate the impedance of the antenna.

Compare the values of R and X between the input and load columns in the rows where the frequencies are close to 3.782 MHz, say from 3.770 to 3.800 MHz. Notice how closely they agree, as they should do near the frequency where the feed line is ½ λ. However, the reason the resistance values at the line input are slightly lower than at the antenna is because of line attenuation, which is 0.259 dB at 3.782 MHz (attenuation A times the square root of 3.782). When a mismatched transmission line has attenuation, the value of the input resistance is always closer to the characteristic impedance Z_C than it is when the line is lossless.

Now make a similar comparison of the values of line input and load around 3.5 MHz and 4.0 MHz. The differences between the input and output impedances on the line at these frequencies are seen to be very large, indicating that the impedance repeatability of a ½-λ line is valid only at the frequency where it is precisely a half wave in length.

The second point of interest is in the R_I and X_I columns in Table 15-4. At frequencies where this feed line is close to ¼-λ at 3.797 MHz, we find impedance values that demonstrate the impedance-*inverting* action of the ¼-wave line. For example, at f = 3.775, R_I = 37.6 and R_L = 64.46. Because of the impedance-inverting action of the ¼-λ line, these two values of resistance lie equidistant geometrically on either side of the line impedance, Z_C. The proof is that the square root of the product of these two numbers is 49.2, while Z_C was found with TRANSCON to be 49 Ω. This is one way of *determining* the characteristic impedance Z_C of the feed line if it is ¼ λ long. However, since Z_C in the feed-line calibration data is 49, and since the X_I and X_L values are nearly exact negatives of each other, the quarter-wave length of the line and the characteristic impedance Z_C are both verified by the measurements.

In summary, the values of the antenna impedances appearing near the frequency at which the feed line is ¼ λ are inverted from those measured at the input terminals of the feed line. This brings us to the point of recognizing that both the impedance and the SWR of an antenna can be measured with a *feed line of any length*, as long as the electrical length is known and used in the impedance-transformation equation as described in the discussion of the LINTRNSM and LINTRNSA computer programs. And finally, observe the large differences in the values of R_I and X_I obtained using a ¼-λ line (Table 15-4) in contrast to those obtained using a ½-λ line (Table 15-5), with both lines terminated by the same antenna impedance.

Program 1

BASIC Listing of T.BAS

This program is available on ARRL diskette. See information on an early page of this book.

```
0 ' ****************************************************************************
10 '*                         "TEE" NETWORK CALCULATIONS                          *
15 '*                                                                           *
20 '*  CALCULATIONS EXECUTED USING STANDARD FORMULAS LISTED IN CHAPTER 14  *
25 '*                                                                           *
30 '*  BY WALTER MAXWELL          FILE NAME "T"               W2DU  1-05-89  *
40 '****************************************************************************
45 '
50 PRINT "Ri ="; : INPUT RI '   INPUT RESISTANCE, Ri
60 PRINT "RL ="; : INPUT RL '   LOAD RESISTANCE, RL
70 PRINT "QL ="; : INPUT QL '   Q OF LOAD SECTION, QL
80 '
90 PRINT TAB(4)"Rp"TAB(15)"Xs in"TAB(25)"Xp in"TAB(36)"Q In"TAB(46)"Xs Load";
100 PRINT TAB(59)"Xp Load"TAB(71)"Xp Total"
110 '
120 RP = RL*(QL^2+1)     '     Rp, VIRTUAL RESISTANCE
130 XSL = RL*QL          '     Xs Load, SERIES REACTANCE, LOAD SECTION
140 XPL = RP/QL          '     Xp LOAD, PARALLEL REACTANCE, LOAD SECTION
150 QI = SQR((RP/RI)-1)  '     QI, Q OF INPUT SECTION
160 XSI = RI*QI          '     Xs In, SERIES REACTANCE, INPUT SECTION
170 XPI = RP/QI          '     Xp In, PARALLEL REACTANCE, INPUT SECTION
180 XPT = RP/(QI+QL)     '     Xp Total, TOTAL PARALLEL REACTANCE, TEE NETWORK
190 '                          QI + QL = TOTAL Q OF NETWORK
200 '
210 M$="#####.##    ####.##   ####.##   ####.##      ####.##      ####.##      ####.##"
220 PRINT USING M$;RP,XSI,XPI,QI,XSL,XPL,XPT
230 PRINT : RUN
```

Program 2

BASIC Listing of PI.BAS

This program is available on ARRL diskette. See information on an early page of this book.

```
0 ' ****************************************************************************
10 '*                       "PI" NETWORK CALCULATIONS                          *
15 '*                                                                        *
20 '*  CALCULATIONS EXECUTED USING STANDARD FORMULAS LISTED IN CHAPTER 14     *
25 '*                                                                        *
30 '*  BY WALTER MAXWELL          FILE NAME "PI"            W2DU  2-21-89     *
40 '****************************************************************************
45 '
50 PRINT "Ri ="; : INPUT RI '   INPUT RESISTANCE, Ri
60 PRINT "RL ="; : INPUT RL '   LOAD RESISTANCE, RL
70 PRINT "QI ="; : INPUT QI '   Q OF INPUT SECTION, QI
80 '
90 PRINT TAB(3)"Rs"TAB(12)"Xs in"TAB(23)"Xp in"TAB(35)"Q Out"TAB(47)"Xs Load";
100 PRINT TAB(61)"Xp Load"TAB(72)"Xs Total"
110 '
120 RS = RI/(QI^2+1)    '      Rs, VIRTUAL RESISTANCE
130 XPI = RI/QI         '      Xp In, PARALLEL REACTANCE, INPUT SECTION
140 XSI = RS*QI         '      Xs In, SERIES REACTANCE, INPUT SECTION
150 QL = SQR((RL/RS)-1) '      QL, Q OF LOAD SECTION
160 XSL = RS*QL         '      Xs Load, SERIES REACTANCE, LOAD SECTION
170 XPL = RL/QL         '      Xp Load, PARALLEL REACTANCE, LOAD SECTION
180 XST = XSI + XSL     '      Xs Total, TOTAL SERIES REACTANCE, PI NETWORK
190 '                          QI + QL = TOTAL Q OF NETWORK
200 '
210 M$="####.##   ####.##   ####.##    ####.##     ####.##     ####.##    ####.##
220 PRINT USING M$;RS,XSI,XPI,QL,XSL,XPL,XST
230 PRINT : RUN
```

Program 3

BASIC Listing of T-CLC.BAS

This program is available on ARRL diskette. See information on an early page of this book.

```
0 ' ***********************************************************************
10 '*       T-NETWORK INPUT IMPEDANCE AND RESPONSE VS FREQUENCY            *
20 '*                                                                      *
30 '*                    SERIES C, PARALLEL L  (C-L-C)                     *
40 '*                                                                      *
50 '*                          FILENAME "T-CLC"             W2DU  1-25-89  *
60 '***********************************************************************
65 '
70 PRINT "INPUT R =";: INPUT RI : PRINT "LOAD R =";: INPUT RL:
80 PRINT "INPUT Q ="; :  INPUT Q1
90 C1 = -RI*Q1  : RP = RI*(1+Q1^2) : L1 = Q1/RP : Q2 = ABS(RP/RL-1)^.5 :
100 L2 = Q2/RP : L = 1/(L1+L2) : C2 = -Q2*RL
110 L1 = 1/L1 : L2 = 1/L2
120 PRINT
130 PRINT "INPUT R =";RI"      LOAD R =";RL"     Q1 =";Q1"   Q2 = ";Q2
140 PRINT "L1 = ";L1;" OHMS   L = ";L;" OHMS    L2 = ";L2;" OHMS
150 PRINT "C1 = ";C1;" OHMS       RP = ";RP;"         C2 = ";C2;"  OHMS"
160 PRINT : PRINT : PRINT "FREQUENCY": PRINT "FACTOR"; TAB(23);"INPUT Z";
170 PRINT TAB(48);"ATTEN, dB)" : PRINT
180 FOR F = .2 TO 5.5  STEP .2 : A = C1/F : B = F*L : C = C2/F : J = RL:
190 K = C : GOSUB 270
200 K = N+1/B : GOSUB 270
210 K = A+N : DB = ((J+50)^2+K^2)/(4*J*50) : DB = 4.34295*LOG(DB)
220 PRINT USING "#.#";F;: PRINT TAB(16)J;:IF K>=0 THEN PRINT "+J";: GOTO 240
230 PRINT "-J";
240 K = ABS (K):
250 PRINT USING "####.#";K;: PRINT TAB(48)DB
260 NEXT F
265 END
270 I = J^2+K^2 : M = J/I : N = K/I : J = M: RETURN
```

Program 4

BASIC Listing of T-LCL.BAS

This program is available on ARRL diskette. See information on an early page of this book.

```
Ø ' ***********************************************************************
1Ø '*        T-NETWORK INPUT IMPEDANCE AND RESPONSE VS FREQUENCY          *
2Ø '*                                                                     *
3Ø '*                   SERIES L, PARALLEL C  (L-C-L)                     *
4Ø '*                                                                     *
5Ø '*                        FILENAME "T-LCL"           W2DU  1-3Ø-89     *
6Ø '***********************************************************************
65 '
7Ø PRINT "INPUT R =";: INPUT RI : PRINT "LOAD R =";: INPUT RL:
8Ø PRINT "INPUT Q ="; :  INPUT Q1
9Ø  L1 = RI*Q1  : RP = RI*(1+Q1^2) :  C1 = -Q1/RP : Q2 = ABS(RP/RL-1)^.5 :
1ØØ  C2 = -Q2/RP : C = 1/(C1+C2) : L2 =  Q2*RL
11Ø  C1 = 1/C1 : C2 = 1/C2
12Ø PRINT
13Ø PRINT "INPUT R =";RI"      LOAD R =";RL"     Q1 =";Q1"   Q2 = ";Q2
14Ø PRINT "L1 = ";L1;" OHMS    RP = ";RP;"       L2 = ";L2;" OHMS
15Ø PRINT "C1 = ";C1;" OHMS   C = ";C;"  OHMS    C2 = ";C2;"  OHMS"
16Ø PRINT : PRINT : PRINT "FREQUENCY": PRINT "FACTOR"; TAB(23);"INPUT Z";
17Ø PRINT TAB(48);"ATTEN, dB)" : PRINT
18Ø FOR F = .2 TO 5.5  STEP .2 : A = L1*F : B = C/F : L = L2*F : J = RL:
19Ø K = L : GOSUB 27Ø
2ØØ K = N+1/B : GOSUB 27Ø
21Ø K = A+N : DB = ((J+5Ø)^2+K^2)/(4*J*5Ø) : DB = 4.34295*LOG(DB)
22Ø PRINT USING "#.#";F;: PRINT TAB(16)J;:IF K>=Ø THEN PRINT "+J";: GOTO 24Ø
23Ø PRINT "-J";
24Ø K = ABS (K):
25Ø PRINT USING "####.#";K;: PRINT TAB(48)DB
26Ø NEXT F
265 END
27Ø I = J^2+K^2 : M = J/I : N = K/I : J = M: RETURN
```

Program 5

BASIC Listing of W5FD.BAS

This program is available on ARRL diskette. See information on an early page of this book.

```
0   '*************************************************************************
10  ' *              Qo-BASED PI-NETWORK CALCULATIONS BY W5FD                 *
20  ' *                                                                       *
30  ' *  CALCULATING PI-NETWORK REACTANCES USING W5FD EQUATIONS BASED         *
40  ' *              ON ACTUAL CIRCUIT OPERATING Q (Qo).                      *
50  ' *                        SEE CHAPTER 14                                 *
60  ' *  FILE NAME: W5FD                          Edited by W2DU 2-21-89      *
70  ' *                                                                       *
80  ' *  SEE W5FD'S QST ARTICLE, AUGUST 1983, P23.  SEE ALSO CORRECTIONS,     *
90  ' *                 "FEEDBACK," QST, JANUARY 1984.                        *
100 '*************************************************************************
110 '
120 X1=0:X2=0:XL=0:R1=0:R2=0:Q1=0:Q2=0
130 INPUT "R1";R1
140 INPUT "R2";R2
150 INPUT "Qo";QO
160 IF (R1*R2*QO)<=0 THEN PRINT "NOT A PI NETWORK" : GOTO 310
170 '  SPECIAL CASE WHERE R1 = R2
180 IF R1 = R2 THEN X1  = (2*R1)/QO : X2 = (2*R2)/QO : Q1 = QO/2
190 IF R1 = R2 THEN XL = (R1*QO)/(Q1*Q1+1) : GOTO 290
200 IF ABS((QO*QO+1)-(R1/R2))<.01 THEN PRINT "L NETWORK" : GOTO 310
210 IF ABS((QO*QO+1)-(R2/R1))<.01 THEN PRINT "L NETWORK" : GOTO 310
220 IF R1/R2>QO*QO+1 THEN PRINT "NO SOLUTION" : GOTO 310
230 IF R2/R1>QO*QO+1 THEN PRINT "NO SOLUTION" : GOTO 310
240 Q1=(R1*QO-SQR(R1*R2*QO^2-(R1-R2)^2))/(R1-R2)
250 Q2=QO-Q1
260 X1=R1/Q1
270 X2=R2/Q2
280 XL=(R1*QO)/(Q1*Q1+1)
290 PRINT TAB(3)"R1"TAB(14)"R2"TAB(23)"QO"TAB(33)"Xc1"TAB(45)"Xc2"TAB(57)"XL"
300 M$="####.#    ####.##     ##.#     ####.##    ####.##     ####.##"
310 PRINT USING M$;R1,R2,QO,X1,X2,XL
320 PRINT : RUN
```

Program 6

BASIC Listing of TRANSCON.BAS

This program is available on ARRL diskette. See information on an early page of this book.

```
10 '  **********************************************************************
20 '  *                        W2DU PROGRAM                                *
30 '  *                                                                    *
40 '  *           OBTAINING TRANSMISSION-LINE CONSTANTS                    *
50 '  *                                                                    *
60 '  *     FROM MEASUREMENTS OF OPEN- AND SHORT-CIRCUITED LINES           *
70 '  *                                                                    *
80 '  *           1. COMPLEX CHARACTERISTIC IMPEDANCE                      *
90 '  *           2. ELECTRICAL LENGTH                                     *
100 ' *           3. ELECTRICAL LENGTH PER MEGAHERTZ                       *
110 ' *           4. ATTENUATION                                           *
120 ' *           5. ATTENUATION PER SQUARE ROOT OF MEGAHERTZ              *
130 ' *                                                                    *
140 ' * REFERENCE: "Theory and Problems of Transmission Lines,"            *
150 ' *            ROBERT A. CHIPMAN, PH.D.                                *
160 ' *            SCHAUM's Outline Series, McGraw-Hill, NY.               *
170 ' *                                                                    *
180 ' *  BY WALTER MAXWELL      Filename:  TRANSCON        9-24-88         *
190 ' **********************************************************************
200 '
210 PRINT
220 A=B=C=D=0
230 PRINT "FREQUENCY ="; : INPUT F 'FREQUENCY AT WHICH MEASUREMENTS WERE TAKEN
240 PRINT "Rsc ="; : INPUT A 'LINE-INPUT RESISTANCE WITH LINE SHORT CIRCUITED
250 PRINT "Xsc ="; : INPUT B 'LINE-INPUT REACTANCE WITH LINE SHORT CIRCUITED
260 PRINT "Roc ="; : INPUT C 'LINE-INPUT RESISTANCE WITH LINE OPEN CIRCUITED
270 PRINT "Xoc ="; : INPUT D 'LINE-INPUT REACTANCE WITH LINE OPEN CIRCUITED
280 RAD = 57.2957795# 'NUMBER OF DEGREES IN ONE RADIAN
290 ZSC = SQR(A*A+B*B) 'ABSOLUTE VALUE OF SHORT-CIRCUITED LINE-INPUT IMPEDANCE
300 SCA = ATN(B/A)*RAD 'ANGLE OF SHORT-CIRCUITED LINE-INPUT IMPEDANCE
310 ZOC = SQR(C*C+D*D) 'ABSOLUTE VALUE OF OPEN-CIRCUITED LINE-INPUT IMPEDANCE
320 OCA = ATN(D/C)*RAD 'ANGLE OF OPEN-CIRCUITED LINE-INPUT IMPEDANCE
330 ZC = SQR(ZSC*ZOC) 'CHARACTERISTIC IMPEDANCE, ZC, OF MEASURED LINE
340 ANG = (SCA+OCA)/2 'ANGLE OF COMPLEX ZC OF LINE
350 ' INPUT ---> (A+JB)/(C+JD)
360 ' OUTPUT ---> R = REAL PART OF RESULT, I = IMAGINARY PART OF RESULT
370 R = (A*C+B*D)/(C*C+D*D)
380 I = (B*C-A*D)/(C*C+D*D)
390 M = SQR(R*R+I*I)
400 P = ATN(I/R)*57.29577950000001#
410 M = SQR(M)
420 P = P/2
430 IF P<0 THEN P=90+P ELSE P=-(90-P)
440 ' CONVERT POLAR TO RECTANGULAR COORDINATES
450 I = M*SIN(P/57.29577950000003#)
460 R = M*COS(P/57.29577950000002#)
470 ' DO COMPLEX NUMBER DIVISION
480 RN = 1+R
490 IN = I
500 RD = 1-R
```

```
510 ID = -I
520 R2 = (RN*RD+IN*ID)/(RD*RD+ID*ID)
530 I2 = (IN*RD-RN*ID)/(RD*RD+ID*ID)
540 'CONVERT RECTANGULAR TO POLAR COORDINATES
550 M2 = SQR(R2*R2+I2*I2)
560 LN = LOG(M2)/2
570 DB = LN*8.685889638#
580 P2 =(ATN(IN/RN)-ATN(ID/RD))*57.2957795#
590 IF RD<0 GOTO 600 ELSE GOTO 610
600 P2 = P2-180 : GOTO 610
610 P3 = P2/2
620 IF P3<0 GOTO 630 ELSE GOTO 640
630 P4 = P3 + 180 : GOTO 650
640 P4 = P3
650 P5 = P4/F
660 DBF = DB/SQR(F)
670 PRINT TAB(11)"CHARACTERISTIC"TAB(28)"ELEC.LINE"TAB(41)"LINE"TAB(49)"LENGTH";
680 PRINT TAB(59)"ATTENUATION"
690 PRINT TAB(13)"IMPEDANCE"TAB(29)"LENGTH"TAB(40)"ATTEN."TAB(49)"PER MHZ";
700 PRINT TAB(59)"PER SQR FREQ."
710 PRINT TAB(3)"FREQ."TAB(13)"ZC"TAB(20)"ANGLE"TAB(29)"DEGREES"TAB(42)"DB";
720 PRINT TAB(51)"PHI"TAB(61)"DB   A"
730 PRINT TAB(1)"-----------------------------------------------------------------------------"
740 M$="###.###  ###.###  ##.###  ###.#####  ##.#####  ###.#####   ##.#####"
750 PRINT USING M$;F,ZC,ANG,P4,DB,P5,DBF
760 GOTO 210
```

Program 7

BASIC Listing of LINTRNSM.BAS

This program is available on ARRL diskette. See information on an early page of this book.

```
10 '   **********************************************************************
20 '   *                              W 2 D U                               *
30 '   *          TRANSMISSION-LINE IMPEDANCE-TRANSFORMATION PROGRAM        *
40 '   *              Transforming Zi at line input to ZL at load           *
50 '   *                           (or vice versa)                          *
60 '   *                                                                    *
70 '   *                           Zi - Zc TANH (A + jB)                    *
80 '   *                  ZL = Zc  ---------------------                    *
90 '   *                           Zc - Zi TANH (A + jB)                    *
100 '  *                                                                    *
110 '  *    To transform ZL to Zi answer "N" to prompt (ZL and Zi are       *
120 '  *    interchanged and (-) is changed to (+) in the equation).        *
130 '  *                                                                    *
140 '  *  WALTER MAXWELL      File Name: "LINTRNSM"  Rev. M     9-26-88     *
150 '  **********************************************************************
160 PRINT "PHI =";: INPUT P : PRINT "A =";: INPUT A : PRINT "Zc =";: INPUT Z1
170 PRINT "LINE-LENGTH CONSTANT IN DEGREES/MHz, PHI ="; P
180 LPRINT "LINE-LENGTH CONSTANT IN DEGREES/MHz, PHI ="; P
190 PRINT "LINE ATTENUATION CONSTANT IN DB/SQR(MHz), A ="; A
200 LPRINT "LINE ATTENUATION CONSTANT IN DB/SQR(MHz), A ="; A
210 PRINT "LINE IMPEDANCE, Zc ="; Z1
220 LPRINT "LINE IMPEDANCE, Zc ="; Z1
230 LPRINT
240 INPUT "TRANSFORMING FROM INPUT TO LOAD (Y,N) --> ";C$
250 IF C$="Y" OR C$="y" THEN Q=1 : C$="Y"  ELSE Q=-1 : C$="N"
260 IF C$="Y" THEN T$ = "TRANSFORMING Zi = Ri+jXi MEASURED AT LINE INPUT TO ZL = RL+jXL AT LOAD"
270 IF C$="N" THEN T$ = "TRANSFORMING ZL = RL+jXL AT LOAD TO Zi = Ri+jXi AT LINE INPUT"
280 LPRINT T$
290 LPRINT
300 P = P/57.29577910000001# : A = A/8.685889636000001#
310 IF C$="N" THEN GOTO 350
320 LPRINT TAB(4)"F"TAB(12)"Ri"TAB(21)"Xi"TAB(29)"RHOi"TAB(37)"SWRi";
330 LPRINT TAB(45)"RHO L"TAB(54)"SWR L"TAB(65)"RL"TAB(74)"XL"
340 GOTO 370
350 LPRINT TAB(4)"F"TAB(12)"RL"TAB(21)"XL"TAB(28)"RHO L"TAB(36)" SWR L";
360 LPRINT TAB(46)"RHOi"TAB(54)"SWR i"TAB(65)"Ri"TAB(74)"Xi"
370 LPRINT "--------------------------------------------------------------";
380 LPRINT "-------------"
390 R1=X1=R=X=0
400 PRINT "FREQUENCY ="; : INPUT F
410 PRINT "Ri MEASURED ="; : INPUT RI
420 PRINT "Xi MEASURED ="; : INPUT XI
430 S = RI : I = XI
440 S = S/Z1 : I = I/Z1 : B = Q*P*F : A1 = Q*A*SQR(F) : T = TAN(B) : R1 = S*(1+T^2)
450 R2 = 2*I+(I^2)*T+(S^2)*T : R1 = R1/(1+R2*T) : I2 = S^2+I^2-I*T-1
460 X1 = I+I2*T : X1 = X1/(1+R2*T) : H = (EXP(A1)-EXP(-A1))/(EXP(A1)+EXP(-A1))
470 R2 = R1*H-R1^2-X1^2-1 : R3 = 1-2*R1*H+H^2*(R1^2+X1^2) : R = (R1+H*R2)/R3
480 X = X1*(1-H^2)/R3 : R2 = R*Z1 : X2 = X*Z1 : R5 = R1*Z1 : X3 = X1*Z1
490 R4 = SQR((R-1)^2+X^2)/SQR((R+1)^2+X^2) : S3 = (1+R4)/(1-R4)
500 R6 = SQR((R1-1)^2+X1^2)/SQR((R1+1)^2+X1^2) : S4 = (1+R6)/(1-R6)
510  M$="##.### ####.##  ####.###   .####  ###.##   .####   ##.###   ###.##   ####.##
520 IF C$="N" THEN GOTO 560
530 PRINT TAB(4)"F"TAB(12)"Ri"TAB(21)"Xi"TAB(29)"RHOi"TAB(37)"SWRi";
540 PRINT TAB(45)"RHO L"TAB(54)"SWR L"TAB(65)"RL"TAB(74)"XL"
550 GOTO 580
560 PRINT TAB(4)"F"TAB(12)"RL"TAB(21)"XL"TAB(28)"RHO L"TAB(36)" SWR L";
570 PRINT TAB(46)"RHOi"TAB(54)"SWR i"TAB(65)"Ri"TAB(74)"Xi"
580 PRINT USING M$;F,RI,XI,R6,S4,R4,S3,R2,X2
590 LPRINT USING M$;F,RI,XI,R6,S4,R4,S3,R2,X2
600 GOTO 390
```

Program 8

BASIC Listing of LINTRNSA.BAS

This program is available on ARRL diskette. See information on an early page of this book.

```
10 '  *********************************************************************
20 '  *                             W 2 D U                               *
30 '  *        TRANSMISSION-LINE IMPEDANCE-TRANSFORMATION PROGRAM         *
40 '  *            Transforming Zi at line input to ZL at load            *
50 '  *                          (or vice versa)                          *
60 '  *                                                                   *
70 '  *                          Zi - Zc TANH (A + jB)                    *
80 '  *                ZL = Zc  ---------------------                     *
90 '  *                          Zc - Zi TANH (A + jB)                    *
100 '  *                                                                  *
110 '  *    To transform ZL to Zi answer "N" to prompt (ZL and Zi are    *
120 '  *    interchanged and (-) is changed to (+) in the equation).      *
130 '  *                                                                  *
140 '  *  WALTER MAXWELL      File Name: "LINTRNSA" Rev. N     9-27-88   *
150 '  ********************************************************************
160 INPUT "TYPE NAME OF DATA FILE --> ";A$
170 OPEN "I",1,A$
180 LINE INPUT #1,B$
190 FOR I = 1 TO LEN (B$)
200 MID$(B$,I,1) = CHR$(ASC(MID$(B$,I,1)) AND 127)
210 NEXT I
220 LPRINT "FILE --> ";A$;" <-- ";B$ : LPRINT
230 INPUT "TRANSFORMING FROM INPUT TO LOAD (Y,N) --> ";C$
240 IF C$="Y" OR C$="y" THEN Q=1 : C$="Y"  ELSE Q=-1 : C$="N"
250 IF C$="Y" THEN T$ = "TRANSFORMING Zi = Ri+jXi MEASURED AT LINE INPUT TO ZL = RL+jXL AT LOAD"
260 IF C$="N" THEN T$ = "TRANSFORMING ZL = RL+jXL AT LOAD TO Zi = Ri+jXi AT LINE INPUT"
270 LPRINT T$
280 INPUT #1,P,A,Z1
290 LPRINT
300 LPRINT "FEEDLINE CALIBRATION:"
310 LPRINT "LINE-LENGTH IN DEGREES/MHz, PHI ="; P
320 LPRINT "LINE ATTENUATION IN DB/SQR(MHz), A ="; A
330 LPRINT "LINE IMPEDANCE, Zc ="; Z1
340 LPRINT
350 P = P/57.29577910000001# : A = A/8.685889636000001#
360 IF C$="N" THEN GOTO 420
370 PRINT TAB(4)"F"TAB(12)"Ri"TAB(21)"Xi"TAB(29)"RHOi"TAB(37)"SWRi";
380 LPRINT TAB(4)"F"TAB(12)"Ri"TAB(21)"Xi"TAB(29)"RHOi"TAB(37)"SWRi";
390 PRINT TAB(46)"RHO L"TAB(54)"SWR L"TAB(65)"RL"TAB(74)"XL"
400 LPRINT TAB(46)"RHO L"TAB(54)"SWR L"TAB(65)"RL"TAB(74)"XL"
410 GOTO 460
420 PRINT TAB(4)"F"TAB(12)"RL"TAB(21)"XL"TAB(28)"RHO L"TAB(36)" SWR L";
430 LPRINT TAB(4)"F"TAB(12)"RL"TAB(21)"XL"TAB(28)"RHO L"TAB(36)" SWR L";
440 PRINT TAB(46)"RHOi"TAB(54)"SWR i"TAB(64)"Ri"TAB(74)"Xi"
450 LPRINT TAB(46)"RHOi"TAB(54)"SWR i"TAB(64)"Ri"TAB(74)"Xi"
460 LPRINT "-----------------------------------------------------------------";
470 LPRINT "-------------"
480 R1=X1=R=X=0
490 INPUT #1,F,RI,XI
500 IF F=0 THEN END
510 S = RI : I = XI
520 S = S/Z1 : I = I/Z1 : B = Q*P*F : A1 = Q*A*SQR(F) : T = TAN(B) : R1 = S*(1+T^2)
530 R2 = 2*I+(I^2)*T+(S^2)*T : R1 = R1/(1+R2*T) : I2 = S^2+I^2-I*T-1
540 X1 = I+I2*T : X1 = X1/(1+R2*T) : H = (EXP(A1)-EXP(-A1))/(EXP(A1)+EXP(-A1))
550 R2 = R1*H-R1^2-X1^2-1 : R3 = 1-2*R1*H+H^2*(R1^2+X1^2) : R = (R1+H*R2)/R3
560 X = X1*(1-H^2)/R3 : R2 = R*Z1 : X2 = X*Z1 : R5 = R1*Z1 : X3 = X1*Z1
570 R4 = SQR((R-1)^2+X^2)/SQR((R+1)^2+X^2) : S3 = (1+R4)/(1-R4)
580 R6 = SQR((R1-1)^2+X1^2)/SQR((R1+1)^2+X1^2) : S4 = (1+R6)/(1-R6)
590 M$="##.### ####.## ####.###   .####  ###.##    .####   ##.###   ###.## ####.##
600 PRINT USING M$;F,RI,XI,R6,S4,R4,S3,R2,X2
610 LPRINT USING M$;F,RI,XI,R6,S4,R4,S3,R2,X2
620 GOTO 480
```

Program 9

BASIC Listing of WRITE.BAS

This program is available on ARRL diskette. See information on an early page of this book.

```
10 '   **********************************************************************
20 '   *                              W 2 D U                                *
30 '   *        TRANSMISSION-LINE IMPEDANCE-TRANSFORMATION PROGRAM           *
40 '   *           Transforming Zi at line input to ZL at load               *
50 '   *                          (or vice versa)                            *
60 '   *                                                                     *
70 '   *                        Zi - Zc TANH (A + jB)                        *
80 '   *               ZL = Zc  ---------------------                        *
90 '   *                        Zc - Zi TANH (A + jB)                        *
100 '  *                                                                     *
110 '  *    To transform ZL to Zi answer "N" to prompt (ZL and Zi are        *
120 '  *    interchanged and (-) is changed to (+) in the equation).         *
130 '  *                                                                     *
140 '  *  WALTER MAXWELL       File Name: "WRITE"              4-28-89       *
145 '  *        To write calculated impedance data to disk file.             *
150 '  **********************************************************************
160 INPUT "TYPE NAME OF INPUT FILE ---> ";A$
165 INPUT "TYPE NAME OF OUTPUT FILE ---> ";D$
170 OPEN "I",1,A$
180 LINE INPUT #1,B$
190 FOR I = 1 TO LEN (B$)
200 MID$(B$,I,1) = CHR$(ASC(MID$(B$,I,1)) AND 127)
210 NEXT I
215 OPEN D$ FOR OUTPUT AS #2 LEN=3500
220 PRINT #2,A$
225 PRINT #2,B$
230 INPUT "TRANSFORMING FROM INPUT TO LOAD (Y,N) --> ";C$
240 IF C$="Y" OR C$="y" THEN Q=1 : C$="Y"  ELSE Q=-1 : C$="N"
250 IF C$="Y" THEN T$ = "TRANSFORMING Zi = Ri+jXi MEASURED AT LINE INPUT TO ZL = RL+jXL AT LOAD"
260 IF C$="N" THEN T$ = "TRANSFORMING ZL = RL+jXL AT LOAD TO Zi = Ri+jXi AT LINE INPUT"
265 WRITE #2,
270 PRINT T$
275 PRINT #2,T$
280 INPUT #1,P,A,Z1
290 WRITE #2,
295 PRINT #2,"FEED LINE CALIBRATION:"
310 PRINT #2,"LINE-LENGTH IN DEGREES/MHZ, PHI =",P
320 PRINT #2,"LINE ATTENUATION IN DB/SQR(MHZ), A =",A
330 PRINT #2,"LINE IMPEDANCE, ZC =",Z1
340 WRITE #2,
350 P = P/57.29577910000001# : A = A/8.685889636000001#
360 IF C$="N" THEN GOTO 420
370 PRINT TAB(4)"F"TAB(12)"Ri"TAB(21)"Xi"TAB(29)"RHOi"TAB(37)"SWRi";
375 PRINT #2,"    F        Ri        Xi      RHOi     SWRi      RHO L   SWR L      RL         XL"
380 PRINT #2,"-----------------------------------------------------------------------------------------"
390 PRINT TAB(46)"RHO L"TAB(54)"SWR L"TAB(65)"RL"TAB(74)"XL"
410 GOTO 480
420 PRINT TAB(4)"F"TAB(12)"RL"TAB(21)"XL"TAB(28)"RHO L"TAB(36)" SWR L";
440 PRINT TAB(46)"RHOi"TAB(54)"SWR i"TAB(64)"Ri"TAB(74)"Xi"
480 R1=X1=R=X=0
490 INPUT #1,F,RI,XI
500 IF F=0 THEN END
510 S = RI : I = XI
520 S = S/Z1 : I = I/Z1 : B = Q*P*F : A1 = Q*A*SQR(F) : T = TAN(B) : R1 = S*(1+T^2)
530 R2 = 2*I+(I^2)*T+(S^2)*T : R1 = R1/(1+R2*T) : I2 = S^2+I^2-I*T-1
540 X1 = I+I2*T : X1 = X1/(1+R2*T) : H = (EXP(A1)-EXP(-A1))/(EXP(A1)+EXP(-A1))
550 R2 = R1*H-R1^2-X1^2-1 : R3 = 1-2*R1*H+H^2*(R1^2+X1^2) : R = (R1+H*R2)/R3
560 X = X1*(1-H^2)/R3 : R2 = R*Z1 : X2 = X*Z1 : R5 = R1*Z1 : X3 = X1*Z1
570 R4 = SQR((R-1)^2+X^2)/SQR((R+1)^2+X^2) : S3 = (1+R4)/(1-R4)
580 R6 = SQR((R1-1)^2+X1^2)/SQR((R1+1)^2+X1^2) : S4 = (1+R6)/(1-R6)
590 M$="##.### ####.## ####.###   .####  ###.##    .####   ##.###   ###.##   ####.##"
600 PRINT USING M$;F,RI,XI,R6,S4,R4,S3,R2,X2
605 PRINT #2,USING M$;F,RI,XI,R6,S4,R4,S3,R2,X2
620 GOTO 480
```

Program HP1

Program Listing for TRANSMISSION-LINE CONSTANTS

This program is for hand-held calculators using reverse Polish notation. The program yields constants len°, α in dB, and Z_C from line-input impedance measurements of Z_{SC} and Z_{OC}. To find the attenuation factor A for use with the calculator impedance transformation program (Program HP2), divide α by $\sqrt{f(MHz)}$, the frequency of the line-input impedance measurements of Z_{SC} and Z_{OC}. Consult the text on how to find PHI from len°. *(W2DU, 1-1-79)*

Line	*Key Entry*	*Comments*
1	H Label A	
2	RCL 2	
3	RCL 1	
4	g → P	
5	STO 1	|Z| of Z_{SC}
6	x ↔ y	
7	STO 2	Angle of Z_{SC}
8	RCL 4	
9	RCL 3	
10	g → P	
11	STO 3	|Z| of Z_{OC}
12	x ↔ y	
13	STO 4	Angle of Z_{OC}
14	RCL 1	
15	RCL 3	
16	×	
17	f → √	
18	STO 5	$\|Z_C\|$
19	R/S	
20	RCL 2	
21	RCL 4	
22	+	
23	2	
24	÷	
25	STO 6	Angle of Z_C
26	RCL 1	
27	RCL 3	
28	÷	
29	f → √	

Store measured data into registers:

Data	*Register*
R_{SC}	R_1
X_{SC}	R_2
R_{OC}	R_3
X_{OC}	R_4

Run program

Algorithm:

$\tanh(\alpha + j\beta)\text{len}° = \sqrt{Z_{SC}/Z_{OC}}$

Read registers at end of program:

R_1	$\|Z_{SC}\|$
R_2	Angle of Z_{SC}
R_3	$\|Z_{OC}\|$
R_4	Angle of Z_{OC}
R_5	$\|Z_C\|$
R_6	Angle of Z_C
$R_{.2}$	α in dB
$R_{.3}$	len°
$R_{.4}$	len° + 180°

Line	Key Entry	Comments
30	STO 0	
31	RCL 2	
32	RCL 4	
33	–	
34	2	
35	÷	
36	RCL 0	
37	f → R	
38	STO 0	
39	x ↔ y	
40	STO 7	
41	x ↔ y	
42	1	
43	x ↔ y	
44	–	
45	STO 8	
46	RCL 7	
47	RCL 0	
48	1	
49	+	
50	g → P	
51	STO 9	
52	x ↔ y	
53	STO .0	
54	RCL 7	
55	CHS	
56	RCL 8	
57	g → P	
58	STO 0	
59	x ↔ y	
60	STO .1	
61	RCL 9	
62	RCL 0	
63	÷	
64	f → LN	
65	2	
66	÷	α in nepers
67	1	
68	g → e^x	
69	f → LOG	
70	2	
71	0	
72	×	
73	×	
74	STO .2	α in dB (line attenuation)

len° corrections for line lengths:

Actual Line Length	*From Calculator*
0° to 90°	len° as is ($R_{.3}$)
90° to 180°	len° + 180° ($R_{.4}$)
180° to 270°	180° – len°.

Line	Key Entry	Comments
75	R/S	
76	RCL .0	
77	RCL .1	
78	–	
79	2	
80	÷	
81	STO .3	len° (line length, see text)
82	R/S	
83	1	
84	8	
85	0	
86	+	
87	STO .4	len° + 180° (see text)
88	H Return	

Program HP2

Program Listing for TRANSMISSION-LINE IMPEDANCE TRANSFORMATION

This program is for hand-held calculators using reverse Polish notation. The program transfers a known impedance at one end of a line to the unknown impedance at the other end. *(W2DU, 7-20-80)*

Line	Key Entry	Comments
1	H Label A	
2	RCL 0	
3	÷	
4	STO 3	x (normalized)
5	x ↔ y	
6	RCL 0	
7	÷	
8	STO 2	r (normalized)
9	1	
10	+	
11	g → P	
12	STO 6	
13	x ↔ y	
14	STO 7	

Store line calibration and Z measurement frequency data into registers:

Register	*Data*
R_0	Z_C characteristic impedance
$R_{.0}$	A attenuation factor
R_1	f_{MHz} (for each measurement frequency)
$R_{.1}$	±PHI Use +PHI if going input to load, –PHI if going load to input.

To run:

1) Key in R of R + *j*X
2) Press ENTER
3) Key in X of R + *j*X
4) Run

Line	*Key Entry*	*Comments*
15	RCL 3	
16	RCL 2	
17	1	
18	−	
19	g → P	
20	RCL 6	
21	÷	
22	STO 4	ρ at known end
23	x ↔ y	
24	RCL 7	
25	−	
26	STO 5	θ at known end
27	RCL 1	
28	f → √	
29	RCL .0	
30	×	
31	CHS	
32	1	
33	0	
34	÷	
35	g → 10^x	
36	RCL 4	
37	x ↔ y	
38	÷ or ×	
(Use ÷ if going input to load, × if going load to input.)		
39	STO 6	ρ at unknown end
40	RCL .1	
41	RCL 1	
42	×	
43	STO .2	
44	2	
45	×	
46	RCL 5	
47	+	
48	STO 7	θ at unknown end
49	RCL 6	
50	g → x^2	
51	STO 2	
52	RCL 7	
53	f cos	
54	RCL 6	
55	×	
56	2	

Algorithm:

Step 1) Z = R +*j*X to $\rho \angle\theta$

Step 2) $\rho \angle\theta_{IN}$ to $\rho \angle\theta_{LOAD}$ (or vice versa)

Step 3) $\rho \angle\theta$ to Z = R +*j*X

Read registers at end of program:

Register	*Data*
R_4	ρ at known end
R_5	θ at known end
R_6	ρ at unknown end
R_7	θ at unknown end
R_8	Resistance at unknown end
R_9	Reactance at unknown end
$R_{.2}$	Line len° at f of measurement
$R_{.3}$	SWR at input
$R_{.4}$	SWR at load

Line	*Key* *Entry*	*Comments*
57	×	
58	–	
59	1	
60	+	
61	STO 3	
62	1	
63	RCL 2	
64	–	
65	RCL 3	
66	÷	
67	RCL 0	
68	×	
69	STO 8	Resistance answer
70	RCL 7	
71	f sin	
72	RCL 6	
73	×	
74	2	
75	×	
76	RCL 3	
77	÷	
78	RCL 0	
79	×	
80	STO 9	Reactance answer
81	RCL 4	
82	1	
83	+	
84	RCL 4	
85	1	
86	x ↔ y	
87	–	
88	÷	
89	STO .3	Input SWR
90	RCL 6	
91	1	
92	+	
93	RCL 6	
94	1	
95	x ↔ y	
96	–	
97	÷	
98	STO .4	Load SWR
99	H Return	

Chapter 16

Reflections in Attenuators, Filters and Matching Networks

(Adapted from Technical Correspondence, *QST*, November 1981)

Sec 16.1 Introduction

Network-type filters and attenuators have one thing in common—they both attenuate energy. However, this is just about the extent of their similarity; they operate on quite different principles and should not be confused with one another. Network-type filters attenuate harmonic and subharmonic energy. Network-type attenuators usually comprise resistive elements to absorb and dissipate power equally at *all* frequencies. See Fig 16-1 as an example of a network type attenuator. The elements are well matched to the system impedance so as not to reflect any power back to the source.

Filters, on the other hand, comprise low-loss reactive elements. One such circuit is shown in Fig 16-2. Low-loss elements are used so as to absorb or dissipate a minimum of power, thereby conserving power at the frequencies to be passed through the filter. (An ideal filter would have lossless elements, and thus dissipate no power.) The reactive elements in the filter are arranged so that the filter is well matched for minimum reflection at the frequencies to be passed, but mismatched for maximum reflection at all frequencies to be rejected. The rejection is accomplished by selectively reflecting back to the source all the power appearing at the unwanted frequencies, where, on return, the reflected power causes the source to be mismatched to the line at only the

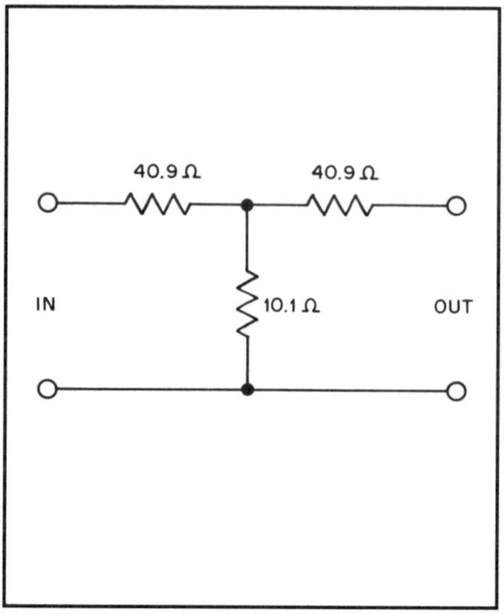

Fig 16-1—A network type of attenuator. The values indicated provide 20 dB of attenuation in a 50-Ω system. When terminated in 50 Ω, the input impedance of the network is 50 + *j*0, and the output energy is attenuated by 20 dB from that at the input, no matter what the input frequency, assuming the resistors have no reactance. Networks such as this may also be designed for unequal input and output impedances.

unwanted frequencies. Thus, attenuators attenuate all frequencies by absorption and dissipation, while filters attenuate only the *unwanted* frequencies by *selective reflection.*

The primary purpose of the matching network is to couple circuits that have different impedance levels, such as a power source and a load, so that the source will make its maximum power available to the load at the fundamental frequency. Low-loss reactive elements, as in the filter, are used to perform the matching function. See Figs 16-2 and 16-3. The distributed elements of transmission-line transformer and stub configurations may be used instead of lumped reactances, but the principles are the same.

The elements of the matching network are chosen so that it produces a mismatch at the fundamental frequency that is *complementary* to the mismatch that would arise if the source and the load were connected together directly without the matching network. Complementary mismatches produce complementary reflections, each mismatch producing a reflected voltage (and current) that is equal in magnitude, but of opposite phase, to that produced by the other.

When the matching network is inserted between the source and load of different impedances, complementary mismatches are produced, one between the source and the network, and its complement between the network and the load. The reflected voltages (and currents) of equal

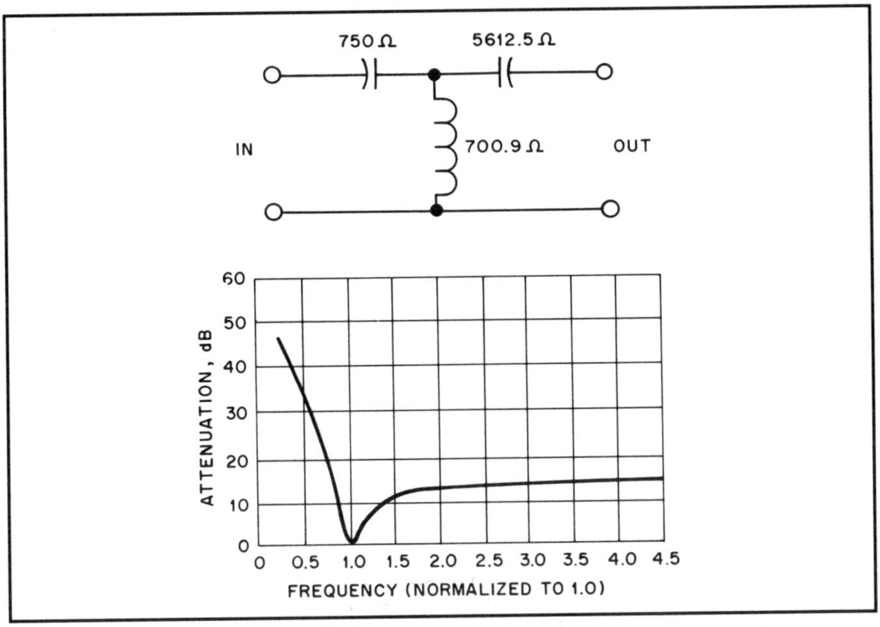

Fig 16-2—A T-network matching circuit of L and C elements, and a curve showing its attenuation versus frequency. The component reactances shown here provide a match for a 5000-Ω load to a 50-Ω source, with an unloaded input-section Q of 15. The attenuation curve is calculated for a 5000-Ω load of pure resistance at all frequencies, but the impedance of loads such as antenna systems are seldom if ever constant across such a frequency range. The input impedance versus frequency for the fixed 5000-Ω load is presented in Table 16-1. By changing the Q and component reactances in the circuit, many other combinations will provide the same 5000- to 50-Ω match, but the attenuation curve will assume a different shape.

magnitude and opposite phase resulting from the mismatches combine in the network to produce a resultant voltage (and current) of zero phase relative to the voltage (and current) of the source wave. Thus, the resultants of the reflected voltages and currents add in phase to those of the source wave. These in-phase additions to the source wave have the effect of re-reflecting the reflected voltages and currents into the forward direction, resulting in the cancellation of all rearward-traveling waves. Thus, a 1:1 conjugate match is formed, and because the source now has its desired or optimum impedance for a load, it makes its maximum power available at the actual load of a different impedance.

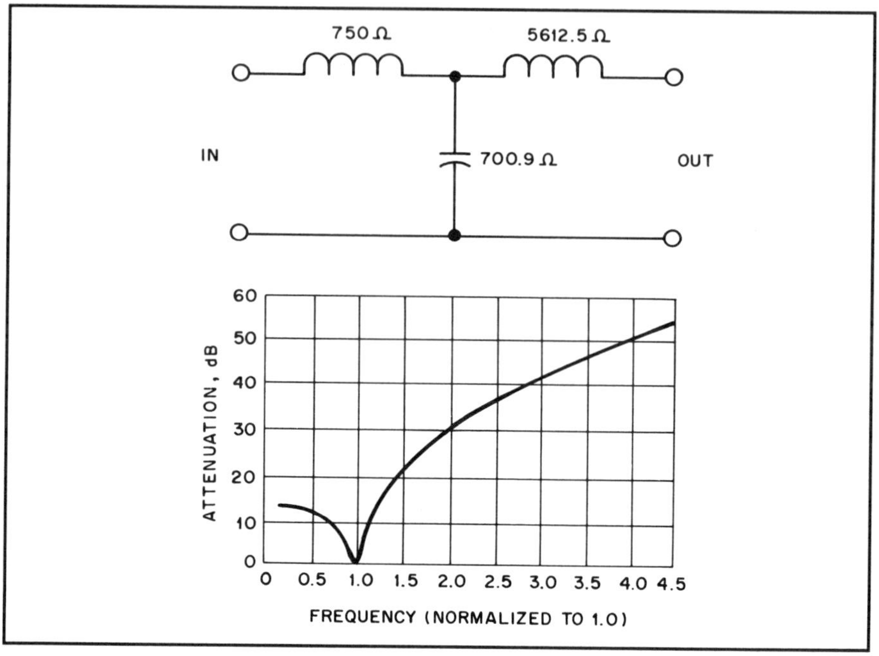

Fig 16-3—Another T-network matching circuit of L and C elements and its attenuation curve. The circuit is identical to that of Fig 16-2 except that inductances have replaced capacitances and vice versa; the reactance values remain the same for similar locations in the circuit. This network, too, will match a 5000-Ω load to a 50-Ω source. The input impedance versus frequency for a fixed 5000-Ω load is given in Table 16-2.

Keep in mind that the "load" now seen by the source is the combination of the actual Z_C-mismatched load and the matching network. Although it may seem difficult to appreciate that an additional mismatch is deliberately introduced to obtain a match (such as with a stub on a line), the mismatches described above must exist as Z_C mismatches to produce the controlled complementary reflections that are required to develop the conjugate match. However, no power is lost from the reflections, and, except for the small I^2R and E^2/R losses in the low-loss elements of the matching network, the maximum source power is transferred to the Z_C-mismatched, but conjugate-matched load. Details of the wave actions in developing the conjugate match are presented in Chapter 4.

As stated earlier, the principles of network and stub matching are identical. To emphasize the general concept that all conjugate matching

Table 16-1

Frequency Versus Network Input Impedance for the C-L-C Circuit of Fig 16-2

A fixed load of 5000 Ω and lossless circuit elements are assumed for all frequencies. The frequency is normalized to 1.0 for the fundamental.

Frequency	*Input Z, Ohms*	*Frequency*	*Input Z, Ohms*
0.4	1.8 – *j*1589.6	3.0	882.3 + *j*1811.7
0.6	8.4 – *j*814.4	3.2	996.5 + *j*1911.0
0.8	23.6 – *j*346.4	3.4	1111.9 + *j*1999.6
1.0	50.0 + *j*0.0	3.6	1227.6 + *j*2078.1
1.2	89.1 + *j*284.4	3.8	1343.1 + *j*2147.3
1.4	140.9 + *j*530.8	4.0	1457.6 + *j*2207.8
1.6	204.9 + *j*750.4	4.2	1570.8 + *j*2260.2
1.8	279.8 + *j*948.8	4.4	1682.1 + *j*2305.1
2.0	364.3 + *j*1129.1	4.6	1791.2 + *j*2343.1
2.2	456.9 + *j*1293.3	4.8	1897.9 + *j*2374.8
2.4	556.3 + *j*1442.7	5.0	2001.9 + *j*2400.8
2.6	661.2 + *j*1578.3	5.2	2103.0 + *j*2421.4
2.8	770.2 + *j*1701.1	5.4	2201.2 + *j*2437.2

Table 16-2

Frequency Versus Network Input Impedance for the L-C-L Circuit of Fig 16-3

A fixed load of 5000 Ω and lossless circuit elements are assumed for all frequencies. The frequency is normalized to 1.0 for the fundamental.

Frequency	*Input Z, Ohms*	*Frequency*	*Input Z, Ohms*
0.4	608.1 – *j*1512.1	2.8	1.2 + *j*1846.0
0.6	228.7 – *j*818.7	3.0	0.9 + *j*2013.4
0.8	100.8 – *j*349.0	3.2	0.7 + *j*2178.5
1.0	50.0 + *j*0.0	3.4	0.6 + *j*2341.8
1.2	27.1 + *j*282.5	3.6	0.4 + *j*2503.5
1.4	15.8 + *j*526.1	3.8	0.4 + *j*2664.0
1.6	9.8 + *j*745.2	4.0	0.3 + *j*2823.5
1.8	6.4 + *j*948.3	4.2	0.2 + *j*2982.0
2.0	4.3 + *j*1140.2	4.4	0.2 + *j*3139.7
2.2	3.0 + *j*1324.2	4.6	0.2 + *j*3296.8
2.4	2.1 + *j*1502.3	4.8	0.1 + *j*3453.2
2.6	1.6 + *j*1675.9	5.0	0.1 + *j*3609.1

is achieved by canceling one mismatch reflection with another, a few words concerning stub matching will be helpful. In stub matching, the stub introduces the deliberate mismatch, and thereby produces the complementary reflections that accomplish the conjugate match on the line. The stub is of the correct length and impedance to produce amplitudes of reflected voltage and current equal to those reflected by the load mismatch. The stub is placed on the line where the voltages (and currents) reflected by the load mismatch are of opposite phase to those produced by the stub. The line is thus matched (no reflections) between the stub point and the source. By comparison, the network generates the same complementary-mismatch reflections as the stub, and simultaneously adjusts the effective electrical distance from the mismatched load so that its complementary reflections are placed at the same electrical position on the line as a correctly placed stub.

Sec 16.2 Harmonic Rejection in Matching Networks

Let us now consider the harmonic rejection capability of matching networks. The key to the rejection is *mismatch*. As we know, the percentage of power that the source makes available to a load is proportional to the quality of the impedance match to the load. Conversely, and crucial to harmonic rejection, the amount of power the source *withholds*, relative to its maximum available power at a given frequency (such as a harmonic), is proportional to the impedance mismatch at that frequency.

Single-frequency matching networks are designed to provide *only one* complementary mismatch, compensating for the load mismatch appearing only at the fundamental frequency. Thus, the beauty of the network is that it inherently fails to provide complementary mismatches to compensate for the mismatches presented by the load at the harmonic frequencies. Consequently, as in the filters described earlier, the source remains mismatched at the undesired subharmonic and/or harmonic frequencies, and delivers reduced power at these frequencies in the amount that the degree of match (or mismatch) dictates. Tables 16-1 and 16-2 indicate this effect by showing the input impedances for the networks of Figs 16-2 and 16-3, respectively, for a range of frequencies and a fixed load. Only at the fundamental frequency does a match occur.

The most common amateur use of a matching network is the antenna tuner, or Transmatch. The network increases operating flexibility by providing the conjugate match to compensate for the mismatch between the feed line and the antenna that arises when a coax-fed antenna is operated off resonance, or when the antenna is fed with open-wire line. In these cases, the load terminating the matching network is the impedance appearing at the input terminals of the feed line. As we know, when the antenna impedance is mismatched to the Z_C impedance of the feed line, reflections from the mismatch cause the input impedance of the line to change from its resistive characteristic value Z_C (usually 50 or 75 Ω) to some new complex value determined by the antenna impedance and the length of the feed line. (See Chapters 7, 9 and 12.) It is this new line-input impedance (which changes as we change frequency) that we match to the source with the Transmatch at only the fundamental frequency.

Obviously the antenna presents a drastically different impedance to the feed line at harmonic frequencies compared with that at its fundamental frequency, and consequently the input impedances to the feed line are vastly different at the harmonic frequencies than what the matching network sees as a load at the fundamental frequency. These changes in load impedance with frequency, as well as the effects described earlier in the paragraph on filter principles, create reflections from the uncompensated mismatches at the harmonic frequencies that are responsible for the harmonic rejection of the matching network.

There are many arrangements and configurations of reactance elements that can be used equally well to generate a complementary mismatch for obtaining a conjugate match. However, there are vast differences among these various configurations concerning the severity of their noncomplementary mismatches that are required to reject harmonics and other unwanted frequencies. Figs 16-2 and 16-3 illustrate this simply. Note that both circuits are T networks having identical reactance values in similar positions of the circuits; only the type of reactance differs, capacitive versus inductive. Both circuits will match a 5000-Ω load to a 50-Ω line at the fundamental frequency. The C-L-C circuit of Fig 16-2 is commonly called a high-pass configuration, and the L-C-L circuit of Fig 16-3 a low-pass configuration. Both circuits, however, attenuate frequencies both lower and higher than the fundamental, as shown by the frequency-response curves. Of course, as frequencies of the harmonic energy are higher than the fundamental, a matching network of low-pass configuration would be the most desirable, a circuit with series inductance and shunt capacitance arms. For a more detailed explanation of the role played by reflections in the technique of impedance matching and harmonic rejection, see Chapters 3 and 4.

As a parting thought concerning the operational adjustment of Transmatches, the amateur should be aware of the trade-off between optimum efficiency (minimum power lost in the network) and optimum harmonic rejection. There are usually several settings of the network that will yield a 50-Ω impedance at the input of the Transmatch. As far as the transmitter is concerned, all such settings are equivalent—the same as if you were feeding a 50-Ω dummy load. However, the 50-Ω input obtained with the lowest loaded Q has the optimum efficiency. At the other extreme, the 50-Ω inputs obtained with the higher loaded Q values offer increased rejection of unwanted frequencies, but with some increase in insertion loss. In either case, the loss is usually small and is well worth the improved operation flexibility afforded by the Transmatch. In the circuits of Figs 16-2 and 16-3, higher Q values are obtained with greater reactances in the series-input arms (lower input capacitance in Fig 16-2, higher input inductance in Fig 16-3). This subject is also touched on in the final paragraph of Sec 14.2.

In conclusion, I would further point out that Transmatch settings which yield a 50-Ω impedance at the input to the Transmatch do not mean the SWR is 1:1 on the feed line between the Transmatch and the antenna. It means only that a 1:1 SWR exists on a 50-Ω line preceding the Transmatch. You have not "brought the SWR down" on the antenna feed line by adjusting the matching network; you simply have a 50-Ω impedance looking into the input of the Transmatch.

The frequency response and input impedance data presented in this chapter for the two T networks were obtained by computer, using the BASIC language. Two different computer programs were used, one for the C-L-C network of Fig. 16-2, and one for the L-C-L network of Fig 16-3. Listings for these two computer programs are presented in Chapter 15, along with an explanation of how to use them for determining the frequency response and input impedance of these two network configurations in general applications.

Chapter 17

How Does A Transmatch Work?

(Adapted from Technical Correspondence, *QST*, March 1985)

Sec 17.1 Introduction

In the Technical Correspondence column of *QST* for October 1984, Clifford Ward, WA5LVG, tells us there is confusion concerning the use of a Transmatch (*Ref 128*).[1] Comments from amateurs I have encountered during my lectures on this subject verify that he is correct. Unfortunately, his attempt to dispel the confusion fails because most of the remarks that follow his initial statement are simply not true. He tells us that, with a Transmatch:

1) Only feed-line tuning is accomplished. (This is not true; the antenna is also tuned.)

2) Antenna system performance is increased only when the Transmatch is located at the antenna feed point. (Not true.)

3) We are providing a proper load only for the transmitter, and not improving the antenna system efficiency. (Also not true.)

Many amateurs avoid using a Transmatch because they misunderstand these points. Let us examine some aspects of impedance matching to understand why Ward's statements are incorrect, and learn the true advantage of using a Transmatch.

It is appropriate to quote here two well-known network theorems, which are pertinent to the subject of impedance matching. They are the Maximum Power Transfer Theorem, and the Conjugate Matching Theorem. The Maximum Power Transfer Theorem states: "The maximum power will be absorbed by one network from another joined to it at two terminals if the impedances looking into the two networks at the

1 All references appear in Chapter 24.

junction are conjugates of each other." (From *Ref 17, p 49*, and *Ref 69, p 403*.) The Conjugate-Matching Theorem states: "If a group of four-terminal networks containing only pure reactances (or lossless lines) are arranged in tandem to connect a generator to its load, then if at any junction there is a conjugate match of impedances, there will be a conjugate match of impedances at every other junction in the system." (From *Ref 17, p 243*, and *Ref 69, p 407*.)

The same principles of transmission-line impedance matching apply, and the same wave actions and reflections occur, whether a match is achieved with a tuned stub or with a network of inductors and capacitors. (See Chapter 4.) In either case, the impedance match achieved is a conjugate match. So let us discuss the Conjugate-Matching Theorem as a basis for an evaluation of Mr. Ward's statements.

Sec 17.2 The Conjugate Match

The term "conjugate match" identifies a condition where the impedances on opposite sides of a junction have identical resistive components, and reactive components that are equal in magnitude but opposite in sign. For example, a conjugate match exists when a source impedance of $50 + j10\ \Omega$ feeds a load impedance of $50 - j10\ \Omega$. When a conjugate match is accomplished at any of the junctions in a system, *all* reactances in the system are canceled, *including any reactance in the load.* This reactance cancellation establishes resonance in the entire system, and the generator delivers its maximum available power to the load. This means that *a nonresonant antenna* as the load is *tuned to resonance by the conjugate match.* Because the Conjugate Matching Theorem states that maximum power transfer occurs equally well with the matching operation performed at *any* junction in the system, it is clear that Ward's statements violate the theorem and are therefore incorrect.

I can illustrate the principles of the Conjugate Matching Theorem with an example. Imagine an ideal system in which a Transmatch and transmission line that feed an antenna are both lossless. (It is customary to assume lossless lines and components while discussing the principles, and then treat the effects of attenuation and loss later.) We'll assume that a Z_0 mismatch exists at the junction of the feed line and a nonresonant antenna, meaning that the line impedance, Z_C, does not match the input-terminal impedance, $R + jX$, of the antenna. (Z_C, and sometimes Z_0, are abbreviations for the characteristic impedance of a transmission line.) We'll also assume that the RF power source, or generator, is adjusted to deliver its maximum available power into the line impedance Z_C.

Case 1—Mismatched Antenna, No Matching

First, we'll consider the case where the generator is connected directly to the input terminals of the mismatched line with no Transmatch. In this case, there are two mismatches at the junction of the antenna and the feed line—both a Z_0 mismatch and a conjugate mismatch. The Z_0 mismatch causes a reflection loss that is transferred back along the line to the generator, presenting the same magnitude of Z_0 mismatch at the line input (*Ref 19, p 5*). This mismatch causes the generator to deliver less than its maximum available power by an amount equal to the reflection loss. Considering that we've specified lossless line in this ideal case, there is no power lost to dissipation in the line—only a reduction in power delivered by the generator (*Refs 19* and *70, p 204*). (A point to remember: All power delivered to a lossless line is absorbed in the load, regardless of the degree of load mismatch. See Fig 6-1 and Eq 6-2.) Hence, the nonresonant antenna absorbs 100% of the reduced power delivered by the generator. I will presently show why the nonresonant antenna absorbs the delivered power with 100% efficiency.

Case 2—Mismatched Antenna, Matched at Antenna Feed Point

Next, we'll consider the case where a matching network placed at the antenna feed point is adjusted to obtain a conjugate match. By the Conjugate-Matching Theorem, an antenna input impedance of $R + jX$ is matched by a Transmatch output impedance of $R - jX$. Thus, the initially reactive nonresonant antenna is now tuned to resonance, because the $+jX$ reactance of the antenna is canceled by the equal, but opposite $-jX$ reactance at the output of the Transmatch. The input impedance of the Transmatch is Z_0, which now provides the transmission line with a Z_0 match: The reflection loss that previously was referred along the line to the generator is now zero; hence the generator sees both a conjugate match and a Z_0 match at the line input, and delivers its maximum available power into the line. As before, the antenna absorbs 100% of the power delivered by the generator. However, the antenna current now flowing through the *resonant-antenna* resistance has increased to a maximum, in proportion to the square root of the increase in power delivered by the generator.

Case 3—Mismatched Antenna, Matched at Line Input

What happens when the matching network (antenna tuner or Transmatch) is placed at the input of the feed line, with the feed-line output connected directly to the mismatched antenna terminals? If the length of the feed line is a multiple of an electrical half wavelength, the antenna impedance, R + *j*X, is repeated at the input of the line. The impedance looking back into the Transmatch is R – *j*X as before, and the Transmatch needs no readjustment to again provide a conjugate match.

However, a line of any length other than a half wave (or multiple thereof) transforms the antenna input impedance to a new value, R′ + *j*X′, at the input terminals of the line. Therefore, the Transmatch must now be readjusted to again obtain a conjugate match. The result of readjustment: Because the antenna impedance is R + *j*X, the impedance looking back into the output terminals of the feed line is now R – *j*X. However, where the +*j*X antenna reactance was previously canceled by the –*j*X reactance at the output of the Transmatch, the antenna reactance is now canceled by the –*j*X reactance at the output of the feed line. Thus, the antenna is again tuned to resonance. The output impedance of the Transmatch is now R′ – *j*X′, the conjugate of the line-input impedance. The input impedance of the Transmatch is again Z_0, the generator delivers the same maximum available power into the Transmatch as it previously delivered into the matched line with the Transmatch at the antenna feed point, and the Transmatch delivers this power into the line. Inasmuch as the line and Transmatch have no losses (ideal case yet, remember?), the antenna still absorbs 100% of the power delivered by the generator. Hence, the same antenna current flows as when the Transmatch was placed at the antenna feed point.

How can this be? Even though we have a conjugate match at the junction of the feed line and the antenna, don't we again have a Z_0 mismatch at that junction, producing the same reflection loss as without the Transmatch? Indeed we do, so to obtain a better understanding of how the conjugate match transfers the maximum available power from the generator to the load, despite a Z_0 mismatch in the path, we must examine the wave actions and reflections resulting from Z_0 mismatches.

Sec 17.3 Wave Actions and Reflections with a Z_0 Mismatch

Whether a tuned stub or a lumped-constant network is used to perform the matching, the matching device itself (such as a Transmatch) presents a mismatch to the line. When adjusted to achieve the conjugate match, the Transmatch generates a new reflected wave having the same magnitude, but of opposite phase as that of the reflected wave arriving from the antenna mismatch. These two waves combine in the Transmatch to produce a resultant wave that is in phase with the voltage and current components of the source wave delivered by the generator. All of the power reflected by the Z_0 mismatch at the antenna *is now added to the source power at the matching point in the Transmatch*, to be subtracted later when it again reaches the antenna. (See Chapters 4, 5 and 6).

The re-reflection of power in the Transmatch results in what is called reflection *gain*, which equals and cancels the reflection *loss* of the Z_0 mismatch. (See *Ref 19* and Chapter 6). This addition of the reflected power to the source power results in forward power between the Transmatch and the antenna that exceeds the source power. When the reflected power is again subtracted by reflection at the antenna mismatch, the remainder, which is absorbed (radiated) by the antenna, equals the source power exactly. (See Chapters 4 and 16.) It may come as a surprise, but these same wave actions and reflection-loss cancellations also occurred in the matching network when it was placed at the antenna terminals. Indeed, it is the multiple reflections produced by the controlled mismatch of any matching device that form the basis for all impedance-matching operations. (See Chapters 4 and 5, Eqs 6-1 and 6-2, the final paragraphs of Sec 9.1, and Chapter 16.)

Sec 17.4 Attenuation and Power Loss

Now let's consider the effects of attenuation and power loss. In lines having attenuation, all power entering the line is absorbed in the load, except for that which is lost through line attenuation, regardless of the mismatch or SWR. (See Chapters 5 and 6.) Of course, the attenuation is greater when there is a load mismatch, because in addition to the attenuation of the forward power, the reflected power is also attenuated during its return to the source. This is the *only significant* difference in performance between placing the Transmatch at either the input or the antenna end of the feed line—except for the obvious difference in (in)convenience in tuning the Transmatch.

The power lost to attenuation increases as the SWR on the line increases when the Transmatch is at the line input. However, when the matched-line attenuation is low, as it is in typical amateur installations, the additional loss from a Z_0 mismatch is so small that the difference in radiated power cannot be discerned by a receiving station. The increase in line attenuation resulting from SWR can be determined from the graphs of Figs 1-1 and 6-1, or it may be calculated precisely from Eq 6-1. A study of these references shows that the logical place for a Transmatch is on the operating desk, not at the antenna.

Chapter 18

The Broadband Double-Bazooka Antenna—How Broad Is It?

(Adapted from Technical Correspondence, *QST*, September 1976)

Sec 18.1 Introduction

The increasing interest in using the coaxial dipole (sometimes called the Double-Bazooka antenna) for its questionable increased bandwidth is disturbing, especially in view of the results of an analysis and some experiments I performed and published several years ago. The antenna, with its inner section of coax and its end sections of open-wire transmission line (ladder line), was popularized for amateur use by Charles Whysall, W8TV, with an article in July 1968 *QST* (*Ref 129*).[1] It appears in several editions of *The ARRL Handbook*, although it has never appeared in *The ARRL Antenna Book*. The results of my analysis and experiments indicate that the coaxial stubs in the coaxial-dipole configuration in general use by amateurs cannot provide the degree of bandwidth that users of the coaxial dipole appear to be measuring. Thus, it appears that features other than the shunt-compensating reactance provided by the coaxial stubs within the dipole must be responsible for achieving the bandwidth credited to the coaxial feature.

Here's why the coaxial feature cannot make any significant contribution to bandwidth when the feed-line impedance is the usual

---

1 All references appear in Chapter 24.

$Z_C = 50\ \Omega$. Depending on the height above ground, the input impedance of the average 40-meter or 80-meter amateur dipole generally runs from 50 to 80 Ω of resistance at resonance. Thus, at resonance, the mismatch on a 50-Ω line is generally quite low, from less than 1.1 to around 1.6 at worst. On either side of resonance, the mismatch increases rapidly because of the reactance appearing in the dipole impedance. With the addition of the coaxial, reactance-compensating shunt stubs, the dipole reactance is either canceled, at best, or at worst, reduced somewhat by the opposite shunt reactance provided by the stubs.

Sec 18.2 Reactance Cancellation

Although you'll see later why it can't be, let's first assume hypothetically that complete cancellation of the reactance can be achieved by the shunt reactance of the coaxial stubs. This means the cancellation is obtained by a *parallel*-connected reactance, which raises the series resistance of the dipole impedance to its equivalent parallel-circuit value, which is *much higher*. And here is the crucial point. When you use a feed line having an impedance which already matches the dipole terminal resistance rather well at resonance, the higher mismatch off resonance caused by the dipole reactance will not be significantly different, whether it is caused by the reactance of the uncompensated dipole or by the increased resistance received in exchange for the canceled reactance.

To illustrate with an example, consider an 80-meter dipole at a height which yields a resonant terminal impedance of $55 + j0\ \Omega$ at 3.75 MHz. The mismatch is 1.1 referred to 50 Ω. At 3.55 MHz, 200 kHz below resonance, the series impedance of the dipole is approximately $50 - j90\ \Omega$, which yields a 5.04:1 mismatch. Now a 90-Ω inductive reactance placed in series with the dipole terminals *would* cancel the 90-Ω capacitive dipole reactance, and would leave the terminal resistance at 50 Ω. And we would indeed have a perfect match.

Unfortunately, the reactance provided by the stubs in the coaxial dipole is in *parallel* with the dipole terminal impedance, *not in series*. And what a difference this makes! The values of the equivalent parallel-circuit components of the dipole impedance ($50 - j90\ \Omega$) are $R_P = 212\ \Omega$, and $X_P = -117.8\ \Omega$, as shown in Fig 18-1A. When the 117.8 Ω of capacitive dipole reactance is completely canceled by an equal inductive reactance in parallel with the dipole impedance, the resulting impedance is $212 + j0\ \Omega$. Bad news! The resonant resistance

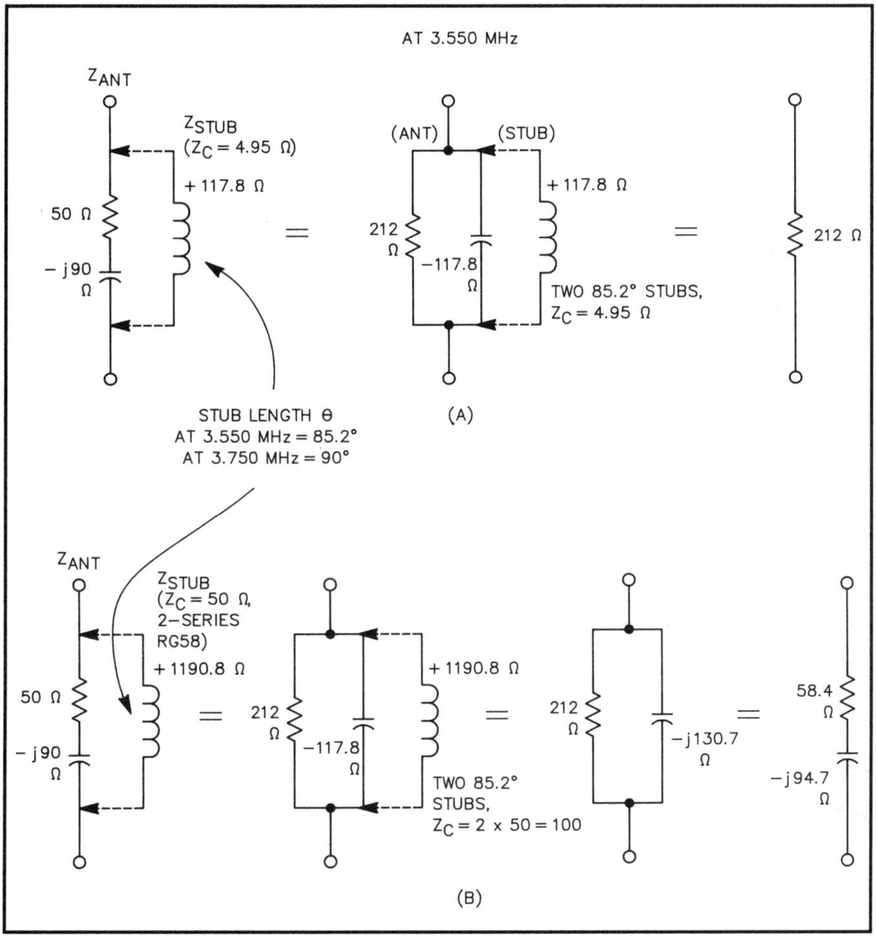

Fig 18-1—Showing the impedance transformation for (A) 4.95-Ω center coaxial section, and (B) 50-Ω center section. At both A and B, the SWR without coaxial stubs is 5.04:1; with stubs at A it becomes 4.24:1, and in the practical case at B, 4.9:1.

of the dipole is now 212 Ω, raised from 50 Ω! Canceling the dipole reactance with *parallel* circuitry has raised the resistive component by a factor of 4.24 from 50 Ω up to 212 Ω, and coincidentally, the mismatch is now 212/50 = 4.24:1. This is the lowest mismatch obtainable with any type of parallel compensation, because even though the dipole reactance has been completely canceled, we're still stuck with the 212-Ω dipole load resistance.

Obviously, a reduction in mismatch from 5.04:1 to 4.24:1 is hardly worthwhile, even if it could be accomplished. But it can't. Why?

Because a canceling reactance of 117.8 Ω would require a coax having a characteristic impedance Z_C of only 4.95-Ω for the stubs—impractical to build. Here's more bad news: Stubs made from 50-Ω coax yield a reactance ten times too high—useless. What about 75-Ω coax?—1.5 times worse. Incredible, you say? Example continues: A short-circuited stub, ¼-λ resonant at 3.75 MHz made from 50-Ω coax, yields an inductive reactance of 595.4 Ω at 3.55 MHz, again 200 kHz below resonance. These values are calculated from the equations below at 3.55 MHz.

$$\text{Stub length } \theta = \frac{3.55 \times 90^\circ}{3.75} = 85.2^\circ \qquad \text{(Eq 18-1)}$$

$$\text{Shunting reactance } X = Z_C \tan\theta \qquad \text{(Eq 18-2)}$$

$$\text{Stub impedance } Z_C = X \cot\theta \qquad \text{(Eq 18-3)}$$

The stubs in each dipole half are connected in series with each other through their center conductors, so the total inductive reactance of the series combination is twice the value of the single stub, or 1190.8 Ω. This is the value appearing in parallel with the dipole impedance when using 50-Ω stubs. (Stubs of 75 Ω would yield 1786.3 Ω.) The combined parallel components of the dipole impedance and shunt-stub reactance (R_P = 212 Ω and X_P = −117.8 Ω in parallel with stubs of +1190.8 Ω) yield total parallel-circuit component values of R_P = 212 Ω and X_P = −130.7 Ω. The series-equivalent dipole input-terminal impedance is now 58.4 – *j*94.7 Ω, also shown in Fig 18-1 at B. The result? A whopping big reduction in mismatch from 5.04 without stubs, all the way down to 4.9:1 with stubs! Conclusion? Isn't it obvious that the stubs are ineffective? And shouldn't it be disturbing?

So you ask what other features can be responsible for the lower mismatch values that appear to be measured by many coaxial-dipole users. I'll give you a number of possibilities.

First, the mismatch values shown here are those which appear at the junction of the feed line and the antenna, while values measured at the input of the feed line will be somewhat lower because of line attenuation.

Second, increased radiator thickness, especially when the stubs are constructed from RG-8, reduces the dipole characteristic impedance, resulting in less reactance than with the thinner wire dipole for the same frequency excursion away from resonance.

Third, the extensions for building out from the shortened, short-circuited ends of the coax stubs to obtain an *external* half wave are usually of multiwire construction such as ladder line, which further increases the effective radiator thickness. Such is especially helpful at the outer ends of the dipole, where the voltage and the electric field are high. This reduces the off-resonance reactance still further.

Fourth, the external dielectric material covering the stub coax increases both dipole capacitance (increasing the electrical length) and effective diameter of the radiator. However, indications under investigation suggest this increase in effective diameter is also accompanied by increased ohmic loss in the external dielectric, which decreases the Q, and thus increases the bandwidth at the expense of efficiency.

And fifth, in the range above 3:1, many SWR indicators show readings considerably lower than the true value. If you are interested in pursuing the subject further, I invite you to read my paper entitled, "A Revealing Analysis of the Coaxial Dipole Antenna," appearing on page 46 of *Ham Radio* for August, 1976 (*Ref 62*).

Sec 18.3 Resistive Losses

Since I wrote the Technical Correspondence item on which the above information is based, the true reason for the increased dipole bandwidth obtained with the Double-Bazooka has been discovered. But the reason is not a happy one. Frank Witt, AI1H (ex-W1DTY), with the aid of a computer, has discovered that the increased bandwidth of the Double-Bazooka obtained by many amateurs actually arises from the previously undetermined resistive loss in the coaxial cable used to form the stubs, and not by reactance cancellation from parallel-connected coaxial stubs (*Ref 122*). In other words, the reduction in SWR obtained by those who use the Double-Bazooka is from resistive loading, and not from reactance cancellation. Unfortunately, the resistive loading results in a reduction in radiated power, power lost in heating the stubs.

This turn of events is ironic for two reasons. First, as I showed earlier, the reactance available in the coaxial stubs in the Double-Bazooka is insufficient to obtain any practical amount of reduction in the SWR-producing antenna-terminal reactance, much less total cancellation. And second, even if the stubs could provide sufficient reactance to obtain complete cancellation, the improvement in bandwidth would still have been inconsequential as a result of the cancellation, as I proved mathematically.

Earlier in this chapter, and in my coaxial dipole analysis (*Ref 62*), I pointed out the reason no bandwidth improvement is possible with parallel-circuit reactance cancellation when the feed line impedance is 50 Ω. You'll remember, this is because the parallel-circuit form of the stub connection used in the cancellation raises the effective terminal resistance as the reactance is lowered, resulting in a negligible reduction of mismatch. I also pointed out in the analysis that an improvement in bandwidth could be achieved with parallel-circuit reactance cancellation by using a feed line having an impedance higher than the resonant terminal resistance of the dipole, and then accepting a corresponding increase in mismatch at the resonant frequency of the dipole. I showed that by using a 144-Ω feed line to feed a dipole having a resonant terminal resistance of 72 Ω (yielding a 2:1 SWR at resonance), the 2:1 SWR bandwidth could be improved from 165 kHz (on 50-Ω line with no stubs) to 565 kHz by canceling the off-frequency dipole reactance with 19.7-Ω stubs. This results in a bandwidth improvement factor of 3.4. This arrangement is particularly appealing for the 80-meter band, because it achieves an SWR of less than 2:1 across the entire band, except at the center frequency where it is exactly 2:1.

In my analysis I offered no concrete suggestion for practicing this arrangement, except to suggest the use of two 75-Ω coaxial lines side by side for the feed line. However, I considered using a lumped-constant capacitor and inductor at the feed point for an impedance transformer to obtain the required increase in feed-point impedance, and for reactance cancellation. But I decided it was impractical to use, so I discarded the idea. On the other hand, Frank Witt came up with the same idea and made it work! (See *Ref 122*.) In addition, he also came up with a very clever method of using a shorted ¼-λ-stub impedance transformer to obtain both the required step-up of impedance and the reactance cancellation (*Refs 123* and *124*). His method uses the larger RG-8, or RG-213 coaxial cable for the ¼-λ stub, ensuring optimum bandwidth improvement by reactance cancellation but with minimal loss in the cable stub. The Snyder dipole (*Ref 130*) uses a somewhat similar method of broadbanding, but with higher loss than with Witt's method, because the Snyder dipole uses the higher-loss RG-141 coax for the reactance cancellation.

For anyone wishing to build a true broadbanded 80-meter dipole, the Witt articles referenced above provide all of the necessary details, plus the explanation of how it works.

Chapter 19

The Conjugate Match and the Z_0 Match

Sec 19.1 Introduction

Material appearing in this chapter was originally written for the 1986 *ARRL Handbook*, and also appears in later editions, as well as the 15th edition of *The ARRL Antenna Book*. This presentation is a modified version.

The primary purpose of matching impedances is to enable a power source or generator, having an optimum load impedance, to deliver its maximum available power to a load having a different impedance. In antenna systems, power is usually delivered through a transmission line having still a third impedance. Various schemes may be used for matching a transmitter to the input of an antenna system. However, the same principles apply, and the same wave actions and reflections occur, whether the matching device is a stub on a transmission line, a ¼-λ line transformer, or a group of lumped reactances such as a pi or T network often found in a Transmatch, or the pi-network tank circuit of a transceiver. The wave actions underlying the matching process are discussed in detail in Chapters 4 and 7. To aid in understanding the principles of matching, we shall first consider transmission lines and the reactance components of networks as lossless elements, and then treat the effects of loss later.

We will be discussing two specific cases of match and mismatch, one on a conjugate basis, and the other on a Z_0 basis.[1] If the power

---

[1] Both Z_C and Z_0 are symbols that represent the characteristic impedance of transmission lines. Although Z_C has been used primarily in other chapters of this book, the symbols may be used interchangeably.

actually delivered to the load is to be compared with the maximum power available from the source, the discussion is on a conjugate basis. If the comparison is made with the power that the source will deliver to a Z_0 load, it is on a Z_0 basis. Considerable confusion has arisen over the use of the terms "match" and "mismatch," as it is not always clear just what basis is intended (*Ref 120*).[2] R. W. Beatty of the National Bureau of Standards (NBS) has set forth a complete set of specific terms and definitions which eliminate this confusion (*Ref 121*). They are as follows:

1) Conjugate match—The condition for maximum power absorption by a load, in which the impedance seen looking toward the load at a point in a transmission line is the complex conjugate of that seen looking toward the source.

2) Conjugate mismatch—The condition in the situation above in which the load impedance is not the conjugate of the source impedance.

3) Conjugate mismatch loss—The loss resulting from a conjugate mismatch.

4) Z_0 match—The condition in which the impedance seen looking into a transmission line is equal to the characteristic impedance of the line.

5) Z_0 mismatch—The condition in which the impedance seen looking into a transmission line is not equal to the transmission-line characteristic impedance Z_0. In general, a conjugate match is a case of Z_0 mismatch.

6) Z_0 mismatch loss—The loss resulting from a Z_0 mismatch (which is canceled by the reflection *gain* obtained with a conjugate match).

7) Conjugate available power—*Maximum* available power.

8) Z_0 available power—The power a source will deliver to a Z_0 load.

Let's now assume we have a classical generator, a load, and an ideal lossless line connecting them, all of which have the same impedance, Z_0. Under these conditions, both a conjugate match and a Z_0 match exist, and the generator delivers its maximum available power to the line. The line transfers it to the load, where it is completely absorbed. I

[2] All references appear in Chapter 24.

will discuss later the effect on the conjugate match when the classical generator is replaced with a transceiver or linear amplifier.

If the Z_0 load is now replaced by a mismatched load, such as an antenna having an impedance $Z_L = R + jX \neq Z_0$, we have two mismatches at the load, a Z_0 mismatch and a conjugate mismatch. The Z_0 mismatch creates a reflection having a magnitude

$$\rho = \frac{Z_L - Z_0}{Z_L + Z_0} \qquad \text{(Eq 19-1)}^3$$

which causes a reflection loss ρ^2 to be referred back along the line to the generator. The reflection causes the generator to see the same magnitude of Z_0 mismatch at the line input, which in turn causes the generator to deliver less than its maximum available power by the amount equal to the reflection loss ρ^2. (See *Refs 19, p 5* and *70, p 204*.) Because the line we are discussing is lossless, there is no power lost in the line—only a reduction of power delivered by the generator, and all of the reduced power delivered is absorbed by the load.

If a matching device is now inserted between the line and the load, the device provides a conjugate match between the two by supplying a reflection gain that cancels the reflection loss. Hence, the line now sees a Z_0 match at the input of the matching device. Because the line is now terminated in a Z_0 match, the generator again sees a Z_0 match at the input to the line, and again delivers its maximum available power into the line, to be completely absorbed by the load.

However, if the matching device is inserted between the generator and the line instead of between the line and load, the device provides the conjugate match at the input of the line, and the generator sees a Z_0 match at the input of the matching device. One feature of the conjugate match is that, once a conjugate match is achieved at any junction of components in the system, there is a conjugate match at all other junctions in the system. Thus, the line now sees a conjugate match at the load, but it still sees a Z_0 mismatch. The Z_0 mismatch still produces a reflection loss that is referred back to the input of the line as before. However, with the matching network at the input of the line, the reflection gain supplied by the network (which again cancels the reflection loss) shields the generator from seeing the referred Z_0 mismatch

[3] This equation is basically the same as Eq 3-1, with Z_0 used in place of Z_C.

which remains at the input of the line, and again, the generator delivers its maximum available power.

As just stated, the reflection gain of the matching network also creates the conjugate match at the load, and thus enables the load to absorb the maximum available power delivered by the generator in spite of the Z_0 mismatch which remains at the load. Inasmuch as the generator sees a Z_0 match at the input of the matching network, it continues delivering its maximum available power into the line through the matching network, and the power is still completely absorbed by the load because of the conjugate match.

The amount of forward power flowing on a lossless line is determined by the expression

$$\frac{1}{1-\rho^2} \times P$$

where

ρ = voltage coefficient of reflection
P = source power

On a real line with attenuation, forward power at the conjugate matching point is

$$\frac{1}{1-\rho^2\varepsilon^{-4\alpha}}$$

where

α = line attenuation in nepers = dB/8.6859
ε = 2.71828 . . ., the base of natural logarithms

The power arriving at the mismatched load with attenuation in the line is

$$\frac{\varepsilon^{-2\alpha}}{1-\rho^2\varepsilon^{-4\alpha}}$$

and the power absorbed in the load is

$$\frac{(1-\rho^2)\,\varepsilon^{-2\alpha}}{1-\rho^2\varepsilon^{-4\alpha}}$$

The power reflected is

$$\frac{\rho^2\, \varepsilon^{-2\alpha}}{1 - \rho^2 \varepsilon^{-4\alpha}}$$

See Eqs 6-1 and 6-2.

Sec 19.2 The Conjugate Matching Theorem

To understand how a conjugate match is created at the load with the matching network at the input of the line, we need to examine the Conjugate Matching Theorem and the wave actions which produce the conjugate match. It is the conjugate match that enables the load to absorb the maximum available power from the generator despite the Z_0 load mismatch.

The Conjugate Matching Theorem was introduced in Chapter 17, but is repeated here for reference. "If a group of four-terminal networks containing only pure reactances (or lossless lines) are arranged in tandem to connect a generator to its load, then if at any junction there is a conjugate match of impedances, there will be a conjugate match of impedances at every other junction in the system." (From *Ref 17, p 243*, and *Ref 69, p 407*.)

To paraphrase from this NBS definition, "conjugate match" means that if in one direction from a junction the impedance has the dimensions R + *j*X, then in the opposite direction the impedance will have the dimensions R – *j*X. And according to the theorem, when a conjugate match is accomplished at any of the junctions in the system, any reactance appearing at any junction is canceled by an equal and opposite reactance, which also includes any reactance appearing in the load, such as a *nonresonant antenna*. This reactance cancellation results in a net system reactance of zero, establishing resonance in the entire system. In this resonant condition the generator delivers its maximum available power to the load. This is why an antenna operated away from its natural resonant frequency is tuned to resonance by a matching network connected at the input to the transmission line.

Conjugate matching is obtained through a controlled wave interference between two sets of reflected waves. One set of waves is reflected by the load mismatch, and the other by a complementary mismatch introduced by the reactances in the matching network. The mismatch introduced by the reactances of the network creates reflected waves of voltage and current at the matching point (at the input of the network), that are equal in magnitude but opposite in phase relative to those arriving from the load mismatch. The two sets of reflected voltage and current waves created by the two mismatches combine at the matching point to produce a resultant voltage wave that is in phase with the forward, or source voltage wave, and a resultant current wave that is 180° out of phase with the forward, or source current wave.

This combination of forward and resultant reflected *voltages in phase*, and forward and resultant reflected *currents 180°* out of phase creates a virtual one-way open circuit to waves traveling toward the source, which totally re-reflects both sets of reflected waves back into the line. This re-reflection prevents the reflected waves from traveling beyond the matching point, which is the reason an SWR indicator connected at the input of the matching network shows zero reflected power at the conclusion of the matching tune-up procedure. On re-reflection at the virtual open circuit, the reflected current undergoes a phase change of 180°, with the result that both the voltage and current components of the reflected waves now emerge in phase with the corresponding components of the source wave. Hence, by superposition, all of the power contained in the reflected waves is added to the power in the source wave. Consequently, with re-reflection, the forward power in the line between the matching network and the load is greater than the source power whenever the line is terminated in a Z_0 mismatch. This is why the forward-power wattmeter connected at the output of the matching network indicates a value higher than the source power, higher by the amount of the reflection gain, which is equal to the reflected power. For more details concerning the phenomenon of re-reflection and the wave actions involved in conjugate matching, see Chapter 4.

This enlargement of the forward power constitutes the "reflection gain" just mentioned above, which cancels the reflection loss and creates the conjugate match that enables the Z_0-mismatched load to absorb the maximum available power delivered by the generator. When the source power enlarged by the reflected power reaches the Z_0-mismatched load, Z_L, the power previously reflected and added to the source power is again subtracted by the Z_0 mismatch reflection, leaving the source power to be completely absorbed by the load. This is why all power entering the line is absorbed in the load, regardless of the mismatch or SWR.

Sec 19.3 Replacing the Classical Generator With an RF Amplifier

So far we have learned that the principal objective of conjugate matching is to re-reflect and conserve the power that would otherwise be lost by reflection from a Z_0 mismatch at the junction of the feed line and the antenna. In other words, although the Z_0 mismatch remains at the junction, the reflection gain obtained with the conjugate match cancels the effect of the Z_0 mismatch and permits the maximum available power to be delivered by the source and to be absorbed in the load—hence, the load is *conjugately matched* to the source, but it is still Z_0 mismatched to the line.

Until now the discussion has concentrated only on obtaining delivery of the maximum available power from a source which we have considered to be a classical generator. Returning for a moment to the definition of the conjugate match given in Beatty's item 1 (Sec 19.1), we see that one of the conditions required for maximum absorption of power in the load is that the load resistance equals the source resistance of the generator. This condition is easily satisfied with the classical generator. However, in this respect we must be aware of a crucial difference between the classical generator and the output circuit of a real-world RF power amplifier. This difference has no effect on our routine matching operations, but it does affect the terminology we use if we are to have a technically accurate description of the matching operations.

The difference is this: In the classical generator, the optimum load resistance required for delivering the maximum available power is always equal to the source resistance of the generator, as required for a conjugate match. However, because of the complex nature of various relationships between voltages and currents in an RF power amplifier, during normal operation the internal source resistance of the amplifier, R_S, is almost never equal to its optimum load resistance, R_L. Hence, because of the different values of resistance existing between the source and the load when we complete a matching operation, we do not have a *true* conjugate match when an RF amplifier is coupled to an antenna tuner or a feed line. For the reader who has been conditioned to believe that the routine matching operation achieves a conjugate match, it may be helpful to understand that, in reality, the tuner or feed line is matched to the *optimum load resistance* of the amplifier, but it is conjugately *mismatched* to the source resistance.

So what was the purpose in my explaining the wave actions involved in the conjugate match as it applies to antenna tuners if a conjugate match is not achieved? First, because the match to the optimum load resistance that we do achieve with an antenna tuner is still based on the *principles* of conjugate matching, even though a conjugate match is not fully achieved. We must keep in mind that with the conjugate match, the source delivers the *maximum available* power to the load. On the other hand, the match we do achieve with an antenna tuner is a conjugate mismatch relative to the delivery of the *conjugate* available power (see Beatty's item 7), but it is a proper match for the delivery of the power made available from an amplifier when it is terminated in its optimum load impedance Z_L. We must keep in mind that when the antenna tuner is adjusted to obtain the match, the load is matched to the *optimum load resistance* R_L of the RF power amplifier, not to the source resistance R_S of the amplifier, which is the reason we cannot call the match a conjugate match as we do with the classical generator.

We must also keep in mind that the objective in using an antenna tuner, or line-matching network, is not to obtain the maximum available power from the transmitter, or RF amplifier, but rather to conserve the power reflected from the Z_0 mismatch at the junction of the feed line and the antenna, and hence to ensure that the output circuit of the transmitter sees its optimum load resistance despite the Z_0 mismatch at the antenna. In other words, all that we require of the antenna tuner is to prevent the power reflected by the Z_0 mismatch from reaching the output of the transmitter. Otherwise, the reflected power would cause the transmitter to become equally mismatched, and would thus reduce its delivered power by the amount of the reflection-loss power ρ^2, relative to what it delivers into a Z_0 match, as described a few paragraphs earlier.

The conjugate mismatch was incurred because of the difference between the source resistance R_S of the RF amplifier and its load resistance R_L. It is important to note that this conjugate mismatch does not affect or impair the cancellation of all reactances in the antenna system obtained by the proper adjustment of the antenna tuner, as described in detail earlier, because the reactance-canceling aspect of conjugate matching principles still applies. The wave actions described in obtaining the re-reflection of reflected voltage and current at the matching point in the matching network are identical, no matter whether the power source is a classical generator or a real-world RF power amplifier. In other words, despite the conjugate *mismatch* at the output terminals of the amplifier, the matching network still cancels the Z_0 mismatch at the junction of the feed line and the antenna, and thus matches the antenna to both the feed line and the amplifier. The matching network also tunes the antenna to resonance when it is operated at frequencies away from its self-resonant frequency.

Sec 19.4 The Effects of Line Attenuation

In lines having attenuation, all power entering the line is absorbed in the load, except for that lost because of line attenuation. When the matching is performed at the line input, the loss from attenuation increases as the Z_0 load mismatch increases. This happens because, in addition to the attenuation of the forward power, the reflected power is also attenuated in the same proportion.

Should this concern us as amateurs? Probably not, if we are using a typical antenna system at HF. When the matched-line attenuation is low, as it is in typical amateur installations, the additional loss because of Z_0 mismatch is small. The additional losses are so low that even with moderate to high SWR values, the difference in power radiated compared to that with a 1:1 SWR is too small to be discerned by the receiving station. The additional power lost versus SWR may be determined from Figs 1-1 or 6-1, or may be calculated using Eq 6-1.

Chapter 20

SWR with Multiband and Nonresonant Antennas

Sec 20.1 Introduction

In Chapter 9, I departed from the study of impedance matching from the viewpoint of wave reflections, and began an in-depth examination of matching directly from the viewpoint of impedance. During that examination I covered several pertinent subject areas, all having a direct bearing on matching antenna terminal impedance to the transceiver.

I began with the assumption that the conventional setup included a transceiver, an SWR indicator, and a feed line running directly to the antenna. Subsequent discussion showed that flexibility in operating bandwidth is vastly improved when the matching capability of the pi-network tank circuit in the final amplifier is extended by the addition of an external matching unit—the antenna tuner, or Transmatch.

Sec 20.2 Multiband Dipoles

So far I have discussed only simple dipoles fed with coaxial transmission line. With the exception of 15-meter operation on a 40-meter dipole, the simple coax-fed dipole is limited to single-band operation, because the excessive antenna to feed-line mismatch precludes its use on frequencies harmonically related to the fundamental frequency for which it is designed.

The simple dipole can be modified in different ways to extend its performance to multiband operation. Let us now examine those different ways.

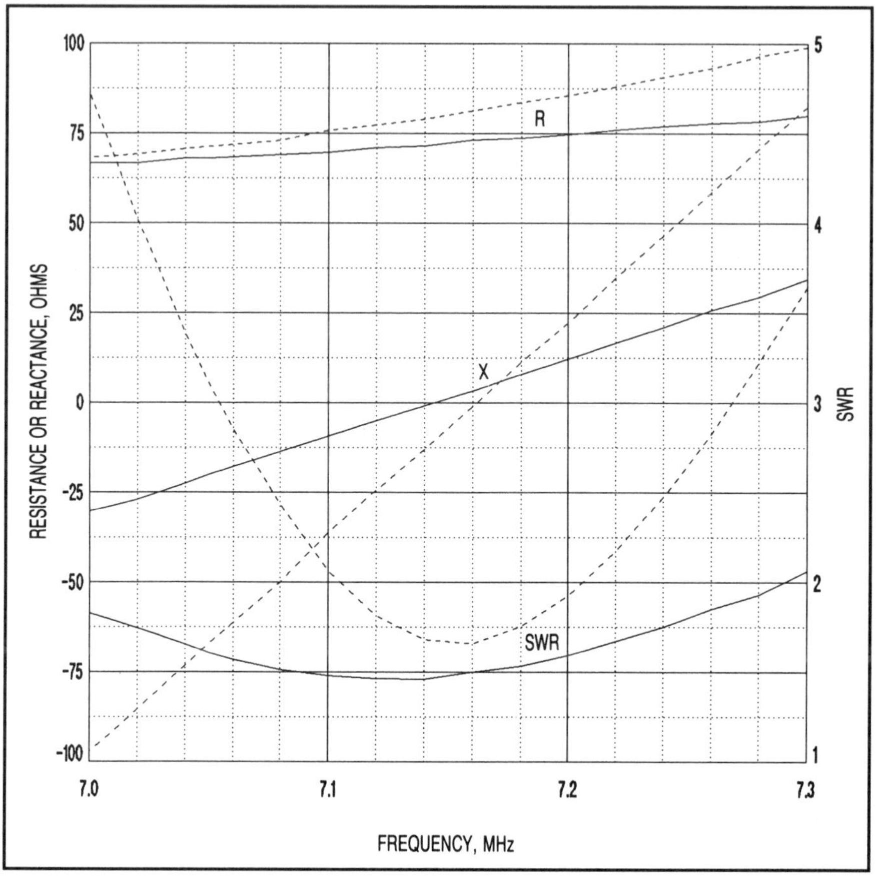

Fig 20-1—Measured feed-point resistance R, reactance X, and SWR versus frequency of 40-meter dipoles at a height of 40 feet. The solid lines show data for a 66-ft, 2-in. radiator with no traps (data from Table 20-1). The broken lines show data for the same antenna but with 40-meter traps and 21-ft 5-in. extensions added at each end to make the antenna usable on 80 meters (data from Table 20-2). ***(Computer-generated plot courtesy of Rick Maxwell, WB4GNR.)***

Sec 20.2.1 Trap Dipoles

One method of achieving multiband operation from a dipole is to insert frequency-selective traps in the antenna. Traps allow operation at the lowest fundamental frequency for which the total electrical length of the dipole is ½-λ, but selectively disconnect dipole portions extending beyond traps designed to present a high impedance in series with the dipole at frequencies higher than the fundamental. Hence, the effective electrical length of the dipole is automatically shortened to

Table 20-1

Measured Antenna Impedance Data Transformed Through Calibrated Feed Line

Measurement data: Z40 (no traps); see Fig 20-1. 66-ft 2-in. 40-meter dipole, height 40 ft, resonant frequency 7.1439 MHz, calibrated RG-214 plus W2DU HF balun, RF bridge General Radio 1606-A, receiver Kenwood TS-530S, signal generator Boonton 250A RX meter.

Feed line calibration:

Line length in degrees/MHz, PHI = 23.7

Line attenuation in dB/$\sqrt{MHz}$, A = 0.082

Line impedance, Z_C = 49 Ω

Freq	R_I	X_I	RHO_I	SWR_I	RHO_L	SWR_L	R_L	X_L
7.000	83.00	–14.500	0.2783	1.77	0.2926	1.827	66.67	–30.19
7.020	80.00	–12.536	0.2580	1.70	0.2712	1.744	66.61	–26.96
7.040	77.50	–8.376	0.2343	1.61	0.2463	1.654	67.96	–22.38
7.050	75.80	–6.667	0.2210	1.57	0.2323	1.605	68.01	–19.98
7.060	74.50	–4.958	0.2102	1.53	0.2210	1.567	68.27	–17.77
7.080	72.50	–1.624	0.1939	1.48	0.2038	1.512	69.04	–13.60
7.100	71.00	0.999	0.1835	1.45	0.1930	1.478	69.61	–10.15
7.120	69.50	5.281	0.1785	1.43	0.1877	1.462	70.95	–5.09
7.140	68.00	8.473	0.1773	1.43	0.1865	1.459	71.45	–0.81
7.145	68.00	9.517	0.1810	1.44	0.1904	1.470	72.04	0.31
7.150	68.00	10.499	0.1848	1.45	0.1944	1.483	72.59	1.38
7.155	68.00	11.321	0.1882	1.46	0.1979	1.493	73.05	2.31
7.160	67.70	12.011	0.1894	1.47	0.1993	1.498	73.12	3.29
7.180	66.50	15.390	0.2000	1.50	0.2104	1.533	73.62	7.88
7.200	66.00	18.750	0.2172	1.55	0.2285	1.592	74.63	12.20
7.220	65.80	22.299	0.2387	1.63	0.2512	1.671	75.88	16.70
7.240	65.70	25.553	0.2598	1.70	0.2733	1.752	76.90	20.94
7.250	65.50	27.586	0.2729	1.75	0.2872	1.806	77.35	23.64
7.260	65.50	29.201	0.2838	1.79	0.2987	1.852	77.79	25.78
7.280	65.50	31.868	0.3019	1.87	0.3177	1.931	78.26	29.43
7.300	66.10	35.890	0.3297	1.98	0.3470	2.063	79.83	34.52

½ λ at the higher frequencies, so that it maintains an acceptable match to a coaxial feed line.

But unfortunately, there is an undesirable penalty to be paid when using the trap method of multiband operation. The penalty is that the operational bandwidth on all the bands on which the trap dipole operates is far narrower than that of separate simple dipoles cut to ½ λ for each band. In other words, the SWR rises much more rapidly with a trap

Table 20-2

Measured Antenna Impedance Data Transformed Through Calibrated Feed Line

Measurement data: Z40 with 40-meter traps; see Fig 20-1. 66-ft 2-in. 40-meter dipole, height 40 ft, KW-40 traps (res. 7.150 and 7.160 MHz), 21-ft 5-in. extensions, calibrated RG-214 plus W2DU balun, RF bridge General Radio 1606-A, receiver Kenwood TS-530S, signal generator Boonton 250A RX meter.

Feed line calibration:

Line length in degrees/MHz, PHI = 23.7
Line attenuation in dB/$\sqrt{MHz}$, A = 0.082
Line impedance, Z_C = 49 Ω

Freq	R_I	X_I	RHO_I	SWR_I	RHO_L	SWR_L	R_L	X_L
7.000	180.00	−68.714	0.6187	4.25	0.6504	4.721	68.18	−97.11
7.020	153.00	−61.396	0.5720	3.67	0.6014	4.017	69.11	−85.27
7.040	132.00	−50.710	0.5175	3.14	0.5440	3.386	70.64	−73.16
7.050	123.00	−45.106	0.4874	2.90	0.5124	3.102	71.27	−66.92
7.060	111.60	−39.518	0.4476	2.62	0.4706	2.778	70.37	−58.89
7.080	103.00	−29.520	0.3975	2.32	0.4179	2.436	72.96	−49.52
7.100	93.20	−16.620	0.3298	1.98	0.3469	2.062	75.73	−36.27
7.120	85.50	−6.461	0.2753	1.76	0.2895	1.815	77.22	−24.25
7.140	80.00	2.661	0.2411	1.64	0.2536	1.680	78.85	−13.09
7.145	79.00	5.178	0.2376	1.62	0.2499	1.666	79.60	−10.16
7.150	78.00	7.203	0.2349	1.61	0.2471	1.656	80.02	−7.59
7.155	77.00	9.790	0.2347	1.61	0.2469	1.656	80.73	−4.45
7.160	75.70	12.430	0.2350	1.61	0.2472	1.657	81.16	−1.00
7.180	72.40	22.006	0.2604	1.70	0.2739	1.754	83.63	11.39
7.200	70.00	28.333	0.2883	1.81	0.3033	1.871	84.43	20.53
7.220	68.30	38.920	0.3515	2.08	0.3698	2.174	88.00	34.81
7.240	67.50	47.238	0.4035	2.35	0.4246	2.476	90.71	46.58
7.250	67.20	51.448	0.4294	2.51	0.4518	2.649	92.01	52.69
7.260	67.00	55.510	0.4538	2.66	0.4775	2.827	93.20	58.66
7.280	67.50	63.874	0.5005	3.00	0.5267	3.225	96.52	70.74
7.300	68.00	71.781	0.5409	3.36	0.5693	3.643	98.94	82.48

dipole than with the simple dipole, as the operating frequency is moved away from the resonant frequency of a trap. Evidence of this sharp rise in SWR is presented in Figs 20-1 to 20-3, and in Tables 20-1 to 20-5. These show antenna-terminal resistance, reactance, and SWR data measured with professional, precision laboratory equipment including a General Radio 1606-A RF bridge. The tables also include the corre-

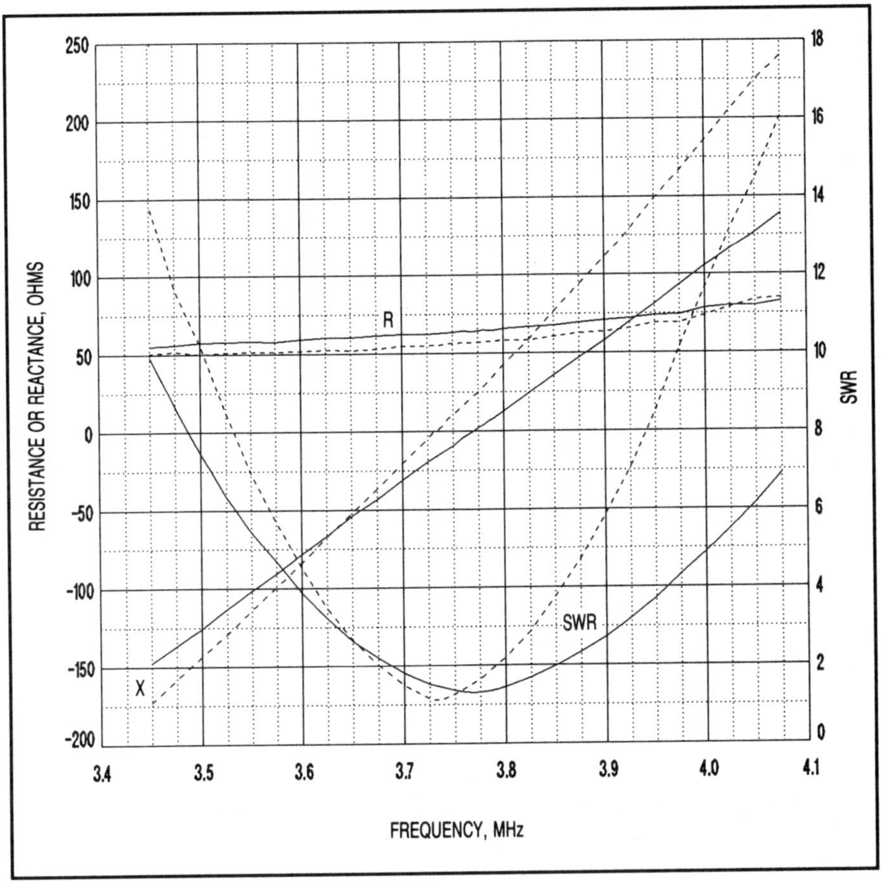

Fig 20-2—Measured feed-point resistance R, reactance X, and SWR versus frequency of 80-meter dipoles at a height of 40 feet. The solid lines show data for a 125-ft radiator with no traps (data from Table 20-3). The broken lines show data for a 109-ft radiator with traps (data from Table 20-4). (This is the 40-meter antenna of Fig 20-1 with traps and extensions added to make the antenna usable on 80 meters.) *(Computer-generated plot courtesy of Rick Maxwell, WB4GNR.)*

sponding reflection coefficient data. A more detailed explanation of trap operation may be found in *Ref 71, p 7-8.*[1]

Sec 20.2.2 Stagger-Tuned Dipoles

There is a second method of obtaining multiband dipole operation that does not have the undesirable bandwidth limitation of the trap

---

1 All references appear in Chapter 24.

Table 20-3

Measured Antenna Impedance Data Transformed Through Calibrated Feed Line

Measurement data: Z80 (no traps); see Figs 20-2 and 20-3. 125-ft 80-meter dipole, height 40 ft, calibrated RG-214 plus W2DU HF balun, RF bridge General Radio 1606-A, receiver Kenwood TS-530S, signal generator Boonton 250A RX meter.

Feed line calibration:

Line length in degrees/MHz, PHI = 23.7
Line attenuation in dB/$\sqrt{MHz}$, A = 0.082
Line impedance, Z_C = 49 Ω

Freq	R_I	X_I	RHO_I	SWR_I	RHO_L	SWR_L	R_L	X_L
3.450	5.90	6.957	0.7889	8.47	0.8171	9.933	55.22	−147.31
3.475	6.70	8.201	0.7653	7.52	0.7927	8.649	56.21	−136.31
3.500	7.70	9.429	0.7370	6.60	0.7635	7.457	57.75	−125.48
3.525	9.00	10.922	0.7026	5.72	0.7279	6.350	58.16	−112.97
3.550	10.50	12.394	0.6655	4.98	0.6896	5.443	58.40	−101.44
3.575	12.25	13.902	0.6256	4.34	0.6483	4.687	58.15	−90.45
3.600	14.80	15.139	0.5704	3.66	0.5912	3.892	59.37	−78.39
3.625	17.90	16.000	0.5084	3.07	0.5271	3.229	60.58	−66.58
3.650	22.00	16.575	0.4345	2.54	0.4505	2.640	60.62	−53.76
3.675	26.00	15.782	0.3640	2.14	0.3774	2.212	61.86	−42.99
3.700	30.90	13.568	0.2791	1.77	0.2894	1.815	62.32	−30.65
3.725	35.00	9.799	0.2021	1.51	0.2096	1.530	62.64	−19.44
3.750	37.20	4.853	0.1478	1.35	0.1533	1.362	63.60	−9.32
3.760	37.50	1.862	0.1346	1.31	0.1397	1.325	64.36	−3.88
3.770	37.80	−0.040	0.1290	1.30	0.1339	1.309	64.14	−0.33
3.775	37.60	−1.457	0.1327	1.31	0.1376	1.319	64.46	2.26
3.780	37.20	−2.566	0.1400	1.33	0.1453	1.340	64.98	4.39
3.785	37.20	−3.699	0.1433	1.33	0.1487	1.349	64.63	6.49
3.790	36.90	−4.749	0.1511	1.36	0.1567	1.372	64.70	8.54
3.800	35.60	−6.711	0.1766	1.43	0.1832	1.449	65.74	12.93
3.825	32.00	−10.850	0.2468	1.66	0.2561	1.688	67.15	24.37
3.850	28.00	−13.507	0.3194	1.94	0.3314	1.992	67.87	35.86
3.875	23.95	−14.452	0.3889	2.27	0.4036	2.353	69.44	47.23
3.900	20.60	−14.359	0.4478	2.62	0.4648	2.737	71.15	57.90
3.925	17.50	−13.885	0.5067	3.05	0.5260	3.220	72.28	69.85
3.950	15.10	−12.911	0.5548	3.49	0.5760	3.717	74.00	81.09
3.975	12.95	−11.950	0.6020	4.02	0.6251	4.334	74.42	93.31
4.000	11.40	−10.375	0.6365	4.50	0.6610	4.899	78.61	105.24
4.025	10.15	−9.242	0.6670	5.01	0.6928	5.510	80.00	116.23
4.050	9.10	−8.272	0.6943	5.54	0.7212	6.175	79.96	126.62
4.075	8.16	−6.994	0.7195	6.13	0.7475	6.920	82.50	139.11

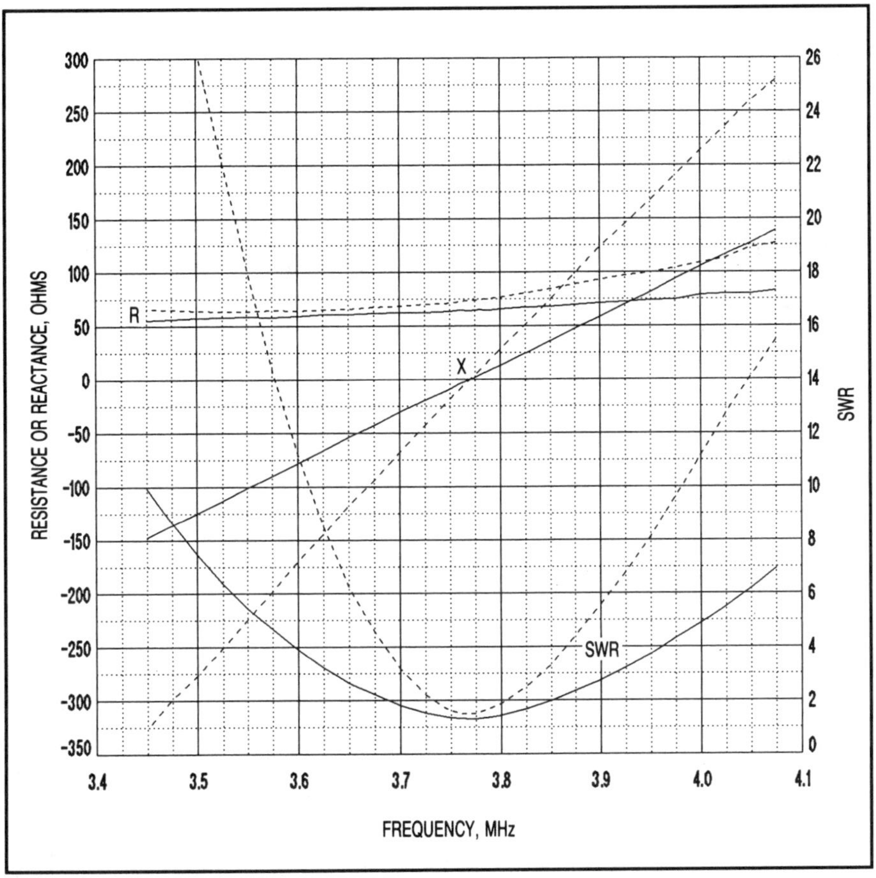

Fig 20-3—Measured feed-point resistance R, reactance X, and SWR versus frequency of 80-meter dipoles at a height of 40 feet. The solid lines show data for a 125-ft radiator with no traps (data from Table 20-3), the same as the solid-line curves in Fig 20-2. The broken lines show data for the same antenna with 80-meter traps and 42-ft extensions added at each end to make the antenna usable on 160 meters (data from Table 20-5). ***(Computer-generated plot courtesy of Rick Maxwell, WB4GNR.)***

dipole. This method uses multiple dipoles fed in parallel from the same feed line, and is called stagger tuning. The length of each separate dipole is ½ λ for each band. Hence, each dipole presents a good impedance match to the feed line on the band for which it is intended, and a poor match on all the others. It is an effective system because none of the dipoles takes power from the feed line except the one which is ½ λ at

Table 20-4

Measured Antenna Impedance Data Transformed Through Calibrated Feed Line

Measurement data: Z80 with 40-meter traps; see Fig 20-2. 66-ft 2-in. 40-meter dipole, height 40 ft, KW-40 traps (res. 7.150 and 7.160 MHz), 24-ft 3-in. extensions, calibrated RG-214 plus W2DU balun, RF bridge General Radio 1606-A, receiver Kenwood TS-530S, signal generator Boonton 250A RX meter.

Feed line calibration:

Line length in degrees/MHz, PHI = 23.7

Line attenuation in dB/$\sqrt{\text{MHz}}$, A = 0.082

Line impedance, Z_C = 49 Ω

Freq	R_I	X_I	RHO_I	SWR_I	RHO_L	SWR_L	R_L	X_L
3.450	4.45	5.507	0.8354	11.15	0.8652	13.840	51.11	−172.70
3.475	5.15	6.907	0.8132	9.71	0.8423	11.683	52.11	−157.90
3.500	5.81	8.571	0.7937	8.70	0.8223	10.252	50.20	−143.30
3.525	6.92	10.213	0.7617	7.39	0.7892	8.489	51.22	−128.76
3.550	8.40	12.113	0.7222	6.20	0.7484	6.948	51.59	−113.43
3.575	10.25	14.098	0.6770	5.19	0.7017	5.704	51.66	−99.06
3.600	13.00	16.250	0.6162	4.21	0.6387	4.536	51.91	−83.68
3.625	17.20	18.207	0.5337	3.29	0.5532	3.477	52.67	−67.38
3.650	22.80	19.589	0.4396	2.57	0.4557	2.674	52.56	−51.84
3.675	30.10	18.640	0.3266	1.97	0.3387	2.024	53.28	−36.53
3.700	37.90	12.973	0.1943	1.48	0.2015	1.505	54.99	−20.50
3.725	43.40	3.222	0.0699	1.15	0.0725	1.156	54.95	−4.63
3.730	43.50	0.268	0.0595	1.13	0.0617	1.132	55.40	−0.74
3.735	43.90	−1.071	0.0561	1.12	0.0582	1.124	54.95	1.07
3.740	43.20	−3.209	0.0718	1.15	0.0745	1.161	55.74	3.95
3.745	42.20	−6.008	0.0993	1.22	0.1030	1.230	56.41	7.97
3.750	41.50	−7.733	0.1186	1.27	0.1230	1.281	56.63	10.60
3.775	35.80	−15.364	0.2350	1.61	0.2438	1.645	57.26	25.32
3.800	27.60	−19.211	0.3641	2.15	0.3778	2.214	58.23	42.60
3.825	21.20	−19.739	0.4675	2.76	0.4851	2.885	58.26	58.56
3.850	15.90	−17.922	0.5591	3.54	0.5802	3.764	60.67	76.79
3.875	12.50	−15.871	0.6266	4.36	0.6504	4.720	62.69	93.92
3.900	10.00	−13.974	0.6833	5.31	0.7092	5.878	63.47	111.27
3.925	8.30	−11.975	0.7247	6.27	0.7524	7.077	66.00	128.79
3.950	6.95	−10.000	0.7605	7.35	0.7896	8.504	69.33	148.59
3.975	5.90	−8.554	0.7908	8.56	0.8212	10.184	69.10	166.25
4.000	5.40	−9.250	0.8077	9.40	0.8388	11.407	53.98	158.40
4.025	4.80	−8.944	0.8269	10.55	0.8588	13.163	46.67	160.29
4.050	4.35	−4.198	0.8380	11.35	0.8705	14.442	83.76	223.92
4.075	4.00	−3.190	0.8496	12.30	0.8827	16.044	85.07	239.80

Table 20-5

Measured Antenna Impedance Data Transformed Through Calibrated Feed Line

Measurement data: Z80 with 80-meter traps; see Fig 20-3. 125-ft 80-meter dipole, height 40 ft, KW-80C traps (res. 3.770 MHz), 42-ft extensions, calibrated RG-214 plus W2DU HF balun, RF bridge General Radio 1606-A, receiver Kenwood TS-530S, signal generator Boonton 250A RX meter.

Feed line calibration:

Line length in degrees/MHz, PHI = 23.7
Line attenuation in dB/$\sqrt{MHz}$, A = 0.082
Line impedance, Z_C = 49 Ω

Freq	R_I	X_I	RHO_I	SWR_I	RHO_L	SWR_L	R_L	X_L
3.450	2.25	0.000	0.9122	21.78	0.9448	35.202	65.33	–325.75
3.475	2.50	1.094	0.9030	19.61	0.9353	29.915	65.11	–298.18
3.500	2.72	2.143	0.8950	18.05	0.9272	26.468	63.88	–276.56
3.525	3.10	3.404	0.8815	15.88	0.9133	22.081	63.65	–250.10
3.550	3.60	4.789	0.8643	13.74	0.8956	18.163	63.48	–224.14
3.575	4.35	6.294	0.8394	11.45	0.8699	14.371	64.88	–198.23
3.600	5.39	8.194	0.8067	9.35	0.8362	11.207	63.96	–170.03
3.625	6.90	10.069	0.7621	7.41	0.7900	8.524	65.36	–144.92
3.650	9.30	12.329	0.6976	5.61	0.7232	6.226	65.72	–117.68
3.675	12.60	14.150	0.6179	4.23	0.6407	4.566	67.39	–94.12
3.700	18.00	15.405	0.5035	3.03	0.5222	3.185	68.36	–68.17
3.725	25.40	13.691	0.3607	2.13	0.3740	2.195	70.05	–42.31
3.745	30.90	9.212	0.2525	1.68	0.2619	1.710	70.76	–23.41
3.750	31.90	7.387	0.2293	1.60	0.2378	1.624	71.33	–18.43
3.755	32.60	5.459	0.2114	1.54	0.2192	1.562	71.95	–13.61
3.760	33.20	3.723	0.1973	1.49	0.2046	1.515	72.03	–9.32
3.765	33.10	1.594	0.1946	1.48	0.2019	1.506	73.29	–4.49
3.775	33.00	–2.172	0.1968	1.49	0.2042	1.513	73.66	4.47
3.800	28.30	–9.921	0.2945	1.84	0.3056	1.880	76.45	28.10
3.825	22.00	–12.811	0.4142	2.41	0.4298	2.508	80.31	50.86
3.850	16.70	–12.987	0.5198	3.17	0.5394	3.343	84.02	74.38
3.875	12.75	–12.258	0.6078	4.10	0.6309	4.418	85.04	98.59
3.900	10.10	–10.205	0.6706	5.07	0.6960	5.580	93.16	123.32
3.925	8.40	–8.764	0.7153	6.03	0.7426	6.770	96.80	145.05
3.950	7.10	–7.418	0.7519	7.06	0.7807	8.120	99.61	167.06
3.975	6.10	–6.038	0.7816	8.16	0.8116	9.614	103.88	190.45
4.000	5.35	–4.750	0.8048	9.25	0.8358	11.178	108.65	213.97
4.025	4.80	–3.603	0.8224	10.26	0.8542	12.717	113.20	236.13
4.050	4.41	–2.395	0.8352	11.14	0.8676	14.103	122.57	260.19
4.075	4.10	–1.472	0.8457	11.96	0.8786	15.469	126.78	279.33

the operating frequency, and thus it is the only one matched to the feed line at that frequency.

As an example, consider a two-dipole arrangement with one dipole cut to ½ λ on 80 meters and the other cut for 40 meters. With this arrangement, ½-λ dipole bandwidth is obtained on both 80 and 40, and in addition, the 40-meter dipole is also a center-fed 3⁄2-λ radiator on 15 meters, which also presents a good match to coaxial feed line. Here is how the frequency selection occurs. On 80 meters, the 80-meter dipole behaves in the conventional manner, but the 40-meter dipole is inactive because its terminal impedance at 80 meters is approximately $14 - j1300\ \Omega$. This impedance results in a mismatch of well over 2400 ~~4000~~:1 on 50-Ω line, assuring its inactivity. On 40 meters, the 40-meter dipole also behaves in the conventional manner, but the 80-meter dipole presents a terminal impedance of approximately 5000 Ω to the feed line at 40 meters, for a mismatch of 100:1. Consequently, this mismatch is also sufficiently severe to result in an inactive 80-meter dipole on 40 meters, especially since the 40-meter dipole is taking practically all the power from the line because of its good match. On 15 meters, the mismatch of the 80-meter dipole is in the same ball park as it is on 40 meters, assuring inactivity on this band.

In practice, I have found that a good match results on 80, 40, and 15 meters when 80- and 40-meter dipoles are prepared for ½-λ operation using the standard length formula

$$\text{len}_{\text{feet}} = \frac{468}{f_{\text{MHz}}} \qquad \text{(Eq 20-1)}$$

When a 20-meter dipole is added to the 80- and 40-meter dipoles, there is a slight degradation in the impedance match of the 40-meter dipole on 15 meters. This results from a reactance component in the 20-meter dipole impedance when operating on 15 meters. There should be no serious problem in adding a 10-meter dipole, except that it may require a little trimming to obtain a satisfactory match at a particular frequency in the band. The bandwidth of each dipole is practically identical with that of a single-fed dipole, and hence, the stagger tuning arrangement provides a multiband antenna system far more satisfactory than a trap dipole. Each dipole in the parallel-fed combination may be supported from different directions if different directions of radiation are desired. Otherwise the shorter dipoles may simply be suspended from the longer ones. Additional information on multiband dipoles in the parallel configuration may be found in *Ref 71, p 7-3*.

With the trap dipole, the conventional pi-network tank circuit is unable to provide an acceptable match to the input impedance of the feed line except perhaps for a few kilohertz on either side of the resonant frequency of the trap. Hence, an antenna tuner is necessary if any wider frequency excursion is to be enjoyed. All solid-state rigs that have no internal antenna tuner require an external tuner in any case, except over very narrow frequency ranges away from dipole resonance where the feed-line input impedance happens to be close to 50 Ω. And even in the case of the simple single dipole, there are some frequencies that will be out of range of pi-network tank circuits when using coaxial feed lines. This is especially true of the 80-meter band. If operation over the entire 80-meter band is desired when using coaxial feed line, the minimum band-edge SWR occurs when the dipole is cut to a length which resonates at 3.750 MHz, the center of the band. This length is usually 125 feet. With this length, the SWR is around 6.5:1 at 3.5 MHz, and 5:1 at 4.0 MHz. Consequently, an external antenna tuner is necessary for operation out to the band edges. The point I'm making is that an antenna tuner is an indispensable part of any amateur station that operates in the HF bands. Anyone operating without one is operating with limited flexibility.

Sec 20.2.3 Random-Length Dipoles

Now I'm coming to the crucial topic concerning multiband operation at HF. I hope I've convinced you of the necessity for an antenna tuner, so why not use it to the fullest advantage and achieve multiband HF operation with the simplest, yet effective and efficient antenna? I'm speaking of a random-length dipole fed with either open-wire line or ladder line. By random length, I mean any length at least ¼-λ long at the lowest frequency of operation.

A center-fed ¼-λ dipole radiates with an effectiveness of approximately 95% relative to a ½-λ dipole, and has a terminal impedance of approximately 7 – *j*1300 Ω. This terminal impedance yields an SWR of 488:1 on 600-Ω line, 600:1 on 450-Ω line, and 848:1 on 300-Ω line. As would be expected, the tuning of the antenna tuner is quite sharp with these values of SWR. However, with the low-loss characteristics of lines of these impedances, the results are good. This is what I use for 160-meter operation—an 80-meter ½-λ dipole, which is ¼ λ on 160 meters, and fed with 300-Ω line. It works very well. I find it more satisfactory than the traditional method of tying the two wires of the parallel feed line together for a single-conductor feed to the resulting flattop. A single wire that center-feeds both dipole halves feeds the halves in a phase relationship that causes current to flow in opposite directions in each half. This opposite current flow results in no radiation from the flattop—the flattop is now only a capacitive-loading top hat, and radiation is only from the feed line!

I'm not suggesting that everyone use a dipole as short as ¼ λ, but I do recommend a length of about ⅜ λ, halfway between ¼ λ and ½ λ, if you can't erect a full ½-λ on 80-meters. For general use, this antenna results in a little higher effectiveness than the ¼-λ dipole, and it does not have such sharp tuning in the antenna tuner. A ⅜-λ dipole has an effectiveness greater than 98% relative to a ½-λ dipole, and has a terminal impedance of approximately 34 – *j*400 Ω. This impedance yields an SWR of 25.5:1 on 600-Ω line, 23.7:1 on 450-Ω line, and 24.6:1 on 300-Ω line, all very reasonable values for open wire or ladder line. A ⅜-λ dipole at 3.5 MHz is approximately 100 feet long, which means that a random length from 90 to 100 feet will make an excellent radiator on all the HF amateur bands, 80 through 10 meters, including the WARC-79 bands. The length of a ⅜-λ dipole may be found using the formula

$$\text{len}_{\text{feet}} = \frac{351}{f_{\text{MHz}}} \qquad \text{(Eq 20-2)}$$

If you don't have room for even 90 feet of straight wire for operation on 80 meters, a 10- to 15-foot portion of each end may be dropped vertically from each end support. There will be no significant change in radiation pattern on 80 and 40 meters. However, there will be a minor change in polarization in the radiation at higher frequencies, but the effect on propagation will be negligible. Although I work 160 meters with a 125-foot dipole, tuning of the antenna tuner would be less critical on 160 meters with a 200-foot dipole, and admittedly, the feed-line loss would be somewhat lower. However, the loss from feed-line attenuation is quite low, even at the high values of SWR. Your attention is invited to the loss versus SWR graph in Fig 6-1 to see just how little loss is incurred with open-wire feeders. The point I'm stressing here is that when using an antenna tuner with open-wire or ladder-line feeders, the dipole length is not critical, because the tuner provides the impedance match throughout the entire antenna system, whatever the dipole length may be.

Sec 20.2.4 The G5RV Antenna

With this background on random-length dipoles behind us, it seems appropriate to make a critical examination of a particular 102-foot dipole that is enjoying a great deal of popularity—Louis Varney's G5RV dipole. In spite of its popularity, its operation is not well understood among many amateurs, so I'll shed a little light on the G5RV. First of all, the reason for the 102-foot length for the G5RV is no secret, but it is not well known. Being unaware of certain antenna principles, many amateurs have come to believe that there is some sort of magic in the 102-foot length, and that their all-band success with this antenna is dependent on this specific length. Nothing could be further from the truth, because, except for 20 meters (as I'll soon explain), any random length of at least 3⁄8 λ at the lowest operating frequency will perform equally well.

What is the significance of the 102-foot length? Unbeknown to many amateurs who use it, Varney designed the antenna to be a resonant 3⁄2-λ radiator on 20 meters—that length is 102 feet. He had two specific reasons for selecting 3⁄2 λ on 20—he wanted a multilobe radiation pattern and a low feed-point impedance. The 3⁄2 λ was a clever choice, because this length yields a multilobe pattern, in addition to a low feed-point impedance that can be line-transformer matched to a 50-Ω line without requiring an antenna tuner. As Varney also intended, this 102-foot length results in a strictly random length on all bands except 20, so except for the 20-meter considerations I just described, there is no magic whatever to this length. The last comment in the previous paragraph should be taken seriously. It should be noted that the 102-foot length of the G5RV is almost exactly the length I recommended above for a random-length antenna, 3⁄8 λ at the lowest frequency of operation, because 100 feet is the length required for 3⁄8 λ at 3.5 MHz.

On 20 meters, the input impedance of the 3⁄2-λ G5RV radiator is low because the feed point is at the center of the central 1⁄2-λ portion. Hence, the impedance (the resonant resistance) is only moderately higher than if the outer 1⁄2-λ sections were eliminated, leaving a single 1⁄2-λ dipole. At midband resonance, the free-space feed-point impedance is approximately 100 + $j0$ Ω, which reduces to around 90 + $j0$ Ω at a convenient height above ground. This results in a mismatch of about

1.8:1 relative to 50 Ω. Varney's choice of the 34-foot matching section, ½ λ on 20 meters, was to make a 1:1 impedance-matching transformer that repeats the $90 + j0$ antenna impedance at its input terminals. Thus, with a suitable choke balun to make a transition from a balanced to an unbalanced line, the low 1.8:1 mismatch makes connecting to a 50-Ω line feasible without requiring an antenna tuner. The SWR on a ½-λ matching section of 300-Ω line is around 3.3:1, while on a 450-Ω line it is about 5:1. Keep in mind that these considerations apply only to 20-meter operation. On all other bands, the G5RV antenna terminal impedance is much higher and reactive, resulting in a higher SWR and making the use of an antenna tuner imperative. Incidentally, the length of a 3/2-λ radiator may be found using the long-wire antenna formula

$$len_{feet} = \frac{492\,(n - 0.05)}{f_{MHz}} \qquad \text{(Eq 20-3)}$$

where n = the number of half wavelengths in the radiator

It should also be mentioned that failure to include a balun between the balanced open wire and the unbalanced coax results in RF radiation in the shack from current flow on the outer surface of the coax shield.

In addition to the misunderstanding concerning the "magical" 102-foot length of the G5RV, there are also other areas of confusion focused on this antenna, some concerning the role of the feed line. There are some who believe that a particular combination of open-wire and coaxial feed line yields a perfect 1:1 match on all bands without a tuner. As stated above, this is true only on 20 meters. Others believe that because the 102-foot dipole length is shorter than ½ λ on 80 meters, a certain length of the feed line is a folded-up portion of the antenna to make up for the difference in length, and that the folded-up portion radiates along with the antenna. Still others believe certain lengths of feed line are to be avoided to prevent "antenna current" from flowing on the feed line because of line resonance. I wish I knew how these myths originate.

My own involvement with the G5RV antenna dates back to the early '70s when I began lecturing on SWR and reflections on transmission lines. My lectures promoted the use of antenna tuners with open-wire feed line on random-length antennas as the best way to achieve all-band operation. I also promoted the concept that the correct length of feed line is that which is required to go from the output terminals of the tuner to the antenna terminals, because, regardless of the length of the feed line, both the feed line and the antenna are made resonant by the conjugate matching action of the tuner. Hence, there is no reason to avoid certain lengths to prevent line resonance, because the tuner makes them resonant anyway.

I first heard of the G5RV when someone in my audience described his 102-foot antenna with open-wire and coax feed line. He claimed it gave him a 1:1 SWR on all bands without a tuner. I told him he must have a lossy coax to get 1:1, because I knew a 1:1 would be impossible with such an arrangement without some exceptionally high resistive loss somewhere in the antenna system. After hearing several more identical claims in later lecture sessions, I analyzed the antenna on all bands, observing it to be $3/2\ \lambda$ on 20 meters, but a random length on all other bands, so I felt confident in rebutting the ridiculous "1:1 on all bands without a tuner" claims. Incidentally, Varney published an update of the G5RV in *The ARRL Antenna Compendium, Volume 1* (*Ref 112*), in which he presented the same specifications for the antenna that I described above, which confirms my earlier observation that his antenna is $3/2\ \lambda$ on 20 meters, and a random length on all other bands.

Let's now examine the other myths and confusion concerning the G5RV that I mentioned earlier. First, we'll consider the feed-line combination believed to yield a 1:1 match on all bands. It has been written that the combination of 33 feet of open-wire line, plus 68 feet of 50-Ω coaxial line will yield such a match. Don't you believe it! A determination of the G5RV antenna-terminal impedance on all bands shows that there is no length of open-wire line of any characteristic impedance Z_C that will transform the antenna impedance Z_A to an impedance that is even close to presenting a match to 50- or 75-Ω coax, except on 20 meters. However, when fairly long lengths of coax follow a length of open wire, the high SWR appearing at the junction of the open wire and the coax will be reduced significantly at the input of the coax because of the attenuation loss in the coax, especially at the higher frequencies. The longer the coax, the lower the input SWR, but remember that this method of lowering the SWR is costly in terms of lost power. Because an antenna tuner is necessary anyway, except on 20 meters, it makes no sense to use any coax at all. Coax performs no useful function in the feed system, and it consumes power unnecessarily because of the high SWR. A more sensible method is to run the open-wire line all the way to the tuner and eliminate the coax entirely.

Second, let's consider the length of feed line believed to be a folded-up portion of the antenna that radiates. Radiation occurs when the electromagnetic field developed by current flow on a conductor is not canceled by an opposing field developed by an equal current flowing in the opposite direction. Hence, radiation occurs as a result of current flowing on an antenna. However, antenna current ceases being antenna current at the antenna terminals, because once it enters the transmission line, the current becomes transmission-line current, with the current in the two conductors flowing in opposite directions. There is no radiation from any portion of the line, because the fields developed by the currents flowing in opposite directions in the two conductors oppose and cancel each other throughout the entire length of the line. Therefore, no portion of the feed line becomes part of the antenna.

Sec 20.3 Antenna Currents on Feed Lines

We know there are times when the feed line does radiate in spite of the statement in the previous paragraph, so let's now examine this concept a little further to determine what conditions must exist to cause the feed line to radiate. Antenna currents and radiation can occur on a two-wire balanced line, but only when the currents in the two wires are unequal. Neither a resonant line nor a resonant combination of line and antenna length will cause the currents to be unequal. Nor will high SWR cause unequal currents or feed-line radiation. Never forget that whatever the length of the feed line (or the antenna), the antenna and feed-line combination is always made resonant by the conjugate matching action of the antenna tuner.

I'll now explain how antenna currents are developed, and it will become clear why feed-line length is not a factor in causing feed-line radiation. As stated earlier, equal transmission-line currents in the two wires of the line flow in opposite directions (push-pull). The electromagnetic fields developed by these two currents are thus 180° out of phase, and oppose each other. Because the separation between the wires is small in terms of a wavelength, the two opposing fields cancel each other, preventing radiation from the feed line. No antenna currents are generated. On the other hand, if the two push-pull transmission-line currents are unequal, their corresponding fields are also unequal, and thus do not completely cancel; hence, an antenna current is developed and radiation results.

The two main causes of transmission-line current imbalance are

(1) asymmetrical positioning of the feed line relative to the antenna, and (2) the antenna itself is unbalanced, either by unequal lengths relative to the feed point (as in the twin-lead fed, misnamed "Windom" antenna), or one portion of the antenna is closer to surrounding objects such as a metal roof. If the feed line is positioned symmetrically relative to the antenna, each half of the antenna induces an equal but opposite current flow on the feed line. This is because each half of the antenna is of opposite polarity. The result is zero induced current, regardless of line length.

If the feed line is *not* positioned symmetrically, the currents induced from each antenna half are not equal. The result is an induced current on the feed line, traveling in the same direction on both wires (push-push). When the push-push induced currents join the push-pull transmission-line currents, the net current is increased in one wire and decreased in the other. The resulting difference in current now flowing in each wire constitutes the antenna current, which does cause the feed line to radiate. Therefore I repeat, if the currents in the two wires of the feed line are equal, antenna current and radiation will be zero, regardless of the length or resonant condition of the feed line. Antenna current flowing on a coaxial (unbalanced) feed line resulting from not using a balun is another subject, and is treated in depth in Chapter 21.

Although it is not pertinent to the G5RV, a strange example of asymmetrical feed-line positioning exists in the misnamed "Windom" antenna. This is a dipole antenna fed off-center with twin lead at the point where a true Windom would be fed with a single-wire feeder. This antenna is not a Windom, and the feed current paths are not the same. In this pseudo Windom, the go and return currents flow on the two wires of the twin lead. In the true Windom, the return path for the feeder current is in the ground-reflected image of the single-wire feeder. And yes, the Windom feeder does radiate, but not for the same reason the pseudo-Windom feeder radiates. The off-center twin-lead feeder of the pseudo-Windom radiates because it carries push-push antenna current in addition to the push-pull transmission-line current, because the currents induced in it from the asymmetrical dipole portions are unequal. There is a legitimate reason for moving the twin-lead feeder away from the center of the antenna, however, as discussed in Sec 21.10. In addition, for ease in matching a ½–λ dipole to a low-impedance line, a feed point ⅓ of its length from one end exhibits low-impedance current loops on the even harmonics of the fundamental frequency, while the center of the dipole exhibits low-impedance current loops on the odd harmonics.

Sec 20.4 Standing Waves and Feed-Line Radiation

It goes without saying that practically every amateur takes SWR measurements on his feed line to determine the status of his antenna system. However, unless the SWR is 1:1, the results of the measurements are often misinterpreted, and the standing wave innocently becomes the culprit in an unfortunate scenario that leads to all sorts of needless effort to get the SWR down to 1:1. Many believe that standing waves produced by an impedance mismatch between the feed line and the antenna result in an energy field surrounding the feed line, causing it to radiate as an antenna. Is this true? No! This is not the case at all, so I'll explain.

A voltage standing wave is simply a variation in the line voltage appearing *between the feed-line conductors* at different positions along the line. And a current standing wave is a corresponding variation in line current flowing on the conductors. Both voltage and current standing waves are illustrated in Fig 9-3. The variations in voltage and current result from the interaction between the forward and reflected waves of voltage and current, as explained in Chapter 9. Let's first consider the conditions prevailing on a matched, open-wire line with no standing waves. Here, there is no variation in either the line voltage or line current as we move in position along the line. And we know from the discussion in the previous section that there is no radiation from the line, because the opposing fields produced by the oppositely flowing currents in the two conductors of the line cancel each other. In other words, there are actually two energy fields surrounding the feed line (open wire, not coax), but they cancel each other because they are of opposite polarity, resulting in no radiation.

Now consider the conditions prevailing on a mismatched line with standing waves. As stated above, we now have a variation in line voltage and current as we move in position along the line—at some points the current is higher and at other points it is lower. The crucial point is that no matter whether the current is higher at one point and lower at another, the current is *equal and opposite* in both conductors at any given point along the line. Therefore, the fields produced by the currents in the two conductors are also equal and of opposite polarity everywhere along the line, and cancel in exactly the same manner as in the line with no standing waves. Hence, there is no radiation from the line, standing waves or not.

So far I have been discussing open-wire lines, where there is nothing to contain the electromagnetic fields produced by the line currents. However, there is an additional reason why there is no radiation from *coaxial* lines because of standing waves. This is because the fields produced by the currents in the conductors of coaxial lines are entirely confined within the coaxial shield surrounding the line. Hence, standing waves on transmission lines do not cause radiation from either open-wire or coaxial transmission lines.

Sec 20.5 The Extended Double Zepp and the ⅝-λ Vertical

There is another dipole type of antenna that is becoming popular with amateurs for its increased gain, called the extended double Zepp, or EDZ for short. It is a collinear dipole, although it can't rightfully be called a dipole because each half of its length is greater than ½-λ (each half is actually 0.64 λ). This means it has a reversal of current flow in each antenna half, and therefore has more than two "poles." Hence, the EDZ rightfully belongs in the realm of the long wire. The EDZ is described in *The ARRL Antenna Book* (*Ref 71, p 8-34*), which shows the current distribution on the radiator and feed line, as well as the radiation pattern.

The reason for my mentioning the EDZ here is that it has a long history relative to AM broadcasting that is not well known, and I believe you will find it very interesting. In the early 1920s, antennas used in AM broadcasting were in their infancy. Many simply used a configuration similar to those used in long-wave, or LF transmission. These antennas generally comprised a top-loaded vertical wire supported by two towers, one at each end of the top-loading wires. The vertical wire was usually much shorter than ¼ λ, and the radiation pattern in elevation was the same as the typical ¼-wave vertical. However, the pattern was not uniform in all azimuth directions because of field distortion caused by the supporting towers. There is no radiation from the top-loading wires, because the currents in each opposing wire are flowing in opposite directions. The wires simply add capacitance to ground, electrically lengthening the short vertical portion.

In 1923, Stuart Ballantine of the Harvard University Physics Department made an important discovery which he published in 1924 (*Ref 115*). He discovered that there is an optimum height for a vertical radiator that produces the maximum radiation in the horizontal, broadside direction. That height is 0.64 λ. The increase in radiation in the horizontal direction is 3.03 dB relative to a ¼-λ vertical radiator. This increase in horizontal radiation arises from a corresponding decrease in the vertical components of the radiation in the major lobe. In other words, the fat half-doughnut pattern of the ¼-λ vertical radiator is squashed down into a thinner doughnut, but one having a greater diameter. However, because this 0.64-λ height is greater than ½ λ, the current in the radiator reverses direction at a point ½ λ from the top, 0.14 λ above the ground. The reversal in current at this point produces a separate radiation field from the bottom 0.14-λ portion, a field that is 180° out of phase relative to the field producing the major lobe of radiation. Consequently, this separate out-of-phase field produces a minor lobe of radiation at a high angle, with a deep null between the major and minor lobes.

I have performed a computer analysis of the 0.64-λ vertical radiator; a pattern integration yields 0.6346 λ (228.45°) as the true optimum height, furnishing a gain of 5.1769 8.1899 dBi, or 3.029 dB over the ¼-λ vertical. The maximum field in the minor lobe, which occurs at 58° in elevation, is 9.13 dB below that of the major lobe. The null between the major and minor lobes occurs in elevation at 35°. You may recognize by now that this vertical radiator is also the well-known ⅝-λ vertical.

Getting back to Ballantine, his discovery formed the basis for the evolution of the vertical tower becoming the radiator in AM broadcasting, replacing the top-loaded vertical wire supported by two towers. There were two principal advantages in using the tower as the radiator. First, the cost for one tower was less than for two. And the second very important advantage was that the single tower eliminated the radiation-pattern distortion caused by the two towers supporting the vertical radiator in the earlier arrangement. In addition, Ballantine's 1923 prediction of increased radiation in the horizontal direction was proved correct when single-tower radiators were erected at his predicted height. However, one unforeseen problem crept in—the radiation from the higher angle minor lobe returned to ground from the ionosphere at night and caused severe fading several miles from the antenna. At the distance where the signal from the ionosphere was the same strength as that of the direct ground wave from the major lobe, wave interference occurred between the two signals and caused the fading. The solution to the problem was the reduction of the tower height to 0.528 λ (190°), which nearly eliminated the minor lobe, while reducing the radiation level by only 1.03 dB. Thus, the fading problem was solved. As a result, many of the AM broadcast stations throughout the world are still using 190° towers.

One other factor that resulted in increased radiation in the horizontal direction was the use of a more efficient ground-radial system that emanated from the experiments of my colleagues, RCA's Brown, Lewis (W2EBS), and Epstein in 1937 (see *Ref 20*). The system they developed is still in use today. If you are interested in more detail concerning the evolution of the vertical radiator and the ground-radial systems in present use, I recommend reading *Radio Antenna Engineering* by Edmund A. Laport, also of RCA (*Ref 57*). Also see Chapters 1 and 5. In addition, Brown's analysis of the ground currents surrounding the base of the vertical antenna also makes interesting reading (*Ref 111*).

Let's now return to the EDZ. What is the connection between the EDZ and the vertical radiator used in AM broadcasting? Simply its length. We must remember that the ground-reflected image of the vertical radiator supplies the missing half of a dipole. Consequently, a dipole having a half-length of 0.64 λ (0.6346 λ if we nit pick) yields the maximum possible radiation in the broadside direction from a single radiator, 3.03 dB greater than that of a ½-λ dipole. With further increases in length, the broadside radiation diminishes rapidly, and becomes zero when the length is 360°. At 360°, the radiation pattern forms two separate, equal lobes resembling a four-leaf clover when viewed on both sides of a pattern representing the radiation.

As stated earlier, the radiation pattern of the EDZ appears on page 8-34 of *The ARRL Antenna Book* (*Ref 71*), and for comparison, the pattern of the ½-λ dipole appears there on page 3-11. The difference in the widths of the lobes is remarkable. However, despite the 3 dBd gain obtained in the broadside direction from the EDZ, the narrower main lobe of the EDZ may work against you at angles away from broadside. Keep in mind that long-distance propagation at HF is via reflection from the ionosphere, so also remember to consider the takeoff angle from the antenna to the ionosphere for the station you're contacting. At angles away from broadside, a dipole may be radiating more energy at the higher wave angles than the EDZ.

I would be remiss if I neglected to discuss the terminal impedance of the EDZ, so for those who are thinking of feeding it with coax, guess again! From interpolating thin-wire antenna impedance data from King

(*Ref 37*), the terminal impedance of a radiator with a half-length of 0.64-λ is 126.18 – *j*658.2 Ω. On a 50-Ω transmission line, the normalized terminal impedance is 2.5236 – *j*13.164, yielding an SWR of 71.57:1. However, on a more realistic approach, here are the values for three higher impedance transmission lines. On 300-Ω line the normalized impedance is 0.4206 – *j*2.194, SWR 14.17:1; on 450-Ω line the normalized impedance is 0.2804 – *j*1.4631, SWR 11.39:1; on 600-Ω line the normalized impedance is 0.2103 – *j*1.0907, SWR 10.59:1. These values of SWR are reasonable for the transmission lines involved, and the balanced output of your antenna tuner should be able to handle these mismatches very well! Just as an exercise, you might want to try calculating these SWR values using the hand-held calculator program listed in Appendix 4 of Chapter 23. I should warn you though, that program is in reverse Polish notation for Hewlett-Packard calculators. Sorry I couldn't include programs for the scientific-notation units.

I mentioned earlier that the 0.64-λ vertical is the same as the ⅝-λ vertical used in many VHF and UHF mobile applications. It is interesting to observe the ease in which this antenna can be matched to a 50-Ω transmission line. The terminal impedance of a vertical radiator over ground, or a ground plane, is half the impedance of the complete dipole having the same half-lengths. Therefore, half of the EDZ terminal impedance of 126.18 – *j*658.2 Ω is 63.09 – *j*329.1 Ω for the ⅝-λ vertical. However, this capacitive reactance of 329.1 Ω was determined for a *thin* radiator, while the whips used in our mobile applications are not thin for lengths and diameters we use at VHF or UHF. So the actual reactance at these frequencies will be considerably less than 300 Ω.

Common practice is to simply insert an inductor having a reactance equal to the capacitive reactance of the antenna in series with the antenna, thus canceling the antenna reactance, and leaving only the 63-Ω resistance component of the antenna terminal impedance to be fed with the 50-Ω line. The result is an SWR of only 1.26:1, which hardly needs any improvement. An alternative method of feeding the ⅝-λ vertical is to connect the inductor between the antenna and ground, and connect the center conductor of the feed line to one of the turns of the inductor. By trimming the length of the antenna and selecting the correct turn to tap onto the coil, an SWR of 1:1 can be achieved. With all of this said about the ⅝-λ antenna, perhaps you are now considering a change to a ⅝-λ vertical from ¼ λ. If so, there is a paper by Donald Reynolds, K7DBA, that is worth studying (*Ref 119*).

Chapter 21

Some Aspects of the Balun Problem

(Adapted from *QST*, March 1983)

Sec 21.1 Introduction

Why all the mystery surrounding baluns? Here's some straight talk to dispel the rumors! The balun—to use, or not to use—is one of today's hottest topics in Amateur Radio. Because certain aspects of the connection between a coaxial feed line and a balanced antenna have been ignored, misunderstanding still exists concerning the function of baluns. Many commercial baluns embody some form of impedance transformer, promoting our tendency to misconstrue them as little more than a matching device, while their primary function is to provide proper current paths between balanced and unbalanced configurations.

To help clarify the misunderstanding, this chapter explains some of the undesirable effects that occur when a balun is not used, and some that occur when baluns employing coupling transformers are used. In many cases, these effects cause significant errors in measurements of antenna terminal impedance and SWR. This chapter also describes a simple and inexpensive method of loading the *outside* of a coaxial feed line with ferrite, which effectively produces a well-balanced, wide-band choke balun. Because this configuration eliminates the coupling transformer (with inherent impedance-transfer ratio errors), the accuracy obtainable in antenna impedance and SWR measurements is greatly improved. In addition, antenna matching networks may be used with this choke balun, because no mismatch limits are imposed by the balun.

Sec 21.2 Transformer Accuracy

Using the General Radio 1606-A precision impedance bridge and the Boonton 250A RX meter, I have made measurements of transformer-type baluns which prove that with a 50-Ω resistive load, the transformers in typical 1:1 or 4:1 baluns *do not* yield a true 1:1 or 4:1 impedance transfer ratio between their input and output terminals. This is because of losses, leakage reactance, and less than optimum coupling. My findings have been substantiated by the work of John Nagle, K4KJ (*Ref 80*).[1] Furthermore, the impedance-transfer ratio of such baluns degrades even further when used with an antenna that is reactive from operation away from its resonant frequency. This degradation of impedance transfer associated with transformer-type baluns poses no serious operational problems. However, SWR curves plotted from measurements of an antenna using such a balun differ significantly from those using a choke balun that has no impedance-transfer error. Thus, when a precision bridge is used to measure antenna impedance (R + *j*X), the data will be erroneous with either a transformer-type balun in the circuit, or with no balun at all.

Sec 21.3 Should SWR Change with Line Length?

We know that the feed-line input impedance changes with line length when the load (antenna) is not matched to the line. Sometimes trimming the length of the feed line helps to obtain a load impedance better suited to match the transmitter. Theoretically the SWR should not change with line length—except for a barely perceptible change because of the corresponding change in line attenuation. Then why does the SWR sometimes change? If the SWR changes significantly with a change in line length, it means that the load impedance terminating the line is also changing. The load impedance can change with line length? *Yes*. If a balun is omitted when you feed a balanced antenna with coaxial cable, the load impedance will change when the line length is changed, and of course the SWR will also change! To explain this commonly experienced phenomenon we must investigate how current flows in an antenna system.

To understand the functions of a balun, it is essential to be familiar with current paths at the feed point of the dipole. These paths are shown

---

[1] All references appear in Chapter 24.

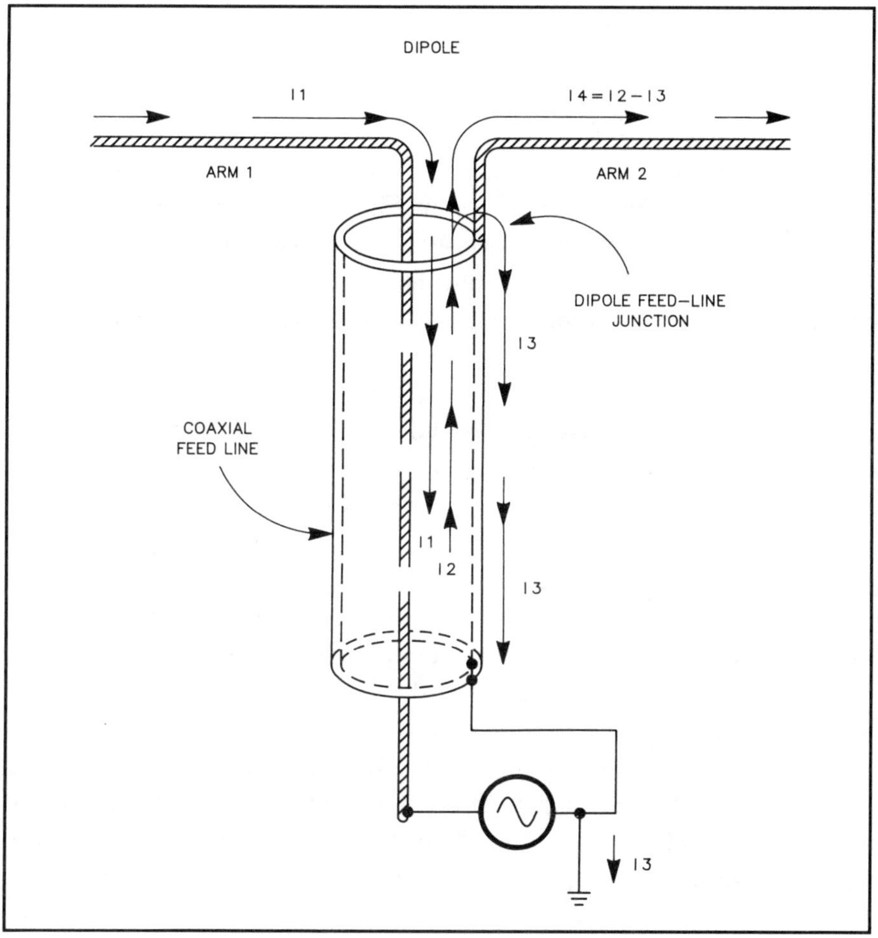

Fig 21-1—Illustration of the various current paths at a dipole feed point.

in Fig 21-1. Because of their symmetrical relationship, the dipole arms couple energy of equal magnitude and opposite phase onto the feed line, thus canceling induced current flow on the outside of the feed line (*Ref 81*). What is disturbing is the discovery that there are *three* paths for current flow in a coaxial feed line, instead of only two. How can there be three current paths in only two conductors? At RF, skin effect isolates current between the inner and outer surfaces of the coaxial shield. This effect, which does not occur significantly at dc or low-frequency ac, prevents currents that flow on the inner braid surface from interacting with those on the outer surface, and vice versa.

As shown in Fig 21-1, while traveling within the transmission line, current I1 flows on the center conductor and I2 flows only on the inner surface of the outer shield. When antenna current is flowing from left to right as shown, I1 flows out of dipole arm 1, downward onto the center conductor, and returns to the generator. Current I2, being of opposite phase and direction, flows upward along the inside surface of the feed-line shield until it reaches the junction of dipole arm 2. At this junction, I2 divides into two separate paths to form I3 and I4. Current I3 flows back down the *outside* surface of the feed line, and I4, which equals I2 – I3, flows to the right onto dipole arm 2. The magnitude of I3 depends on the impedance to ground provided by the outside surface of the coaxial shield.

If the effective path length to RF ground is an odd multiple of ¼ λ, the impedance to ground will be very high, making I3 negligible. In this case, I1 and I4 will be nearly equal. On the other hand, if the RF path to ground is a multiple of ½ λ, the impedance to ground will be fairly low, and current I3 may be substantial. This results in unequal currents in the dipole arms and radiation from the feed line. In many instances, this RF path to ground includes the transceiver line cord and some house wiring, terminating at the power-line ground! Thus, the amplitude of I3 varies with changes in feed-line length because of changes in the impedance between dipole arm 2 and ground.

Keep in mind that transmission-line currents I1 and I2 *cannot* produce radiation, because their fields are not only of equal magnitude and opposite phase, but their fields are also confined within the shield of the coaxial cable. However, the field developed by I3 *does radiate*, and thus, the outer surface of the coaxial braid effectively becomes dipole arm 3, which is connected in parallel with dipole arm 2.

To clarify this equivalent connection of radiators, I've simplified the circuit as shown in Fig 21-2. Since currents I1 and I2 do not interact with any other currents, we may hypothetically place the RF generator directly at the input terminals of the antenna. Now that the coaxial cable is no longer needed to transfer power from the generator to the antenna, the third conductor of the feed line (the outside surface) can be replaced with a single wire connected between arm 2 and RF ground. We have not changed the circuit electrically because I3, which previously flowed on the outside of the coaxial cable, still flows to ground—but now on the single wire shown.

We know that, depending on height, the impedance of a dipole at resonance is usually between 50 and 75 Ω, and is purely resistive. At

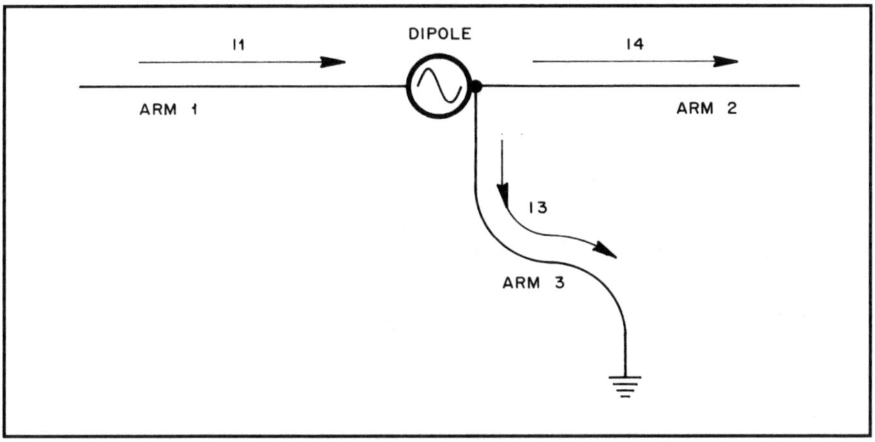

Fig 21-2—Simplified electrical representation of Fig 21-1.

frequencies above resonance, the resistance increases gradually, and series inductive reactance appears; below resonance, the resistance decreases and capacitive reactance appears. The impedance of each dipole arm is half of the total dipole impedance. Since the far end of arm 3 is at RF ground, its impedance behavior follows that of a short-circuited transmission line. Thus, when the length of arm 3 is an odd multiple of ¼ λ, its impedance is a parallel-resonant maximum, a high resistance typically from 2000 to 3000 Ω. This high resistance in parallel with arm 2 has little effect on the total dipole impedance. However, as the effective length of arm 3 departs from ¼ λ (or odd multiples thereof) by changes in either its physical length or the generator frequency, the input resistance of arm 3 decreases, and reactance also appears in series with the resistance. This reactance is inductive when the length decreases and capacitive when the length increases.

If the length of arm 3 is a multiple of ½ λ, the resistance will be a series-resonant minimum value (but not zero, because of arm 3 radiation and ground loss). Thus, when arm 3 departs substantially from an odd multiple of ¼ λ, the net resistive and reactive components of the parallel combination of arms 2 and 3 are different than those of arm 1. Consequently, the dipole impedance is different than if arm 3 was not present.

Returning to Fig 21-1, we can now see that, without a balun, changing the feed-line length also changes the antenna length (arm 3), which in turn affects the impedance at the input end of the feed line.

Therefore, the SWR measured at the input of the transmission line does change with line length when no balun is present to eliminate I3. This phenomenon explains a point that is often puzzling for the amateur who uses no balun, and who believes that he must trim his dipole each time the feed-line length is changed "to keep the SWR down."

It is evident that, in coupling an unbalanced line to a balanced load (such as a dipole), the *primary* function of a balun is to block the current path between the inside and outside surfaces of the coaxial shield. With a balun in the circuit, I2 will not divide at the end of the feed line to form I3, but instead will flow only onto dipole arm 2. Thus, when I3 is zero, I4 equals I1, and the currents flowing on dipole arms 1 and 2 are equal and therefore balanced.

After presenting all the above, let me stress that a balun at the antenna feed point will *not* prevent current from flowing on the outside of the coax shield when the coax is asymmetrically coupled to the antenna. Although I pointed out the above concept to Joe Reisert, W1JR, for use in his balun article (*Ref 82*), he apparently missed my point concerning the source of external current I3. Consequently, his Fig 2 and the associated paragraph do not address the *principal* function of a balun. Contrary to his explanation of his Fig 2, when antenna currents on the feed line are caused by *asymmetrical coupling* to the antenna, a balun *will not* eliminate these currents, but will only change their phase and magnitude.

Sec 21.4 The Effects of Using No Balun

It should now be obvious that obtaining accurate impedance measurements of a dipole antenna is difficult. When a transformer-type balun is used to avoid errors caused by I3, impedance-transfer errors obscure the true impedance at the antenna terminals. If the balun is omitted, the true impedance is obscured by the impedance of arm 3 shunting half of the dipole. Since there is no practical way to determine the impedance of arm 3, the true antenna impedance and SWR cannot be calculated from the measured data (*Ref 83*).

Refer again to Fig 21-1. Bear in mind that for any given physical length of feed line, the *electrical* length of the coaxial braid surface carrying I3 is not the same as that of the inside conductors carrying I1 and I2. This is because the dielectric constants and the propagation velocity factors are different. For example, the velocity factors for polyethylene and Teflon, the inner dielectric of various coaxial cables, are 0.659 and 0.695, respectively. The factor for foamed polyethylene is between 0.75 and 0.81, depending on the amount of air in the foamed material. If the outside surface of the coaxial cable is bare, the velocity factor for the outer shield carrying I3 approaches 0.95. However, the usually thin outer covering of polyvinylchloride (or sometimes Teflon) reduces the antenna-current velocity factor to a value somewhat less than 0.95.

From an operational viewpoint, current I3 itself is usually not detrimental to the performance of simple dipoles for 160 through 40 meters. In addition, I3 alone does not cause TVI, unless the feed line is much closer to the TV antenna than the transmitting antenna. However, radiation from external feed-line current can cause severe distortion in the radiation patterns of directive antennas, such as Yagis and quads. Unless a gamma match or other type of unbalanced input-matching scheme is used, all beam antennas with balanced input terminals require a balun if the optimum performance of the antenna system is to be achieved when fed with coaxial cable. For example, when a balun is not employed, the feed line and tower together become a separate, non-directional vertical antenna. The tower then produces unwanted vertically polarized radiation that fills in the rearward null in the beam pattern, destroying the front-to-back ratio. The tower radiates along with the feed line, because currents flowing on the outside surface of the line are induced in the tower through inductive and capacitive coupling between the feed line and the tower.

Sec 21.5 The Choke Balun

Although many baluns embody some form of coupling transformer, an alternative is to insert an RF choke in the outer conductor of the feed line. This presents a high impedance to I3 without affecting the internal line currents. Advantages of this method are the lack of limitations on either maximum SWR or power handling. In addition, there is no impedance-transfer error that plagues transformer types of baluns (which causes a skewing of SWR and impedance plots), because the choke balun has no coupling transformer—the feed line goes straight through to the antenna terminals!

The simplest choke balun is formed by coiling up a few turns of the feed line, starting where it connects to the antenna terminals. In the frequency range of 14 to 30 MHz, several turns of feed line coiled in a 6- to 8-inch diameter make an inductor with enough series reactance to minimize I3 and practically eliminate feed-line radiation. Unfortunately, this form of choke (with its air core) is not practical below 14 MHz, because too much coiled-up feed line would be required to reduce I3 to an acceptable level.

A word of caution is in order when the choke balun is used on tower-mounted antennas: The choke coil should be placed directly at the feed terminals of the driven element. If the coil is placed away from the feed terminals, any portion of feed line between the terminals and the coil is coupled to the boom or mast, which in turn is coupled to one arm of the driven element. This unbalanced coupling results in an imbalance of currents in the driven-element, causing pattern skewing and tower radiation.

The frequency range of the choke balun can be extended to well below 2.0 MHz by using a core of high-permeability ferrite instead of air. With higher core permeability, the choke inductance increases dramatically, thereby retaining the high reactance needed to minimize I3 at the lower frequencies. Of great importance, no core saturation occurs at high power levels in the choke balun (a serious problem in transformer-type baluns), because the core excitation is low level, produced only by I3, and not by the high internal line current that feeds the antenna.

At my suggestion, Reisert made his choke balun with a Q1 material (μ = 125 to 400) ferrite toroid, winding 9 turns of RG-141 coaxial cable on the core for use from 14 to 30 MHz (*Ref 82*). However, his 12-turn balun appears to provide only marginal performance at 4 MHz. The problem stems from the toroidal winding arrangement. It is difficult to get a sufficiently tight wrap of coaxial cable around the toroid, and the result is a coupling loss that makes it impossible to utilize the full value of the core permeability.

Sec 21.6 The W2DU Balun—Constructed With Ferrite Beads

I have obtained greatly improved choke-balun performance by placing several ferrite beads or sleeves of even higher permeability *around* the coaxial feed line (*Ref 84*). For readers who wish to build this simple coaxial balun, bead materials of various size and RF characteristics are available that dramatically increase both the resistance and reactance of a conductor. (Adding resistance to the reactance in this circuit improves the operational bandwidth of the balun with no increase in loss.) In general, the impedance of the outer coaxial braid surface increases almost proportionately with the number of beads placed over it. A combination of 50-Ω Teflon-dielectric RG-303 cable (or RG-141, with the fabric covering removed) and ferrite beads having an ID of 0.197 inch and a length of 0.190 inch form a superb, compact, wide-band balun. While the two inner conductors of the coaxial cable remain unaffected, the beads introduce a high impedance in series with the outer surface of the braid. This configuration effectively isolates the external output terminal of the feed line from that at the input end.

I made a test balun by slipping 300 no. 73 beads (μ = 2500 to 4000) over a piece of RG-303 coaxial cable. The impedance of the outer conductor of the cable measured 4500 + *j*3800 Ω at 4.0 MHz; 15.6 + *j*13.1 Ω was measured utilizing a single bead. For practical baluns to be used from 1.8 to 30 MHz (less than 12 inches long, including connector), use 50 no. 73 beads (Amidon no. FB-73-2401, or Certified Communications no. 73-W2DU.)[2] For 30 to 250 MHz, use 25 no. 43 beads (μ = 950 to 3000, Amidon no. FB-43-2401, or Certified Communications no. 43-W2DU). No. 64 beads (μ = 250 to 375) are recommended for use above 200 MHz, but I have not yet experimented with them. The coaxial cable need be only long enough to hold the beads and to access the end connectors. The graphs in Fig 21-3 show the measured values of series resistance R, reactance X, and impedance Z versus frequency of the outer braid surface of a choke balun for both the 25- and 50-bead types. With either balun, I3 will be negligible.

At full legal input levels, no power-handling problems will arise using these baluns, because the CW power-handling capability of the cable is 3.5 kW at 50 MHz, and 9 kW at 10 MHz (*Ref 87*). Any suitable connector that will mate with the load end of your feed line can be used at the input of the balun coax, and the balanced-output terminals may simply be pigtails formed by the inner and outer conductors of the balun

2 At this writing, ferrite bead materials are available from Amidon Associates, 12033 Otsego St, N Hollywood, CA 91607, tel 818-760-4429, and "The Wire Man" (Press Jones, N8UG), Certified Communications, 261 Pittman Road, Landrum, SC 29356 (tel 800-433-9473 and 800-727-9473). The Wire Man also has W2DU baluns available both in kit form and fully fabricated, ready for installation. Fully fabricated W2DU baluns are also manufactured by the Unadilla Antenna Manufacturing Company, PO Box 4215 BV, Andover, MA 01810 (tel 508-475-7831), and are available at all Amateur Electronic Supply (AES) outlets. (NOTE: W2AU baluns made by Unadilla are ferrite-core, transformer-type voltage baluns.)

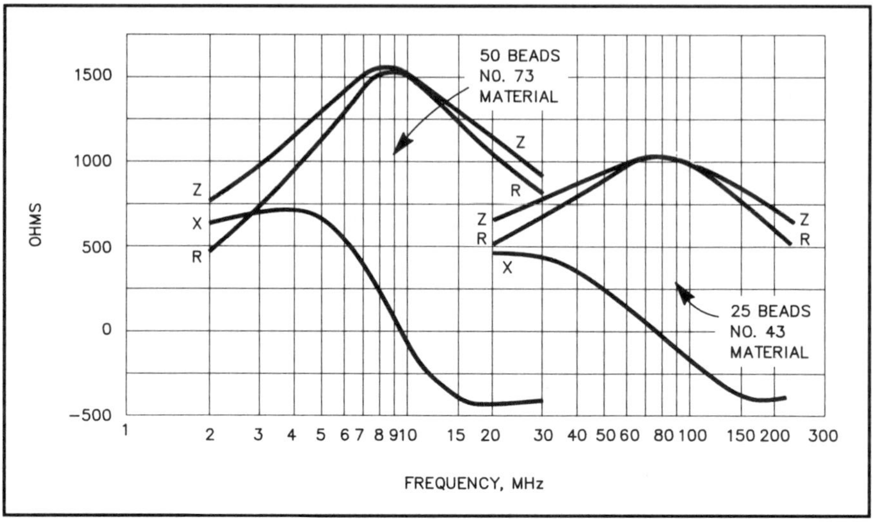

Fig 21-3—Graph of frequency versus series impedance of coaxial-balun shield outer surface.

coax. Methods for connecting the output terminals of the balun to the antenna are left to your ingenuity.

To emphasize simplicity, what VHF antenna buff wouldn't delight in dumping his unwieldy, frequency-sensitive, ½-λ line balun? You can replace it by simply putting some ferrite beads on the last few inches of the coaxial feed line!

Sec 21.7 Analysis of Voltage and Current Baluns

Roy Lewallen, W7EL, has performed extensive analysis and testing of imbalance on different circuit arrangements of both the choke balun and the ferrite-core transformer balun (*Ref 118*). His analysis shows that, while choke baluns are *current* baluns, bifilar-wound 4:1 and trifilar-wound 1:1 transformer baluns are *voltage* baluns. To the best of my knowledge, with the exception of the W2DU ferrite-bead balun (which is a current balun), all 1:1 baluns available on the commercial market are trifilar transformer baluns, and are therefore *voltage* baluns.

Lewallen has determined analytically that current baluns *force* equal currents into opposite halves of a dipole, independent of the impedance of either half. On the other hand, voltage baluns provide only equal voltage to the opposite halves, and thus do not provide equal currents in each half if the impedances of the two halves are not identical. His tests show overwhelmingly that choke-type current baluns provide the best balance of the dipole currents, and the least flow of antenna current on the transmission line. Lewallen's paper is a real eye-opener concerning balun design and use. Tests performed by Dr John (Jack) Belrose, VE2CV, provide corroborating evidence of the validity of Lewallen's analysis and tests, which I describe in Sec 21.10.

Sec 21.8 Verifying Output-Current Balance In Current Baluns

As stated, the ferrite-bead choke balun is a current balun. I use a simple test technique that proves the balun is a current balun, and which also determines the *degree* of current balance between the balanced output terminals. As illustrated in Fig 21-4A, mount the balun on a one-foot square aluminum plate, with the shield side of the unbalanced coaxial input terminal grounded to the plate. Connect a separate resistor between each of the coaxial output terminals and the plate. Fig 21-4B shows the electrical equivalent of this test setup. With an RF voltage applied at the unbalanced input terminals, measure the voltage appearing across each resistor with an RF voltmeter. (I use a Hewlett-Packard 410B.) When both resistors are of the same value, identical values of voltage will appear across each resistor, indicating equal current in each resistor, and signifying a balanced output.

Although equal current flowing in equal resistances signifies a balanced output relative to voltage, it does not yet prove we have a current balun. However, we can prove we have a current balun if we can show that equal currents flow in each terminating resistor when their resistances are *unequal*. Indeed, this is the case, because with this balun the voltages appearing across unequal values of terminating resistances are directly proportional to the values of the resistance, while the currents are equal. For example, say the two terminating resistors are 50 and 100 Ω, and the input voltage is adjusted so that 1 volt appears across the 50-Ω resistor. Then 2 volts will appear across the 100-Ω resistor. Applying Ohm's Law now proves that equal currents are flowing in the two terminations, even though their impedances are different.

The ferrite beads perform their task of isolating the output and input terminals of the coaxial-line outer conductor by inserting a high impedance between them, which permits the balanced output from an unbalanced input. If you have any doubt about that, consider this: If the beads were not inserting a high resistance in series with the outer surface

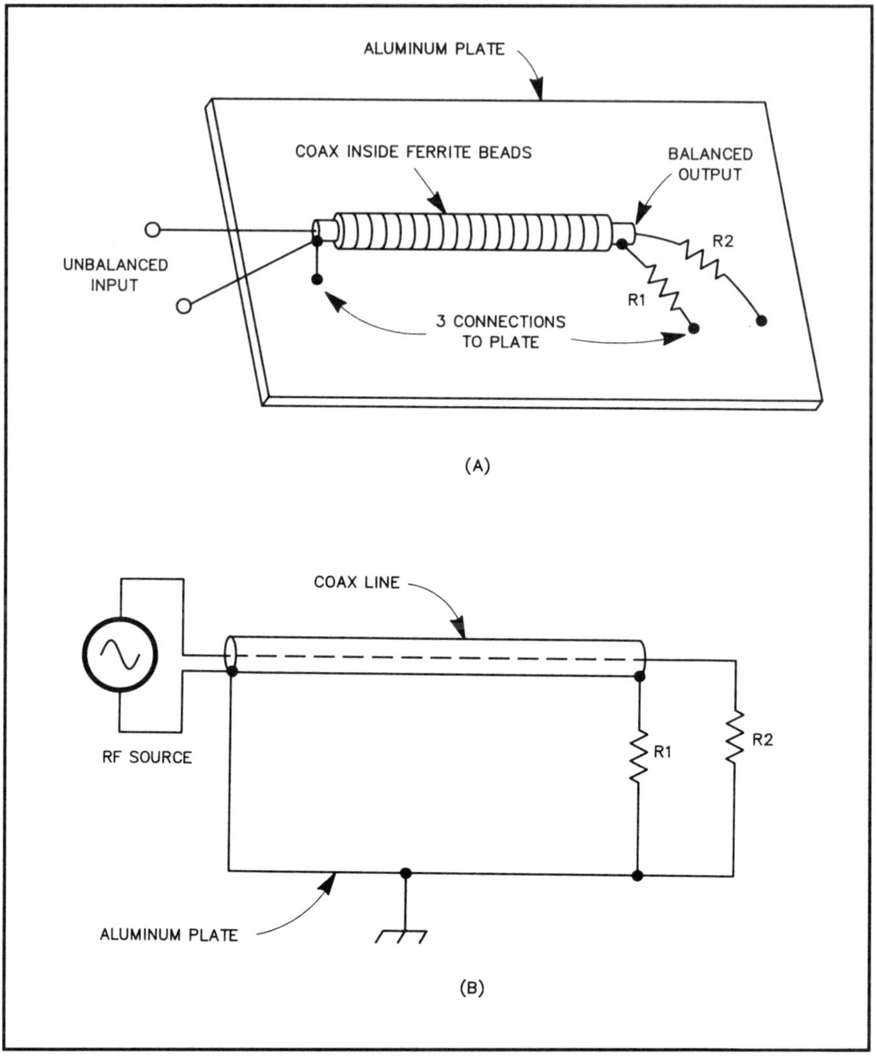

Fig 21-4—The drawing at A shows a test setup for measuring the degree of balance in a current balun, and B shows its electrical equivalent, with an RF source. With a perfect balun and an RF voltage applied at the unbalanced input, the voltage drops across R1 and R2 will be directly proportional to their resistances. This proves that equal current flows in the two resistors, and no current flows on the outside of the coax shield to short out R1, contrary to the way it appears in B.

of the outer conductor, terminating resistor R1 in Fig 21-4 would be short circuited by the low resistance of the outer conductor, and the voltage appearing across R1 would be zero.

Sec 21.9 Baluns With Antenna Tuners

In providing a balanced output for feeding open-wire or ladder-line transmission lines, it is common practice among manufacturers of antenna tuners to place the balun in the output circuit of the tuner. In all manufactured tuners with which I'm familiar, the baluns used are transformer-type, 4:1 voltage baluns wound around ferrite cores, usually toroids. Unfortunately, the output circuit is not necessarily the ideal place for the balun, for reasons I'll explain shortly. And further, the voltage balun is vastly inferior to the current balun in obtaining balanced currents in the feed line. Sec 21.10 explains why it is that when you use balanced feeders, the ideal arrangement for an antenna tuner is to place the balun at the *input* of the tuner, and for the balun to be a choke-type *current* balun such as the W2DU balun.

Let's first examine some of the problems encountered when you use a ferrite-core, transformer-type voltage balun in the output circuit of the tuner. When a transformer balun has a ferrite core, the core is subjected to the total magnetic flux developed by the load current, and the resulting high magnetic flux density can cause the core to saturate. When the core saturates, the RF waveform in the output becomes distorted, creating undesirable harmonic signals. A transformerless choke balun made with coiled-up coax, or a short coax with external ferrite beads, has no core to saturate. And in addition, the external beads are not subjected to the magnetic flux developed by the load current. The beads are subjected only to the flux developed by the vanishingly small current flowing through the high impedance that the beads create on the outside surface of the outer conductor. Hence, no harmonics are generated with the ferrite-bead current balun.

Another problem encountered with the ferrite-core transformer-type voltage balun is the distributed capacitance between its windings, which causes current unbalance between the two output ports that feed the balanced feed line. The input impedance of a balanced feed line can range in value from low to very high, and generally has a reactance component. The higher the line-input impedance, and the higher the operating frequency, the greater is the effect of the distributed capacitance in contributing to output current unbalance. On the other hand, output current unbalance is negligible with the ferrite-bead current balun.

Another undesirable feature of the 4:1 ferrite-core balun placed at the output of the tuner is that it can easily be permanently damaged. This can happen from overloading the balun when you operate with high power into a line having a high SWR that results in a high input impedance containing a large amount of reactance.

A still further undesirable feature of the 4:1 ferrite-core balun is its contribution to power loss. Typical losses with this type of balun range from around 0.5 dB at 2 MHz to 2 dB at 30 MHz. In contrast, the loss in the W2DU balun is minimal, only from 0.1 to 0.2 dB across the band, because the only loss involved is in the attenuation suffered in a 10.5-inch length of coaxial line.

Sec 21.10 Placing the Balun at the Input of the Antenna Tuner

At least three writers have published articles that advocate placing the balun at the *input* of the tuner, rather than at the output, for the same reasons I explained above. Consequently, it is for these reasons I'm predicting that this will be the way antenna tuners will be built in the future. These writers are John Belrose, VE2CV (*Ref 132*), Albert Roehm, W2OBJ (*Ref 127*), and Richard Measures, AG6K (*Ref 133*). Belrose (in 1981) showed a balanced T network driven by a 4:1 voltage balun (but see the paragraphs that follow). Measures uses a balanced L network fed by a 1:1 choke, or current balun comprising a coiled-up section of coaxial line. To cover the range of 1.8 to 30 MHz, his balun is a rather bulky 20 feet of coax coiled on a 5-inch diameter ABS pipe. The bulkiness can be reduced by using the W2DU configuration—a 10.5-inch long coax section covered with external ferrite beads, which is just what Roehm did in his arrangement. However, the results of Roehm's and Measures' placing the balun at the input of the tuner are excellent.

Let me add a few words about the work of John (Jack) Belrose, PhD, VE2CV (see *Ref 132*). Jack is an ARRL Technical Advisor (TA), and is well known in Amateur Radio for his expertise in antennas. He has for some time been the Director of the Radio Communications Lab, Department of Communications, of the Canadian Government. Jack has recently conducted tests on a novel approach to the broadbanding of dipole antennas, and is publishing the results in *QST* (*Ref 134*). His antenna is asymmetrical with respect to the impedance seen relative to each wire of a balanced feed line. Consequently, he discovered that the currents in each wire of the feed line were severely unbalanced when using the 4:1 voltage balun to feed the balanced antenna tuner. He then substituted a commercially manufactured W2DU current balun and repeated the measurements. To his surprise, the feed-line currents were almost perfectly balanced when using the W2DU balun. Jack's feed line comprised two parallel coaxial lines; the outer conductors were connected together and grounded to the tuner, and the center conductors were used as the balanced feeders.

Jack's measurements also revealed that when he used the voltage balun, the current on the outer conductors was high, and varied widely over the frequency range from 2 to 30 MHz, also signifying poor balance. However, when using the W2DU current balun, the outer conductor current was very low, and nearly constant, verifying the good balance. These results of Jack's measurements are exciting for me, because they offer additional verification of my own measurements that prove the W2DU current balun solves many of the problems experienced with transformer baluns that I described earlier.

Jack has prepared graphs that compare input impedance and attenuation versus frequency of a ferrite-core balun and the W2DU balun. The data in the graphs again confirm my measurements. They attest to the remarkable superiority of using ferrite beads around a coaxial line to form a current balun, compared to the ferrite-core transformer-type voltage balun. Jack has also devised a unique method of forming a 4:1 current balun by using two W2DU baluns, connecting them in parallel at the input and in series at the output. This information may appear in a future issue of *QST* as Technical Correspondence.

Chapter 22

The Quadrifilar Helix Antenna

Sec 22.1 Introduction

The quadrifilar helix antenna, shown in Fig 22-1, is seeing use on commercial and military, as well as amateur spacecraft. For amateur applications it first saw use on AMSAT-OSCAR 7 in November 1974. The basic information in this chapter was prepared while I was employed as an antenna engineer at RCA's Astro-Electronics Division in Princeton, New Jersey. Information also appears in Chapter 20 of *The ARRL Antenna Book*, 15th edition (*Ref 71*).[1] While working at RCA, I assisted in the design of the quadrifilar helix. We consider this antenna to be an outstanding contribution to practical spacecraft antenna engineering from the viewpoint of flexibility in design. The quadrifilar helix provides a large range of radiation characteristics from a radiator of extremely small size and weight, from which we can select a pretested design to match the radiation requirements of any particular spacecraft. The quadrifilars shown in Fig 22-1 are flight-model units for operation at 1800 and 2200 MHz. I also assisted in the design of quadrifilars for 121.5, 243, and 1600 MHz, which are flying on TIROS-N weather spacecraft in the worldwide Search and Rescue (SAR) Emergency Locator Transmitter (ELT) service. The service is primarily for downed or stricken aircraft, over either land or sea.

One reason for presenting details of the quadrifilar in this book is that it is an inherently excellent antenna for ground station use in the amateur satellite service. Once you study the following engineering aspects of the quadrifilar, you'll discover its advantages over other circularly polarized antennas, such as crossed Yagis. It is useful at

---

[1] All references appear in Chapter 24.

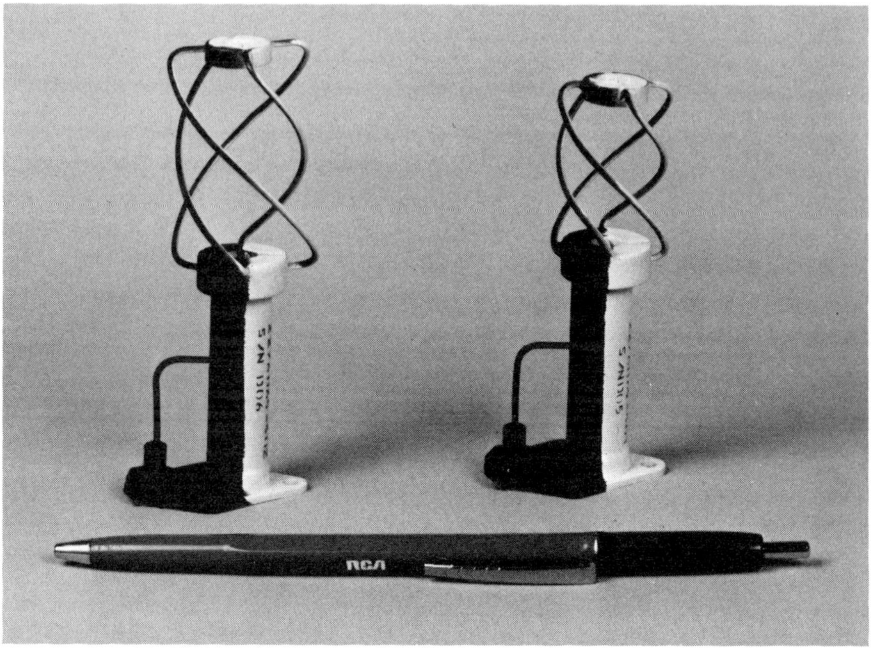

Fig 22-1—Two flight-model quadrifilar helix antennas. The larger is for use around 1800 MHz, and the smaller around 2200 MHz. Radiation from each antenna is circularly polarized and of the same screw sense everywhere throughout its radiation sphere.

144 MHz and up, and all the critical dimensions are included in this chapter. The dimensions are listed in wavelengths, so all you have to do is convert them to the desired frequency. The details of feeding, phasing, and infinite balun construction are given in the drawings and photos, along with sufficient explanation for you to construct your own quadrifilar.

Radiation from the quadrifilar helix antenna is circularly polarized and of the same screw sense everywhere throughout the radiation sphere. The antenna embodies a unique configuration and method of feeding loop elements to produce radiation having a controllable pattern shape. Refer again to Fig 22-1. The quadrifilar antenna comprises two bifilar helical loops oriented in a mutually orthogonal relation on a common axis. The terminals of each loop are fed 180° out of phase, and the currents in the two loops are in phase quadrature (90° out of phase).

By selecting the appropriate configuration of the loops, a wide range of radiation pattern shapes is available, with excellent axial ratio appearing over a large volume of the pattern. The basic form of the quadrifilar antenna was developed by Dr C. C. Kilgus of the Applied Physics Laboratory, Johns Hopkins University, who has published several papers that establish the theoretical basis for its operation. See *Refs 88* through *96*. The Brown and Woodward citation (*Ref 90*) is especially helpful in understanding the radiation characteristics of the quadrifilar. That paper describes radiation from the combination of a dipole and a circular loop sharing a common axis, and fed in phase, which produces radiation characteristics similar to the quadrifilar. This version of the quadrifilar helix uses a novel feed system which includes the infinite balun. The feed system also includes a means for attaining the required 90° current phase relationship without requiring additional components to achieve separate differential-phase excitations.

The quadrifilar antenna is applicable for general use in the frequency range above 30 MHz. It is especially attractive for certain spacecraft applications because it can provide omnidirectional radiation in a single hemisphere (a cardioid volume of revolution) without requiring a ground plane. This no-ground-plane feature affords a dramatic saving in weight and space. In addition, the inherent cardioid radiation characteristic that eliminates the need for the ground plane also affords freedom in the choice of a mounting position on a spacecraft. The profile of the volume beneath the antenna has relatively little effect on the radiation pattern, provided that the quadrifilar is mounted at least ¼ λ above conducting surfaces which would form a reflective plane.

Quadrifilars designed to operate in the frequency range of 1800 and 2200 MHz can be extremely light, weighing only about 0.7 ounce. Developed by RCA Astro-Electronics, several of these antennas are flying on Air Force satellites of the Defense Meteorological Support program, also built by RCA. A slightly smaller version of similar design (donated by RCA Astro-Electronics) is the 2304.1-MHz beacon antenna, which flew on AMSAT-OSCAR 7.

Quadrifilars having a somewhat different configuration from that shown in Fig 22-1 have also been developed by RCA for use on RCA-built NOAA TIROS-N weather satellites. The design differs because the radiation pattern requirements are different. The design of the quadrifilars for the TIROS-N spacecraft is noteworthy because the radiation pattern has been shaped to maintain a nearly constant signal level versus slant-range distance between the spacecraft and the ground station during the in-view portion of the spacecraft orbit.

Sec 22.2 Physical Characteristics

There is a tendency to confuse the quadrifilar with the conventional helical antenna, probably because portions of the quadrifilar are helically shaped. As a result, the radiation characteristics of the quadrifilar antenna are sometimes misunderstood. Although the term quadrifilar is not incorrect, the term itself often fails to conjure up a true picture of the physical characteristics. Therefore, since several characteristics of the quadrifilar antenna differ radically from those of the conventional helical antenna (for example, opposite screw sense of circular polarization), the following paragraphs examine some aspects of both the physical and the radiation characteristics of the quadrifilar to clarify the confusion.

In describing the physical characteristics of the quadrifilar, I will concentrate on the simple half-turn, ½-λ configuration. As I stated earlier, the quadrifilar helix is a combination of two bifilar helicals arranged in a mutually orthogonal relationship along a common axis, and enveloping a common volume.

What is a bifilar helix? One way of visualizing the half-turn ½-λ bifilar helix is to develop it from a continuous square loop of wire one wavelength in perimeter, as shown in Fig 22-2. As in the driven element of the conventional quad antenna, each side of this square loop is ¼ λ, and the feed terminals are formed by opening the loop at the midpoint of the bottom side. We call this square-loop configuration a zero-turn, ½-λ bifilar loop. There is a half wavelength of wire radiator extending away from each feed terminal around the loop to the antipodal point on the opposite side of the square (loop half-length $len_E°$ equals ½ λ). The drawing shows length len° with a script letter L. This loop is a balanced-input device, requiring a balanced, two-wire feed line with push-pull currents.

To visualize the development of the quadrifilar helix, imagine inserting an imaginary cylinder of diameter D = ¼ λ inside the loop. Then, while holding the bottom side of the loop fixed, grasp the top side and give it a half turn of rotation with respect to the bottom. As a result, shown in Fig 22-3, each of the two vertical sides of the square loop becomes a half-turn helix as it curves around the surface of the imaginary cylinder. However, because of the curved paths of the once-straight vertical sides, the distance between the top and the bottom has shrunk, and the axial length, $len_P°$, is less than the ¼-λ diameter.

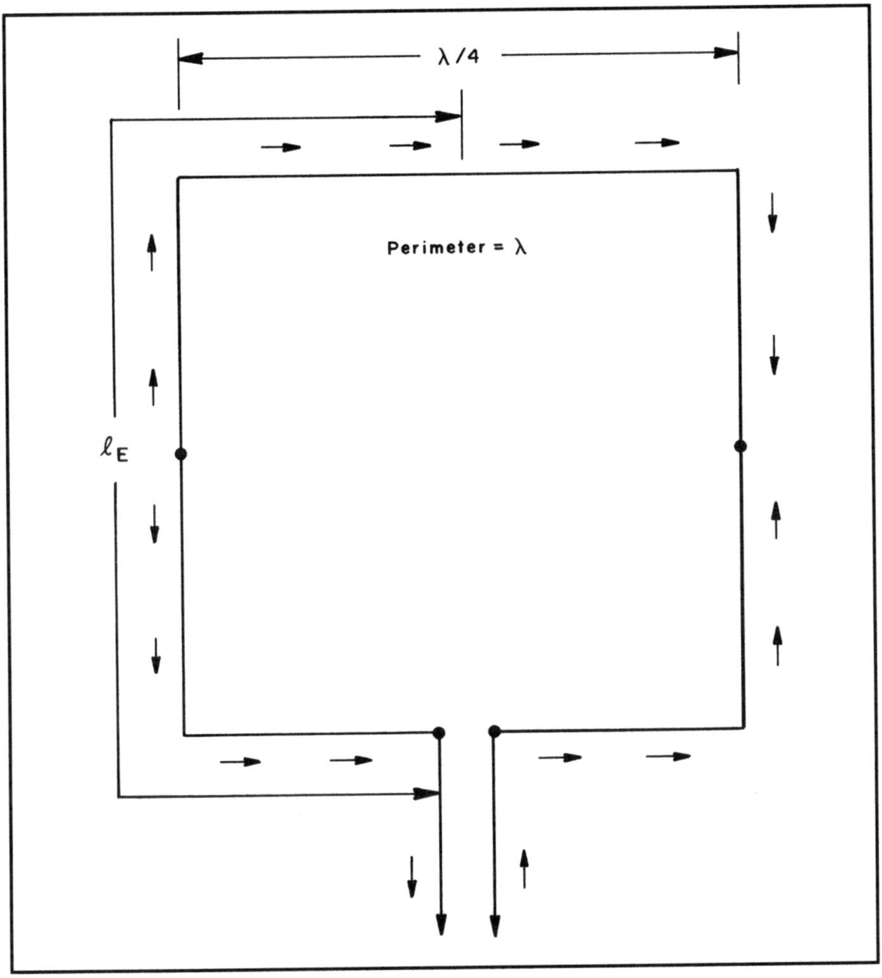

Fig 22-2—Horizontally polarized square loop radiator. The small arrows represent the direction of current flow.

With these particular proportions of the ½-turn, ½-λ bifilar, some of the radiation characteristics are not particularly attractive. However, some physical parameters of the loop may be selected to obtain characteristics that make this antenna especially attractive. Such parameters include the electrical length of the conductors, the number of turns, the cylindrical diameter D, and length $len_P°$. In addition, for a given conductor length, the length-to-diameter ratio, abbreviated len/diam, of the cylinder is also an important variable, controlled by the diameter and number of turns. For example, the antennas shown in Fig 22-1 are

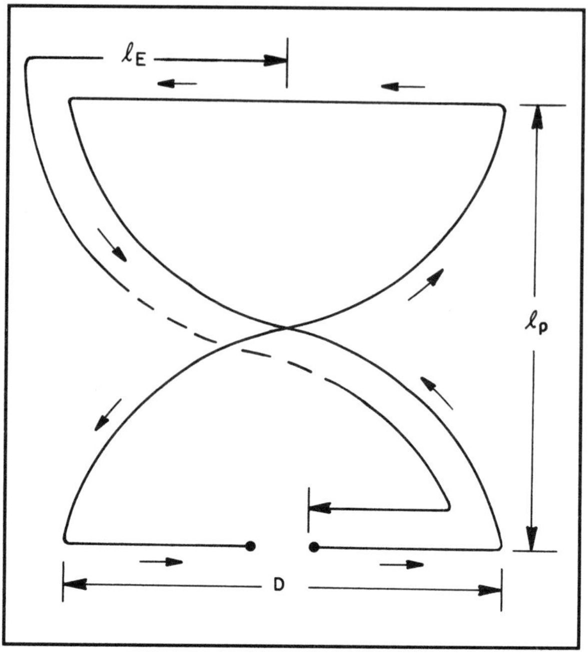

Fig 22-3—Half-turn bifilar helix loop. Distance len$_E$° equals the loop half-length.

½-turn, ½-λ quadrifilars. But by reducing the diameter from 0.25 λ to 0.18 λ for this model, the axial length is inherently increased to 0.27 λ. This results in vastly improved radiation characteristics.

Sec 22.3 Electrical Characteristics

Let's take a short qualitative look at some basic changes in the radiation pattern that result from the ½-turn twist of the square loop. First, recall that the square loop having a perimeter P = 1 λ is basically a broadside array of two dipole elements, with the top element being voltage fed from the bottom element. In this loop the currents in the top and bottom sides flow in the same direction, as shown in Fig 22-2. The horizontally polarized fields produced by the currents in both top and bottom sides are therefore in phase. The two fields add to form the conventional figure-8 broadside radiation pattern. The bidirectional lobes in the pattern are at right angles to the plane of the loop. The null in the pattern appears bidirectionally on a horizontal line that is in the plane of the loop midway between the top and bottom sides.

In the vertical sides of the loop, the current in the top half of each side flows in the direction opposite to that in the corresponding bottom half. Consequently, the vertically polarized fields produced by both halves of each vertical side are mutually out of phase, and add to zero in all directions. This cancellation of the vertical fields results in zero net vertically polarized radiation.

On the other hand, in the bifilar helix loop having the half-turn twist, the currents in the top and bottom sides flow in opposite directions because of the physical half-turn rotation of the top side. This current relationship is shown in Fig 22-3. Thus, the fields produced by the currents in the top and bottom sides are now out of phase with each other, forming an end-fire array relationship.

In the direction broadside to the plane formed by the top and bottom sides, the fields now cancel completely. As in the ordinary end-fire array, this results in zero radiation in the broadside direction, where maximum radiation appeared with the square loop. The conventional lobes of end-fire radiation contributed by the top and bottom sides now occur toward the top and bottom of the antenna as drawn in Fig 22-3. This radiation is horizontally polarized.

The currents flowing in the helical portions of the loop retain the same current-flow pattern as in Fig 22-2. However, the physical positions of each current segment in the twisted vertical wires has now been shifted to a new position and to a new orientation in the helical paths. This results in a corresponding different position and vector direction for each elemental field produced by the helical current elements. As would be expected, the addition of all these elemental fields now results in a composite field consisting of both horizontally and vertically polarized fields.

Sec 22.4 Difference Between Quadrifilar and Conventional Helix

Before considering the radiation characteristics of the bifilar helix further, it may be helpful to mention some critical distinctions between the Kilgus bifilar helix and the conventional helical antenna configurations. In the conventional helical antenna having more than one radiating element, the elements are generally fed in phase. However, Kilgus found that interesting results were obtained with a helical antenna that has two elements spaced radially at 180° when the two elements are fed 180° out of phase. This finding led Kilgus to his further investigation of the bifilar helical antenna having out-of-phase feed, and still further to the quadrifilar configuration. Thus, an alternative way of visualizing the bifilar helix is as a conventional helical antenna with two elements radially spaced at 180°, but with the outer ends radially connected to each other, and with the elements fed out of phase.

Dr. Kilgus' analyses provide a more complete theoretical basis for the functioning of both the bifilar and the quadrifilar helix. He shows the current distribution and radiation characteristics of the half-turn, ½-λ bifilar helix to be similar to those of a loop-dipole combination described by Brown and Woodward (*Ref 90*). Brown and Woodward analyzed a combination horizontal loop and vertical dipole sharing a common axis. They show that while both a loop and a dipole produce identical toroid-shaped radiation patterns, the electric field produced by the dipole current in this arrangement is vertical, and the electric field produced by the loop current is horizontal. When the currents in the loop and dipole are in phase, their fields are in phase quadrature, a requirement for circularly polarized radiation. When the vertical and horizontal fields are also equal in magnitude, the resulting radiation is circularly polarized, with the same screw sense everywhere throughout the radiation sphere.

Experimental data obtained by Kilgus supports his theoretical analysis, and confirms the similarity of radiation characteristics between the half-turn, ½-λ bifilar helix and the loop-dipole combination. For the purpose of obtaining additional data pertinent to the development of quadrifilar antennas for use on the TIROS-N spacecraft, I performed extensive measurements of pattern shape, polarization, and axial ratio on many different bifilar and quadrifilar helix configurations. The data obtained from literally hundreds of my measurements agree with the Kilgus data, adding further validity to the earlier findings as well as extending them.

Sec 22.5 Development of the Quadrifilar Helix Antenna

In this section we examine the intrinsic cardioid, or omnidirectional hemispheric radiation characteristic of the quadrifilar. In doing so, the toroidal radiation pattern of the bifilar helix becomes of primary interest. Recall that in the toroidal radiation pattern of the loop-dipole, maximum radiation occurs broadside to the axis, with the null on the axis. On the other hand, as mentioned earlier, maximum radiation from the bifilar loop occurs bidirectionally along the axis of the loop, while the null appears in the direction perpendicular to the plane formed by the top and bottom sides, or radials of the loop. Except for this 90° rotation of the axes, the radiation characteristics of the two antennas are similar.

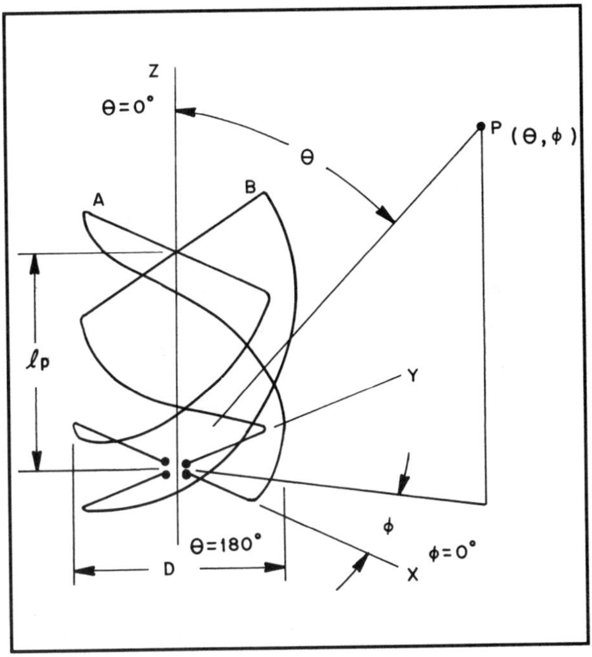

Fig 22-4—Half-turn quadrifilar (two half-turn bifilar loops).

We may develop a quadrifilar helix from the bifilar (which I will call bifilar A) by adding a second bifilar, B, on the same axis and enclosing the same space, but rotated 90° relative to bifilar A. This arrangement is shown in Fig 22-4. The fields radiated by bifilar B are identical to those of bifilar A, except that the entire radiation pattern of bifilar B is rotated 90° relative to that of bifilar A. Consequently, the null of bifilar A and the maximum radiation of bifilar B appear at the same point in space, and vice versa. On the other hand, in the axial directions the radiation fields from both bifilars are equal.

This field relationship is the key to the cardioid radiation characteristic of the quadrifilar. When both bifilars are fed in a 0-180° and 90-270° phase relationship (excited simultaneously with a mutual 90° current-phase relationship), the cardioid radiation pattern appears in the far field. This is because the fields of both bifilars are in phase in one axial direction. In this direction, the two fields add, while in the opposite direction, the fields are out of phase and cancel. As the cancellation is not perfect off axis, the result is a cardioid-shaped pattern of revolution about the axis. In the broadside direction normal to the axis, the respective nulls and maxima of the two individual bifilars compensate each other to produce a uniform omnidirectional radiation pattern around the axis.

Sec 22.6 The Quadrifilar Shape Factor

As I stated earlier, the radiation pattern shape and the axial ratio of the polarization can be controlled. This is done by tailoring both the length of wire and number of turns in the loop, and the length-to-diameter ratio, len/diam, of the formed cylinder. Kilgus' data shows these relationships graphically for certain ranges of the parameter values. However, let us now examine how the Air Force 5D, the amateur OSCAR 7, and the TIROS-N quadrifilar designs were tailored to obtain the radiation characteristics required for their particular missions.

When using bifilars having a diameter D = 0.25 λ, formed by placing a half twist in the square loop of Fig 2, the len/diam ratio is less than one, which yields poor front-to-back and axial ratios. By decreasing the diameter to D = 0.18 λ while retaining the half-turn, ½-λ loop, the axial length increases to $len_P°$ = 0.27 λ. The resulting ratio, len/diam = 1.5, yields a vast improvement in both front-to-back and axial ratios. These are the design values used in the Air Force 5D and OSCAR 7 quadrifilars. The radiation patterns obtained with this design are shown in Figs 22-5 and 22-6. The patterns are presented in the standard IEEE spherical-coordinate system of notation, and the orientation of the antenna relative to the coordinate axes is shown in Fig 22-4.

The θ radiation pattern shown in Fig 22-5 was measured while using circularly polarized illumination of the quadrifilar. This pattern shows an on-axis gain of 5 dBic (decibels relative to isotropic, circular), and a front-to-back ratio greater than 20 dB. The radiation at any angle θ is uniform for all values of ϕ around the Z axis. Thus, the single θ pattern (at any value of ϕ) represents the shape of the envelope of the volume of revolution about the Z axis, which defines the solid radiation pattern in all directions.

The pattern shown in Fig 22-6 was obtained by using a spinning dipole to illuminate the quadrifilar, to obtain axial-ratio information. The periodic ripple appearing in the pattern results from the rotation of the dipole. The maxima and minima, respectively, correspond to the times when the dipole is parallel to the major and minor axes of the polarization ellipse. The axial ratio may be determined from the difference between adjacent maximum and minimum values. Thus, the axial

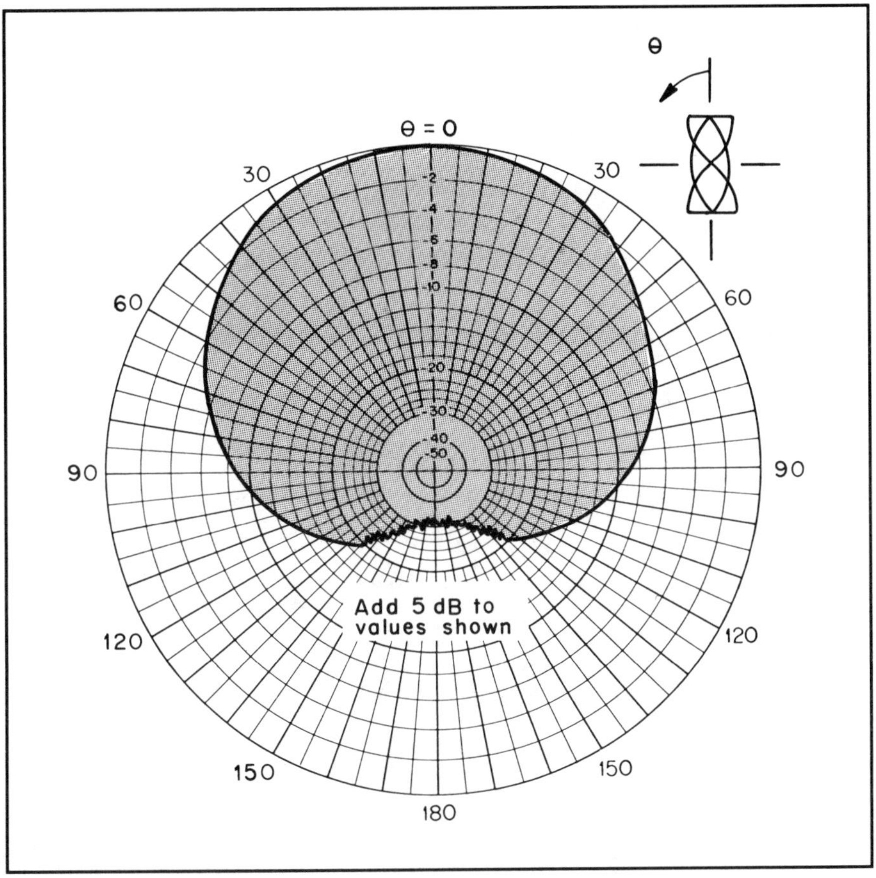

Fig 22-5—Radiation pattern of the half-turn, ½-λ quadrifilar antenna when illuminated with a circularly polarized source.

ratio of this design is seen to be less than 2 dB over a beamwidth of ±30°.

As mentioned earlier, the shape-factor parameters for the TIROS-N configuration were determined empirically from my measurements. I took measurements on many combinations of quadrifilar configurations in which each physical parameter was separately varied in small increments while holding the other variables constant. Hundreds of shape-factor combinations were measured, which provided families of patterns from which to select the parameters for the desired pattern shape. The TIROS-N mission required a radiation-pattern shape which provides a nearly constant signal level to the ground station during in-view time. This pattern shape is shown in Fig 22-7. The bifilar

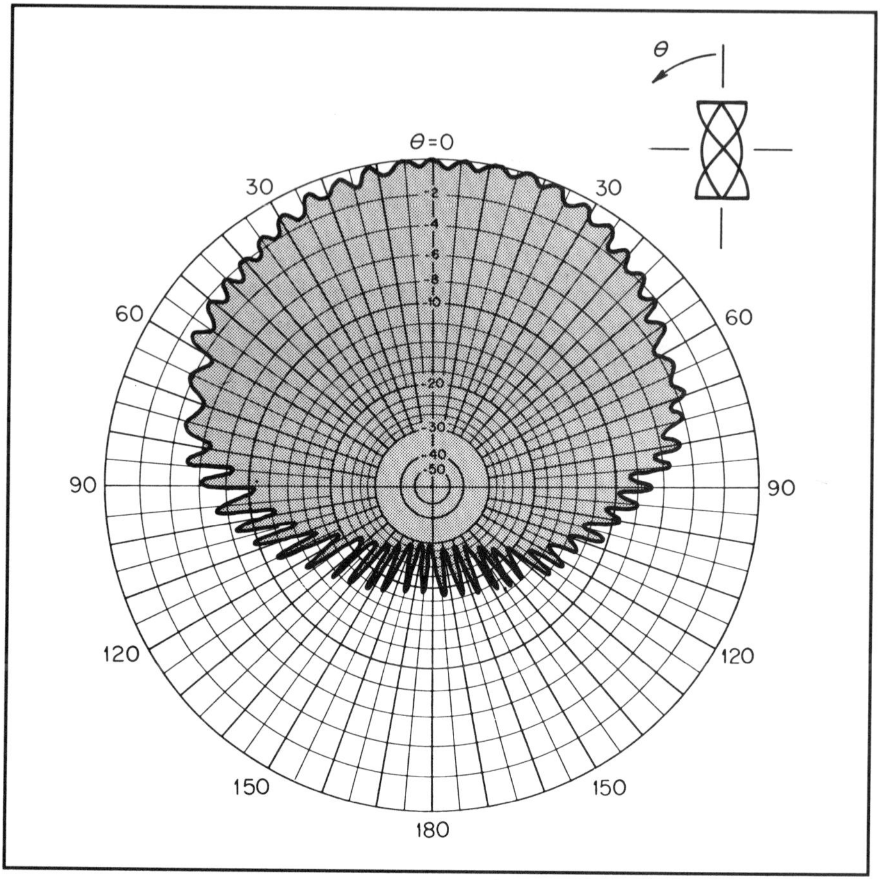

Fig 22-6—Quadrifilar radiation pattern when illuminated with a spinning dipole. This illumination adds axial-ratio data, represented as the difference between adjacent maximum and minimum values of a ripple wave.

parameters which yield this pattern shape were found in the acquired data families to be 1½ turns, 1.25-λ loop, D = 0.1 λ, len_P° = 1.0 λ, and len/diam = 10.0. These quadrifilars, flying on TIROS-N, are used for transmitting weather and other data in the 1600-MHz region, completely separate from the 121.5, 243, and 1600-MHz SAR-ELT quadrifilars flying on the same spacecraft.

Sec 22.7 Methods of Feeding the Quadrifilar Antenna

Feeding the quadrifilar with a single, unbalanced coaxial line requires special attention. Because the individual bifilar loops are bal-

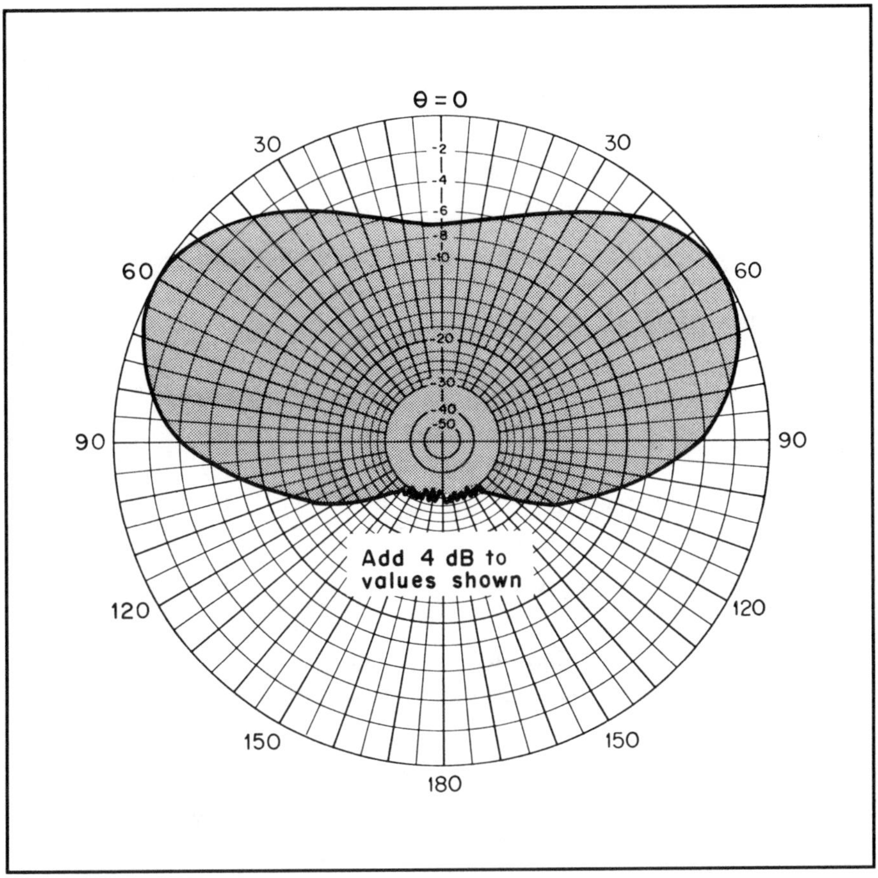

Fig 22-7—Radiation pattern of the 1½ turn, 1.25-λ quadrifilar antenna.

anced-input devices, some form of balun is required to provide balanced push-pull currents to the terminals of each bifilar. In addition, to obtain the unidirectional radiation characteristic of the quadrifilar, the two bifilar loops require separate excitations having a relative phase difference of 90°. Further, the sense of the 90° phase relationship determines from which end the quadrifilar radiates.

Several different balun and quadrature-phase circuit arrangements are available for feeding the quadrifilar, such as the folded balun, the split-sheath balun, or a combination of 90° and 180° hybrids, and so forth, as described by Bricker and Rickert (*Refs 88, 89*). However, to save weight and to effect simplicity in the construction, a unique feeding arrangement is used. This technique features the infinite, or inherent balun, combined with a novel method of self-phasing the two bifilar loops to achieve the 90° differential current relationship between the loops. The constructional simplicity is apparent in Fig 22-1, which illustrates the quadrifilar configuration used on both the TIROS-N and Air Force 5D satellites, and on the amateur satellite, OSCAR 7.

Sec 22.8 The Infinite or Inherent Balun

In the infinite balun, shown in Fig 22-8, we see the coaxial feed line extending into the loop and shaped to form the first half of a bifilar loop. At the end of the coax, the inner conductor is connected to the opposite or second half of the bifilar loop to form the feed point. The other end of the second half of the loop is connected to the outside surface of the coax feed line. Thus, the loop is closed at the antipodal point of the loop, where the feed line enters. The second half of the loop is a solid wire of the same diameter as the outer diameter of the coax.

In operation, current flowing on the inner conductor emerges at the feed point to flow onto the second half of the loop. For current flowing on the inside of the outer conductor of the coax, on its arrival at the end of the coax, the only path for current flow is around the end and onto the outside of the outer conductor. Now such external feed-line current is the desired antenna current, because the outside portion of coax extending from the feed point to the antipodal point is the radiator. Externally, the antipodal point demarks the end of the feed line and the

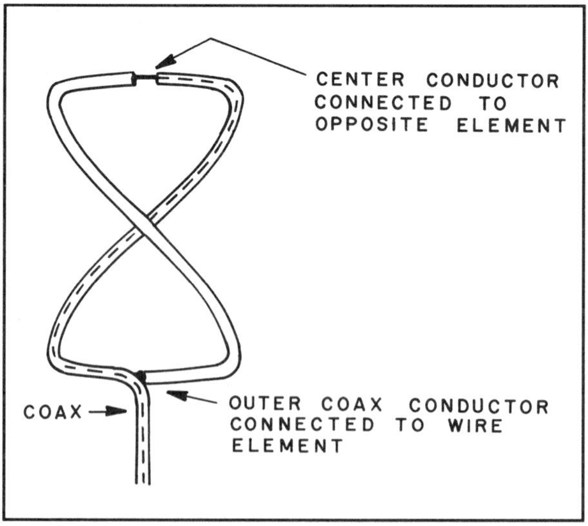

Fig 22-8—Half-turn bifilar loop with infinite balun feed.

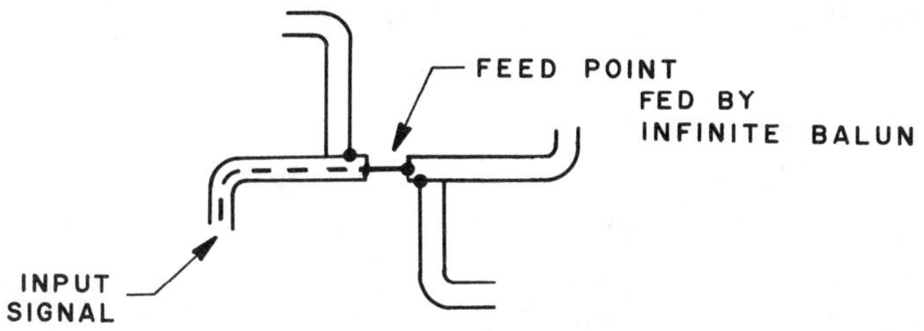

Fig 22-9—Feed arrangement for 90° self-phasing of loops.

beginning of the loop radiator. Because of skin effect, the transmission-line currents flowing inside the coax portion of the loop are completely divorced from the antenna currents flowing externally on the loop. Their only relationship is that the internal transmission-line currents emerge at the feed point, where they become the external antenna currents.

As the feed line is dressed away from the antipodal point symmetrically relative to the loop, currents induced on the feed line because of coupling from each half of the loop are equal and flow in opposite directions. The opposing currents on the line thus cancel each other, decoupling the feed line from the loop. In other words, from the external viewpoint of the loop radiator, the source generator can be considered to exist directly between the two input terminals of the loop at the feed point, and the feed line effectively disappears. Thus, the current-mode transition in the coaxial-line portion of the loop—from an internal unbalanced mode entering at the antipodal point, to an external balanced mode emerging at the feed point—constitutes an inherent balun device. Such a device is called an "infinite balun."

Sec 22.9 Self-Phased Quadrature Feed

As stated, a quadrature-phase current relationship is required between the two bifilar loops of the quadrifilar array. This requirement is met by using the self-phasing method. The orthogonal bifilar loops are designed such that one loop is larger relative to the desired resonant frequency length and therefore inductive, while the other loop is smaller and therefore capacitive. Using this method, the two loops are fed in parallel by connecting the terminals of both loops together at the feed point, as shown in Fig 22-9. This self-phasing method requires only one coaxial feed line, and any of the balun arrangements mentioned earlier (shown by Bricker and Rickert, *Ref 88*) may be used. It is evident that one loop half is the coax feed line, while the other three half loops are simply solid wire.

The larger inductive loop is designed such that, at the operating frequency, the reactive component X_L of the loop terminal impedance is equal to the resistive component, R. Similarly, the smaller capacitive loop is designed so its reactive component X_C equals R at the operating frequency. The $\pm X = R$ relationship is important, because to obtain a relative current phase of 90° between the two loops, the larger loop current must lag by 45° and the smaller must lead by 45°.

For the current phase of the larger loop to lag or the smaller loop to lead by 45°, their phase angles must be 45°, or have an arc tangent of ±1. This occurs only when $\pm X = R$. When the two loops are fed in parallel, the relative currents in the loops differ in phase by 90° without requiring any additional components to obtain separate differential-phase excitations. Dimensions that yield the correct phase relationship with a loop wire diameter of 0.0088 λ are as follows.

Smaller loop:

$D = 0.156\ \lambda$
$len_P° = 0.238\ \lambda$
Perimeter = 1.016 λ

Larger loop:

$D = 0.173\ \lambda$
$len_P° = 0.260\ \lambda$
Perimeter = 1.120 λ

where

D = diameter of imaginary cylinder on which bifilar is wound
$len_P°$ = axial length, as shown in Fig 22-3

The resultant impedance versus frequency response of the parallel-loop combination is shown in the Smith Chart, Fig 22-10. The formation of the cusp in the impedance locus is the design goal that signifies that the 90° phase relationship exists between the loops.

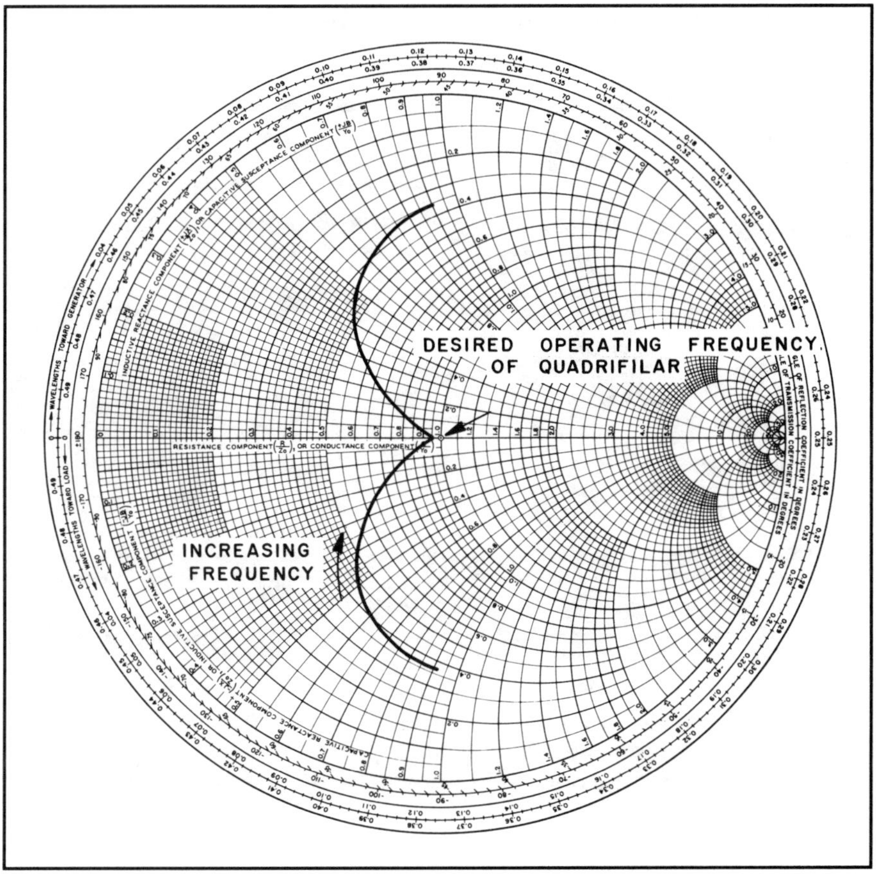

Fig 22-10—Impedance versus frequency plot of self-phased quadrifilar antenna.

Sec 22.10 The Quadrifilar in Space

As I've indicated above, the quadrifilar antenna was developed for spacecraft use. Fig 22-11 shows the comparative size between an experimental S-band quadrifilar and the flight-model system of a circularly polarized crossed-dipole antenna over a ground plane, which the quadrifilar replaced. The two antennas have identical radiation characteristics. Four of the crossed-dipole ground-plane units flew on ITOS (TIROS-M). They were replaced with quadrifilars on the newer TIROS-N to save space and weight, but accomplished the same communications tasks on the same frequencies. Using the same technique

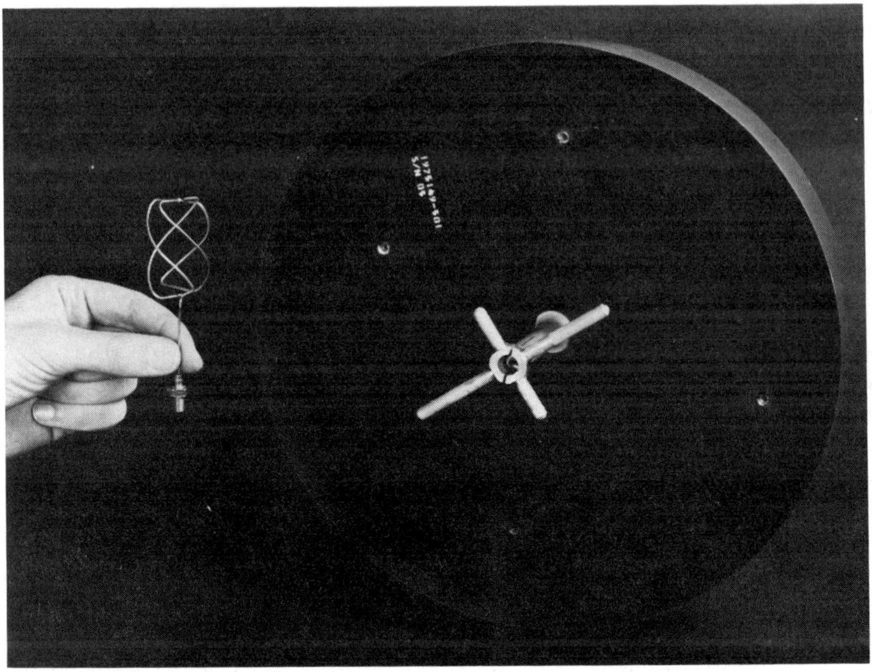

Fig 22-11—Comparing the size of a quadrifilar antenna and a flight-model, circularly polarized crossed dipole over a ground plane system, which it replaced. The two antennas have identical radiation characteristics, but the quadrifilar obviously requires less space and is lighter.

as described in Sec 22.9 for the quadrifilar, the required 90° dipole phasing to obtain circular polarization in the crossed-dipole array was accomplished by using the short and long dipole elements shown in the picture. The short elements are capacitive, causing the dipole currents to lead by 45°, and the long elements are inductive, causing the currents to lag 45°, resulting in a differential phase of 90°. The balun is a split tube.

Before the launch of OSCAR 7, AMSAT publicized the quadrifilar antenna in the *AMSAT Newsletter* for March 1975 (*Ref 131*). The cover picture for that issue is included here as Fig 22-12. The antenna is shown mounted on an OSCAR 7 test model at the RCA Space Center, attached to an ITOS weather satellite (TIROS-M) in the piggyback launch configuration. Fig 22-13 shows another view of the test model, while Figs 22-14 and 22-15 show the quadrifilar mounted on the OSCAR 7 spacecraft as it is being readied for flight. In spite of its small

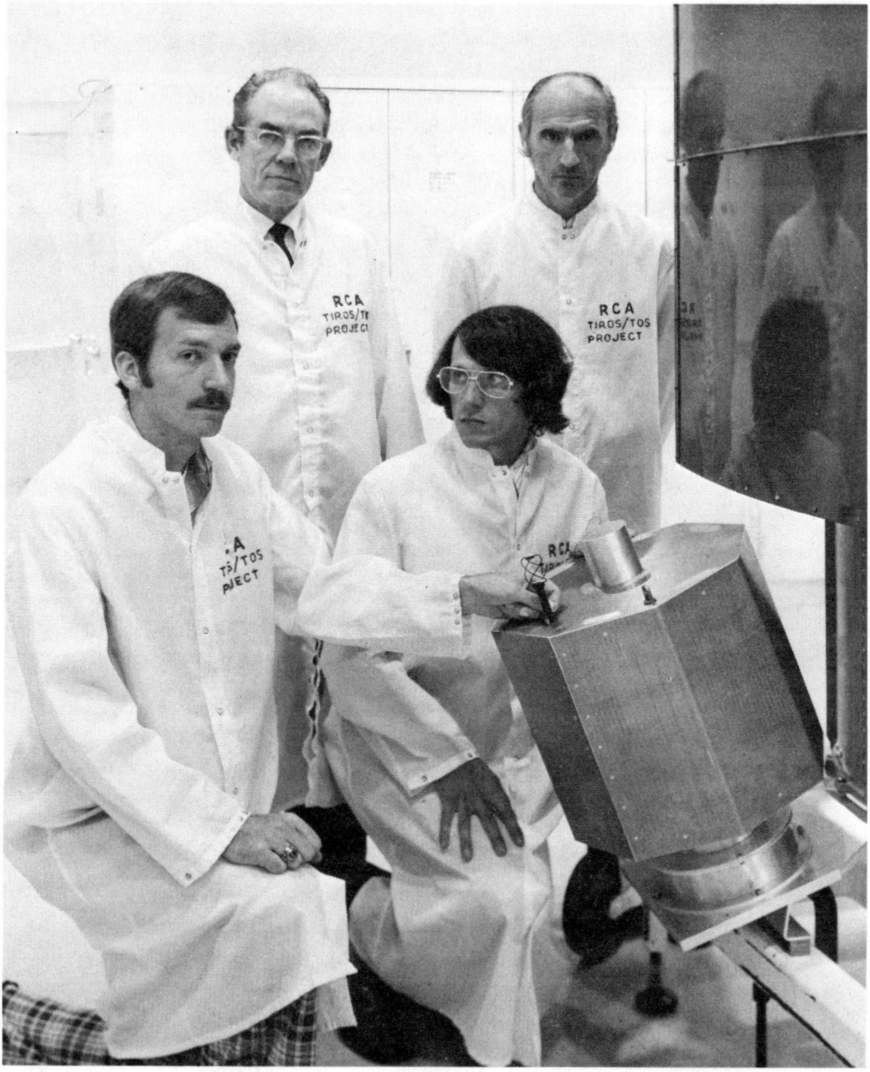

Fig 22-12—This photo was used as the cover picture for the March 1975 issue of the *AMSAT Newsletter* (*Ref 131*). Shown clockwise from the upper left are Walt Maxwell (W2DU), Walt Ozmon (W2WGH), OSCAR 7 Project Manager and NASA engineer Jan King (W3GEY), and Randy Bricker, who is pointing out the radiation characteristics of the 2304-MHz OSCAR 7 beacon antenna.

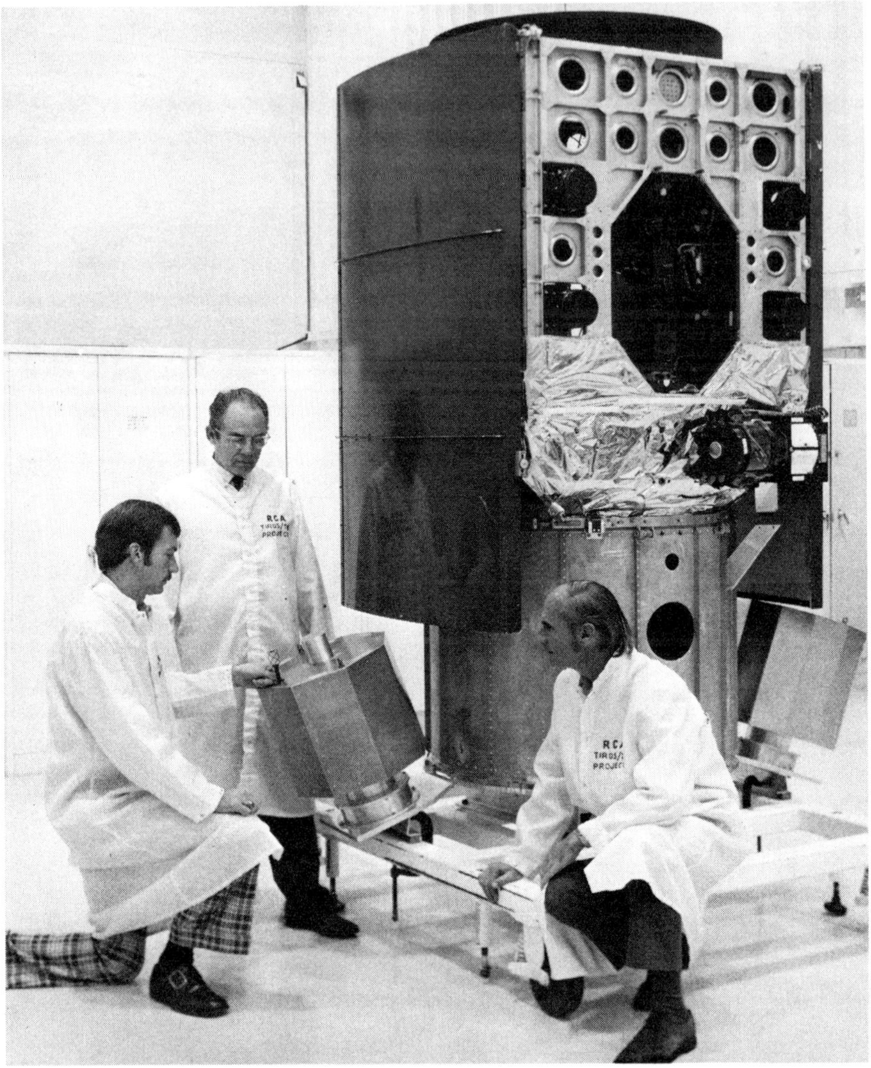

Fig 22-13—The OSCAR 7 quadrifilar on its test model in its piggyback launch configuration alongside the flight-model ITOS spacecraft with which it flew. Left to right are Bricker, Maxwell (W2DU), and Ozmon (W2WGH).

Fig 22-14—The OSCAR 7 quadrifilar, lower right, mounted on the OSCAR 7 spacecraft during prelaunch readiness operations.

Fig 22-15—A close-up view of the OSCAR 7 quadrifilar during prelaunch readiness.

size (0.7 ounce), the quadrifilar antenna boasts tremendous performance for spacecraft use. The original design was developed by RCA Engineers Randy Bricker, Herb Rickert, and myself, for use on the highly successful USAF spacecraft, the Block 5-D Meteorological Satellite, which was built by the RCA Astro-Electronics Division of Princeton, New Jersey.

The OSCAR 7 antenna is circularly polarized, radiates hemispherically without requiring a ground plane, and has a gain of 5 dBic on axis going down smoothly to 0.0 dBic 90° off axis over the entire 360° around the edge of the hemisphere, with good polarization circularity at the hemisphere edges. It is fed with an infinite balun, and has elements phased at 90° with no phasing line. With a modification I devised to operate at the 2304-MHz beacon frequency, this antenna was fabricated under the direction of Bricker and myself specifically for the OSCAR 7 amateur spacecraft, and presented to AMSAT by RCA, which designed and built the TIROS-ESSA-ITOS weather satellite series. I also served as consulting engineer for all antenna systems on AMSAT

OSCAR 7. RCA Technicians Walt Ozmon, W2WGH, and Joe Rovenski, W2HLO, performed many of the impedance and radiation pattern measurements during the final testing of the OSCAR 7 quadrifilar antenna.

Scaled up for lower frequencies, the quadrifilar would have similar performance characteristics. At 146 MHz, the quadrifilar would form a cylinder approximately 13 inches in diameter and 20 inches high, while at 432 MHz, the diameter and height would be 4.4 inches and 6.8 inches. At 29 MHz the dimensions would be 5.5 feet and 8.4 feet, respectively. The circular polarization reduces polarization fading caused by Faraday rotation of the electric field vector, especially on the 10-meter downlink.

In summary, the quadrifilar helix antenna differs significantly from the conventional helical antenna. Control of pattern shape and other radiation characteristics are available by selecting appropriate dimensional parameters. A novel balun for feeding this balanced-input device from coaxial feed line is used, as is a method of self-phasing the loop radiating elements to obtain a quadrature current relationship between the loops. The quadrifilar helix is a durable antenna, one that is seeing use on both amateur and commercial satellites.

Chapter 23

Appendices

APPENDIX 1—MINIMUM-SWR RESISTANCE EQUATION

I have never seen a reference to the minimum-SWR resistance in relation to terminating loads in any of the engineering or academic literature. Therefore, I believe I am the originator of the relationships represented by Eqs 5-1 and 5-2, discussed in Sec 11.3, and the derivation given here.

THEOREM:

When a transmission line of characteristic impedance Z_C is terminated with an impedance $Z = R + jX$, for every value of X there is corresponding value of R which yields the minimum value of SWR. For minimum SWR when $R = Z_C$, X must be zero.

Prove that $r = \sqrt{x^2 + 1}$ (see Fig 23-1)

where

r = normalized resistance component of Z yielding minimum SWR

x = normalized reactance component of Z

$$r' = \text{radius of resistance circle} = \frac{1}{r+1}$$

$$U_C = \text{center of resistance circle} = \frac{r}{r+1}$$

$$\frac{1}{x} = \text{center of reactance circle}$$

ρ = magnitude of reflection coefficient

$$\tan\theta = \frac{1}{1/x} = x = \frac{\rho}{r'}$$

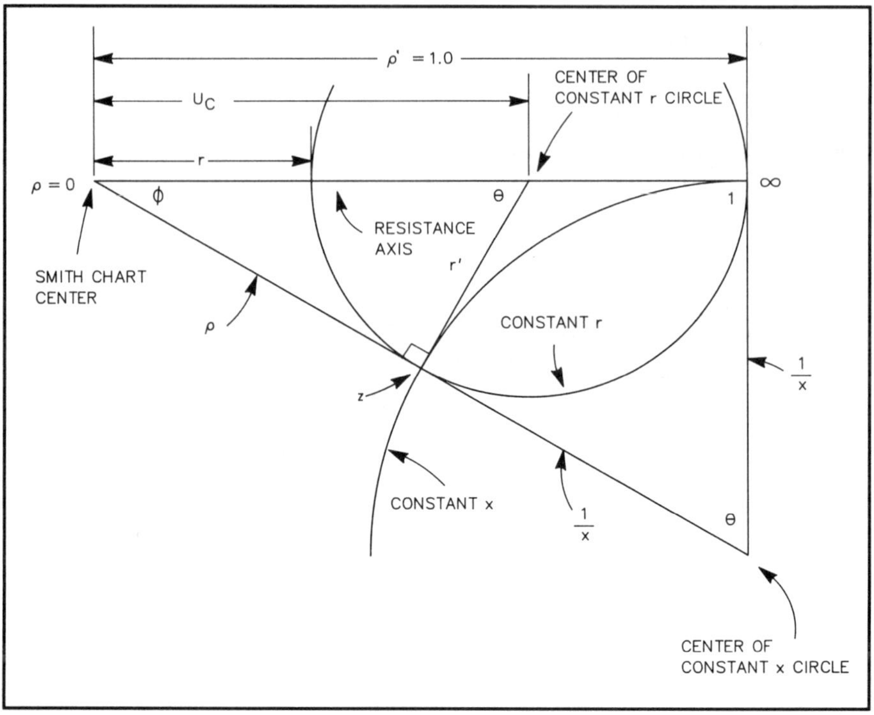

Fig 23-1—Smith Chart geometry for Appendix 1 proof of minimum-SWR equation.

$$\rho + \frac{1}{x} = \sqrt{(1/x)^2 + 1^2} = \frac{\sqrt{x^2 + 1}}{x}$$

$$\rho = \frac{\sqrt{x^2+1} - 1}{x}$$

$$r' = \frac{\rho}{x} = \frac{\sqrt{x^2+1} - 1}{x^2} = \frac{1}{r+1}$$

Inverting,

$$\frac{1}{r'} = \frac{x^2}{\sqrt{x^2+1} - 1} = r + 1$$

Clearing radical in denominator,

$$\frac{x^2}{\sqrt{x^2+1}-1} \times \frac{\sqrt{x^2+1}+1}{\sqrt{x^2+1}+1} = r + 1$$

Multiplying and canceling,

$$\frac{x^2\sqrt{x^2+1}+x^2}{x^2+1-1} = r+1$$

$$\sqrt{x^2+1}+1 = r+1$$

$$\sqrt{x^2+1} = r$$

Therefore the theorem is proved.

APPENDIX 2—DETERMINING SWR FROM R + *j*X, METHOD 1

The SWR resulting from an impedance mismatch may be obtained from the expressions SWR = R/Z_C or SWR = Z_C/R *when the load is a pure resistance.* When the load is a complex impedance Z = R + *j*X, the exact SWR can be determined after finding the magnitude of the reflection coefficient ρ from Eq 3-1, repeated here in modified form.

$$\bar{\rho} = \rho\underline{/\theta} = \frac{Z_L - Z_C}{Z_L + Z_C} = \frac{R - Z_C + jX}{R + Z_C + jX} \qquad \text{(Eq 3-1)}$$

where

Z_L = the complex load impedance, R + *j*X

Z_C = the characteristic impedance of the transmission line

The SWR may now be found from Eq 3-3, also repeated here.

$$SWR = \frac{1 + \rho}{1 - \rho} \qquad \text{(Eq 3-3)}$$

The SWR cannot be determined from any relationship between line impedance Z_C and the magnitude |Z| of the complex load impedance, as shown in Table 23-1. From the table, you can see that different values of load impedance Z_L can have the same magnitude |Z|, but produce different values of SWR on a feed line with a given Z_C, which in this case is 50 Ω.

Table 23-1
Complex Load Impedances Versus SWR

Polar Form		*Rectangular Form*		
|Z| at	*± Angle*	*R*	*± jX*	*SWR*
50	0.0°	50	0	1.00
50	25.84°	45	21.79	1.59
50	36.9°	40	30.0	2.00
50	45.0°	50	50.0	2.618
50	53.1°	30	40.0	3.00
50	66.42°	20	45.83	4.79
50	72.54°	15	47.7	6.51
50	78.46°	10	48.99	9.90
50	84.26°	5	49.79	19.95
50	87.13°	2.5	49.94	39.98
50	88.85°	1	49.99	99.99
50	90.0°	0	50.00	∞

APPENDIX 3—DETERMINING SWR FROM R + *j*X, METHOD 2

I derived this method for calculating the exact value of SWR from R + *j*X from Eq 3-1 after a suggestion from Ken Miller, W2KF.

1) Normalize the load impedance by dividing by the characteristic impedance Z_C of the transmission line.

$$\frac{R + jX}{Z_C} = r + jx$$

2) Find the b term of the quadratic formula

$$\frac{b \pm \sqrt{b^2 - 4ac}}{2a}$$

using values of r and x in the equation

$$b = \left(\frac{x^2 + 1}{r}\right) + r$$

3) Calculate the SWR from the simplified quadratic equation

$$SWR = \frac{b + \sqrt{b^2 - 4}}{2}$$

The a and c terms of the complete quadratic equation reduce to 1 from normalizing the R and X terms of the load impedance. The negative root of the discriminant is ignored.

Let's now try an example by calculating the SWR on a 50-Ω line produced by a load impedance of 40 + *j*30 Ω.

Normalizing: $r + jx = \frac{40 + j30}{50} = 0.8 + j0.6$

Find the b term: $b = \left(\frac{0.6^2 + 1}{0.8}\right) + 0.8 = 2.50$

Calculate the SWR: $\text{SWR} = \dfrac{2.50 + \sqrt{2.50^2 - 4}}{2} = 2.000$

This is an excellent example because the answer is exactly 2:1 with no fractional remainder. To test your arithmetic, other examples to try which also give exact answers when $Z_C = 50\ \Omega$ are $30 + j40$, SWR = 3:1; and $80 + j90$, SWR = 4:1. It is also remarkable to note that the following different load impedances all produce an SWR of 2.618:1 on 50-Ω lines: $25 + j25$, $50 + j50$, $100 + j50$, and $130 + j10$. For a calculator program, see Appendix 4.

APPENDIX 4—PROGRAM FOR EXACT SWR FROM R + *j*X, METHOD 2

Listed below is a calculator program for determining the SWR from R + *j*X, using Method 2 of Appendix 3. This listing is for a programmable Hewlett-Packard calculator or other brand using reverse Polish notation.

Line	*Key Entry*
1	H Label A
2	RCL 0
3	÷
4	$g \rightarrow x^2$
5	1
6	+
7	$x \leftrightarrow y$
8	RCL 0
9	÷
10	STO 1
11	÷
12	RCL 1
13	+
14	ENTER
15	$g \rightarrow x^2$
16	4
17	−
18	$f \rightarrow \sqrt{\ }$
19	+
20	2
21	÷
22	H Return

Instructions:

1. Store line impedance Z_C in Register R_0
2. Key in R value of impedance R + *j*X
3. Press ENTER
4. Key in X value of impedance R + *j*X
5. RUN

APPENDIX 5—DERIVATION OF EQUATIONS 6-1 AND 6-2

I derived these equations to determine the power transmitted through a transmission line with attenuation and terminated in a mismatch (see Fig 23-2).

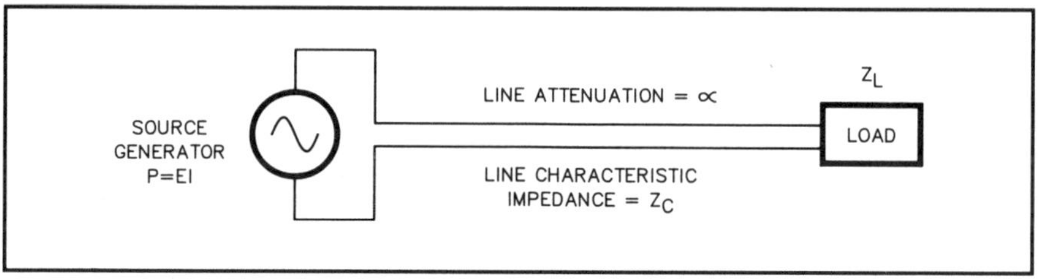

Fig 23-2—Illustrating the transmission line for the Appendix 5 derivation of Eqs 6-1 and 6-2.

ρ_L = voltage reflection coefficient at load

$$= \frac{Z_L - Z_C}{Z_L + Z_C} = \frac{SWR - 1}{SWR + 1} \qquad \text{(Eqs 3-1 and 3-2)}$$

ρ_I at line input $= \rho\, \varepsilon^{-2\alpha}$

ρ_I^2 at line input $= \rho^2\, \varepsilon^{-4\alpha}$

ρ_L^2 = power reflection coefficient at load (equals reflected power)

where

$$\alpha = \text{line attenuation in nepers} = \frac{dB}{20 \log \varepsilon} = \frac{dB}{8.6859}$$

(one neper = 0.115129 decibel)

ε = 2.71828 (the base of natural logarithms)

The forward *voltage* wave encounters a reduction in level by the loss factor of $1/\varepsilon^{\alpha}$, or $\varepsilon^{-\alpha}$ in traveling from the source to the load. The reflected voltage wave encounters the same reduction factor $\varepsilon^{-\alpha}$ in its return to the source. Hence, the voltage reflection coefficient at the line input is $\rho_I = \rho\, \varepsilon^{-2\alpha}$.

The forward *power* encounters a *one-way* attenuation-loss factor $\varepsilon^{-2\alpha}$. Power returning to the source from a load reflection encounters a *two-way* loss—once during the trip to the load, and again during its return to the source. The power reflection coefficient at the line input is therefore $\rho_I^2 = \rho^2 \varepsilon^{-4\alpha}$.

Let power supplied by the source = 1, and let reflected power reaching the input (which adds to the source power) = r. Then,

1 + r = total forward power at line input,

$(1 + r)\,\varepsilon^{-2\alpha}$ = forward power arriving at the load, and

$$\frac{r}{1+r} = \text{input power reflection coefficient} = \rho_I^2 = \rho^2\,\varepsilon^{-4\alpha}.$$

Now let $\frac{r}{1+r}$ (and $\rho^2\,\varepsilon^{-4\alpha}$) = a.

Then, total forward power (1 + r) at the line input $= 1 + \frac{a}{1-a}$

$$= \frac{1 - a + a}{1 - a} = \frac{1}{1 - a}$$

Hence, it follows that

Forward power at line input		*Forward power at load after line atten.*		*Power reflected at load*		*Power absorbed in load*
$1 + r$	$\rightarrow$	$(1+r)\,\varepsilon^{-2\alpha}$	$-$	$\rho^2(1+r)\,\varepsilon^{-2\alpha}$	$=$	$(1-\rho^2)(1+r)\,\varepsilon^{-2\alpha}$

and

$$\frac{1}{1-a} \rightarrow \frac{\varepsilon^{-2\alpha}}{1-a} - \frac{\rho^2\,\varepsilon^{-2\alpha}}{1-a} = \frac{(1-\rho^2)\,\varepsilon^{-2\alpha}}{1-a}$$

But $a = \rho^2\varepsilon^{-4\alpha}$, so by substitution we have Eq 6-1:

$$\frac{1}{1-\rho^2\,\varepsilon^{-4\alpha}} \rightarrow \frac{\varepsilon^{-2\alpha}}{1-\rho^2\,\varepsilon^{-4\alpha}} - \frac{\rho^2\,\varepsilon^{-2\alpha}}{1-\rho^2\,\varepsilon^{-4\alpha}} = \frac{(1-\rho^2)\,\varepsilon^{-2\alpha}}{1-\rho^2\,\varepsilon^{-4\alpha}}$$

where power absorbed = source power times the quantity

$$\frac{(1-\rho^2)\text{ times one–way line–attenuation factor}}{1 - (\rho^2\text{ times two–way line–attenuation factor})}$$

For a lossless line, α becomes zero, making $\varepsilon^{\alpha} = \varepsilon^{0} = 1$, so we have Eq 6-2:

$$\frac{1}{1-\rho^2} \quad \rightarrow \quad \frac{1}{1-\rho^2} \quad - \quad \frac{\rho^2}{1-\rho^2} \quad = \quad 1$$

$$1+r \quad \rightarrow \quad 1+r \quad - \quad r \quad = \quad 1$$

APPENDIX 6—POWER RELATIONSHIPS ON MISMATCHED TRANSMISSION LINES

The power relationships on a mismatched transmission line may be determined from the following equations.

ρ = magnitude of voltage reflection coefficient = $\frac{SWR-1}{SWR+1}$

ρ^2 = magnitude of power reflection coefficient (power reflected by mismatched load)

$(1 - \rho^2)$ = power absorbed by mismatched load

Example 1

Lossless line (attenuation $\alpha = 0$); line with matched load, SWR = 1:1, $\rho = 0$, $\rho^2 = 0$, and $(1 - \rho^2) = 1$. Transmitter matched to line-input impedance, $Z_C = 50\ \Omega$. See Fig 23-3A. All the transmitter power, 100 W, is absorbed in the load.

Example 2

As in example 1, a lossless line (attenuation $\alpha = 0$); line with mismatched load, SWR = 3:1, $\rho = (3 - 1)/(3 + 1) = 0.5$. The 3:1 load mismatch causes a 3:1 transmitter mismatch; *transmitter not rematched* to the new line-input impedance. See Fig 23-3B. Here the transmitter reduces its delivered power by the amount of the reflected power, $\rho^2 = 25$ W. Note that all the transmitter power delivered, 75 W, is absorbed in the load.

Example 3

As in example 2, a lossless line (attenuation $\alpha = 0$); line still mismatched with 3:1 SWR, but *transmitter rematched* (conjugate) to new line-input impedance. See Fig 23-3C. The transmitter again delivers 100 W to the line, and 100 W are now absorbed in the Z_C-mismatched, but conjugate-matched load. How is the 133.33 W of forward power obtained? From Eq 6-2 (also Appendix 5), forward power = $1/(1 - \rho^2) = 1/0.75 = 1.3333$ when the mismatch is 3:1. 100 W $\times$ 1.3333 = 133.33 W.

Example 4

Line with attenuation, $\alpha = 0.5$ dB, such as 87 feet of RG-59 or

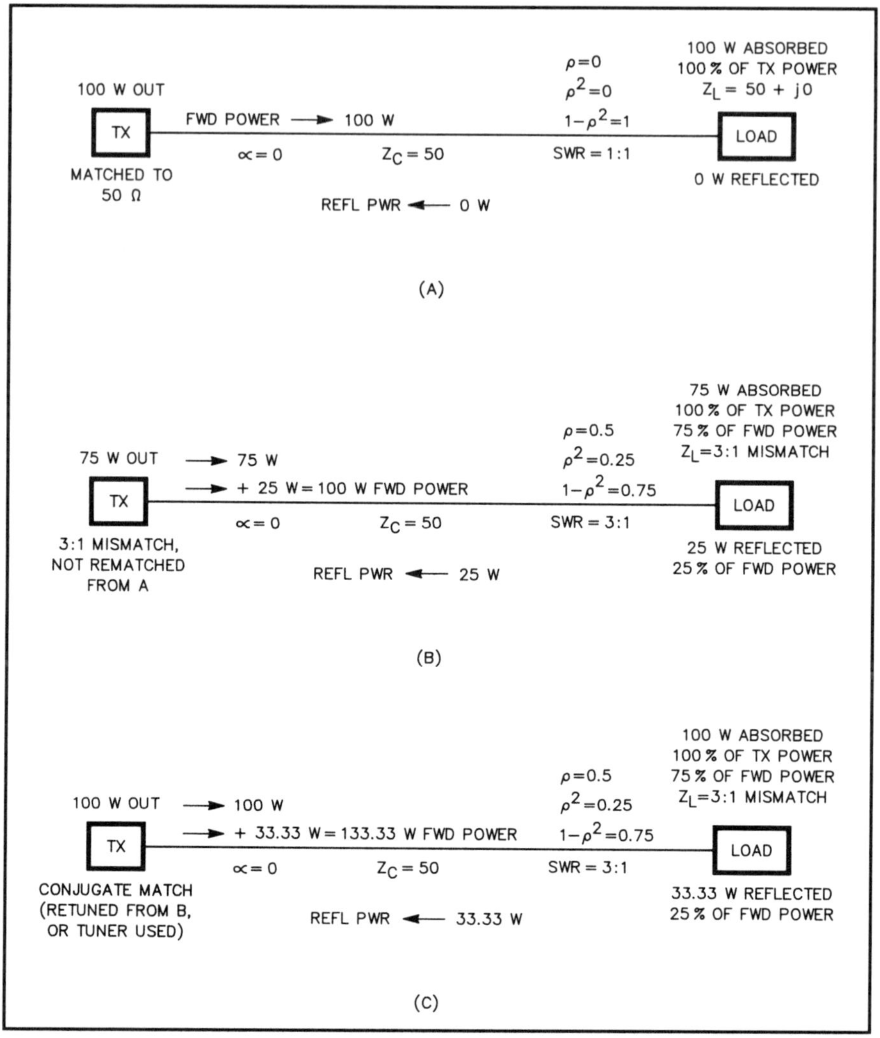

Fig 23-3—Illustrating the transmission-line conditions for a lossless line with different loads and input matching conditions. Drawings A, B and C respectively represent Appendix 6 examples 1, 2 and 3.

175 feet of RG-8 at 4.0 MHz; matched load, as in example 1; SWR = 1:1. See Fig 23-4A. The transmitter is matched to the line-input impedance Z_C = 50 Ω. However, 100 W of forward power leaving the transmitter suffers 0.5 dB line attenuation loss; hence, only 89.125 W reaches the load, where it is all absorbed.

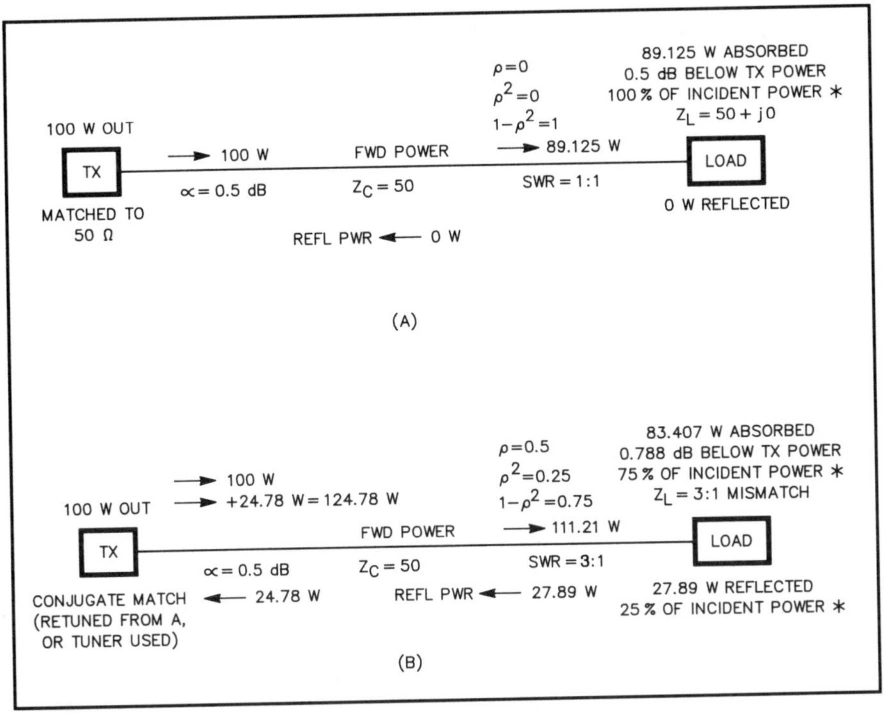

Fig 23-4—Illustrating the transmission-line conditions for a line with 0.5 dB loss having different loads and input matching conditions. Drawings A and B respectively represent Appendix 6 examples 4 and 5. ****Incident* **power is the forward power that reaches the load.**

Example 5

Line with attenuation, α = 0.5 dB, such as 87 feet of RG-59 or 175 feet of RG-8 at 4.0 MHz, as in example 4, but line mismatched with 3:1 SWR at load. See Fig 23-4B. The 3:1 load mismatch referred back to the line input through 0.5 dB of line attenuation results in a 2.61:1 mismatch to the transmitter until matched by retuning, or with a Transmatch. The rematched transmitter delivers 100 W to the line. With 0.5 dB matched-line loss, plus 0.288 dB additional line loss from the 3:1 SWR, the total line-attenuation loss is 0.788 dB. The power lost to line attenuation with the 3:1 mismatch is 16.59 watts, in contrast to 10.87 watts lost when the line is matched (example 4, Fig 23-4A). The forward power, 124.78 W, suffers 0.5 dB loss to 111.21 W during the trip to the load; the reflected power, 27.8 W, suffers 0.5 dB loss to 24.78 W during the return trip to the transmitter, leaving 24.78 W to add to the source power, 100 W. The power absorbed in the load and the power lost in line attenuation account for all the power delivered by the transmitter!

Chapter 24

References

Asterisks (*) indicate material containing misleading or erroneous information, as discussed in earlier chapters.

1 *The ARRL Handbook*, any recent edition (Newington, CT: The American Radio Relay League).

2 *The ARRL Antenna Book*, 10th ed. (Newington, CT: The American Radio Relay League, 1964), pp 70-80 and 90-136.

3 G. Grammer (W1DF), "The Whys of Transmission Lines," *QST*, in three parts: Jan 1965, p 25; Feb 1965, p 24; Mar 1965, p 19.

4 G. Grammer (W1DF), "Simplified Design of Impedance-Matching Networks," *QST*, in three parts: Mar 1957, p 38; Apr 1957, p 32; May 1957, p 29.

5 G. Grammer (W1DF), "Antennas and Feeders," *QST*, in three parts: Oct 1963, p 30; Nov 1963, p 36; Dec 1963, p 53. [EDITOR'S NOTE: Published references to a Part IV of this series are erroneous.]

6 "Standing Waves—Good or Bad?", Technical Topics, *QST*, Dec 1946, p 56.

7 B. Goodman (W1DX), "My Feedline Tunes My Antenna," *QST*, Mar 1956, p 49. Reprint: "My Feed Line Tunes My Antenna," *QST*, Apr 1977, pp 40-42.

8 L. G. McCoy (W1ICP), "Choosing a Transmission Line," *QST*, in two parts: Dec 1959, p 42; Feb 1960, p 40.

9 L. G. McCoy (W1ICP), "Antennas and Transmatches," *QST*, Oct 1964, p 18.

10 L. G. McCoy (W1ICP), "A Versatile Transmatch," *QST*, Jul 1965, p 58.

11 L. G. McCoy (W1ICP), "A Completely Flexible Transmatch for One Watt to 1000," *QST*, Jun 1964, p 39.

[12] L. G. McCoy (W1ICP), "A Transmatch for Balanced and Unbalanced Lines," *QST*, Oct 1966, p 38.

[13] L. G. McCoy (W1ICP), "When is A Feed Line Not A Feed Line?", *QST*, Aug 1965, p 37.

[14] C. C. Drumeller (W5JJ), "The Mismatched RF Transmission Line," *73*, Nov 1969, p 68.

[15] P. H. Smith, "The "Mega-Rule," *QST*, Mar 1969, p 24.

[16] Y. Beers (WØJF), "Match or Not to Match," *QST* Sep 1958, p 13.

[17] W. L. Everitt, *Communication Engineering*, 2nd ed. (New York: McGraw-Hill Book Co, 1937).

[18] W. C. Johnson, *Transmission Lines and Networks*, 1st ed. (New York: McGraw-Hill Book Co, 1950).

[19] P. H. Smith, *Electronic Applications of the Smith Chart in Waveguide, Circuit and Component Analysis* (New York: McGraw-Hill Book Co, 1969, and Malabar, FL: Robert E. Krieger Pub Co, 1983).

[20] G. H. Brown, R. F. Lewis (W2EBS), and J. Epstein, "Ground Systems as a Factor in Antenna Efficiency," *Proc IRE*, Jun 1937, pp 753-787. (Discusses ground resistance versus number of radials—a classic work on which FCC standards for AM broadcast ground-radial systems are based.)

[21] R. M. Smith (W1FTX), "Getting the Most Into Your Antenna," *QST*, Jul 1952, p 21.

[22] "Impedance Matching With an Antenna Tuner," Technical Topics, *QST*, Oct 1946, p 38.

[23] M. Ferber (W1GKX), "Coaxial Cable Attenuation," *QST*, Apr 1959, p 20.

[24] R. Leo (W7LR, ex-K7KOK), "An Impedance-Matching Method," *QST*, Dec 1968, p 24.

[25] G. L. Hall (K1TD, ex-K1PLP), "Smith-Chart Calculations for the Radio Amateur," *QST*, in two parts: Jan 1966, p 22; Feb 1966, p 30.

[26] L. A. Cholewski (K6CRT), "Some Amateur Applications of the Smith Chart," *QST*, Jan 1960, p 28.

[27] P. C. Amis (W7RGL), "Antenna Impedance Matching," *CQ*, in two parts: Nov 1963, p 63; Dec 1963, p 33.

[28] J. R. Fisk (W1HR), "How to Use the Smith Chart," *Ham Radio*, Nov 1970, p 16. (Also see *Ref 29* for corrections.)

[29] W. Maxwell (W2DU), "Correspondence on How to use The Smith Chart," *Ham Radio*, Dec 1971, p 76 (corrections to information of *Ref 28*).

[30] P. H. Smith, "L-Type Impedance Transforming Circuits," *Electronics*, Mar 1942, p 68.

[31] R. E. Gordon (WØKFI), "L Networks for Reactive Loads," *QST*, Sep 1966, p 30.

[32] J. J. Schultz (W4FA), "Broad-Band Balun," *CQ*, Aug 1968, p 66.

[33] *Reference Data for Radio Engineers*, 4th ed. (New York: Federal Telephone and Radio Co, 1956), p 564.

[34] R. C. Fenwick (K5RR, ex-W5KTR), "Antenna and Transmission Line Quiz," *QST*, Questions: Jul 1965, p 19; Answers: Aug 1965, p 55. Reprinted in *QST*, Technical Correspondence, Questions: Jan 1981, p 43; Answers: Feb 1981, p 46.

[35] J. C. Slater, *Microwave Transmission* (New York: McGraw-Hill Book Co, 1942), pp 7-69.

[36] W. I. Orr (W6SAI), "Broad Band Balun For A Buck," *CQ*, Feb 1966, pp 42-45.

[37] R. W. P. King, *Transmission Line Theory* (New York, McGraw-Hill Book Co, 1955, and New York: Dover Publications, Inc, 1965), pp 77-83.

[38] W. B. Bruene (W5OLY, ex-WØTTK), "An Inside Picture of Directional Wattmeters," *QST*, Apr 1959, p 24. (A classic!)

[39] Radio Research Laboratory Staff, Harvard University, *Very High-Frequency Techniques*, Vol I, ed. I (New York: McGraw-Hill Book Co, 1947), pp 10-15.

[40] J. Hall (K1TD, ex-K1PLP), "Accuracy of SWR Measurements," *QST*, Nov 1964, p 50.

[41] L. G. McCoy (W1ICP), "The Ultimate Transmatch," *QST*, Jul 1970, p 24.

[42] W. W. Mumford (W6CU), "Directional Couplers," *Proc IRE*, Feb 1947, p 160.

[43] H. H. Skilling, *Fundamentals of Electric Waves*, 2nd ed. (New York: John Wiley and Sons, 1948), p 123.

[44] S. C. Shallon (W6EL), "The Monimatch and SWR," *QST*, Aug 1964, p 54.

[45] L. G. McCoy, "Is a Balun Required?", *QST*, Dec 1968, p 29.

[46] J. Hall (K1TD, ex-K1PLP), and J. Kaufmann (WA1CQW), "The Macromatcher, An RF Impedance Bridge for Coax Lines," *QST*, Jan 1972, p 14.

[47] G. H. Brown and O. M. Woodward, Jr, "Experimentally Determined Impedance Characteristics of Cylindrical Antennas," *Proc IRE*, Apr 1945, p 257.

[48] R. W. P. King, *Theory of Linear Antennas* (Cambridge, MA: Harvard Univ Press, 1956), pp 169-176.

[49] R. W. P. King and F. G. Blake, "The Self-Impedance of the Symmetrical Antenna," *Proc IRE*, Jul 1942, p 335.

[50] J. Sevick (W2FMI), "The Ground-Image Vertical Antenna," *QST*, Jul 1971, p 16.

[51] J. Sevick (W2FMI), "The W2FMI Ground-Mounted Short Vertical," *QST*, Mar 1973, p 13.

[52] J. D. Kraus (W8JK), *Antennas*, 1st ed. (New York: McGraw-Hill Book Co, 1950); 2nd ed. (New York: McGraw-Hill Book Co, 1988).

[53] E. C. Jordan, *Electromagnetic Waves and Radiating Systems* (New York: Prentice-Hall, Inc, 1950) p 415.

[54] D. L. Fayman (WØGI), "A Simple Computing SWR Meter," *QST*, Jul 1973, p 23.

[55] W. H. Anderson (VE3AAZ), "SWR's Significance," *CQ*, Oct 1970, p 8.

[56] C. E. Smith and E. M. Johnson, "Performance of Short Antennas," *Proc IRE*, Oct 1947, p 1026.

[57] E. A. Laport, *Radio Antenna Engineering* (New York: McGraw-Hill Book Co, 1952).

[58] H. Jasik, *Antenna Engineering Handbook*, 1st ed. (New York: McGraw-Hill, 1961). Also R. C. Johnson and H. Jasik, *Antenna Engineering Handbook*, 2nd ed. (New York: McGraw-Hill, 1984).

[59] W. Schreuer (K1YZW), "Notes on Directional SWR Indicators," *Ham Radio*, Dec 1969, p 65.

[60] J. S. Belrose (VE2CV) "Short Antennas for Mobile Operation," *QST*, Sep 1953, p 30. Information also contained in any edition of *The Mobile Manual for Radio Amateurs* (Newington, CT: The American Radio Relay League), now out of print. (The 4th ed. appeared in 1968.)

[61] R. Leo (W7LR, ex K7KOK), "How To Design L-Networks," *Ham Radio*, Feb 1974, p 26.

[62] W. Maxwell (W2DU), "A Revealing Analysis of the Coaxial Dipole Antenna," *Ham Radio*, Aug 1976, p 46.

[63] L. Gray and R. Graham, *Radio Transmitters* (New York: McGraw-Hill Book Co, 1961), pp 115-133.

[64] F. D. Schottland, "Pi Networks as Coupled Circuits," *Electronics*, Aug 1944.

[65] *J. T. Kroenert (WA1YTC), "What Your Wattmeter Really Reads," *QST*, Feb 1981, p 26.

[66] S. Gibilisco (W1GV), "How Important is Low SWR?", *Ham Radio*, Aug 1981, p 33.

[67] *H. Woods (W9IK), "Exploding the Power Myth," *73*, Dec 1976, p 120.

[68] *C. C. Drumeller (W5JJ), "Logic and Reflected Power," *73*, Jun 1973, p 65.

[69] W. L. Everitt and G. E. Anner, *Communication Engineering*, 3rd ed. (New York: McGraw-Hill Book Co, 1956), p 330.

[70] R. A. Chipman, "Theory and Problems of Transmission Lines," *Schaum's Outline Series* (New York: McGraw-Hill Book Co, 1968).

[71] G. L. Hall (K1TD), Ed., *The ARRL Antenna Book*, 15th ed., (Newington, CT: The American Radio Relay League, 1988), Chapter 28. (This reference is by far the best I've seen for the amateur.)

[72] D. DeMaw (W1FB), "Murch UT-2000-B Transmatch," Product Review, *QST*, Apr 1980, p 50.

[73] L. G. McCoy (W1ICP), "To Use Or Not To Use A Transmatch," *CQ*, Feb 1986, p 13.

[74] E. A. Wingfield (W5FD), "New And Improved Formulas for the Design of Pi and Pi-L Networks," *QST*, Aug 1983, and "Feedback," *QST*, Jan 1984, p 49.

[75] L. G. McCoy (W1ICP), "Set the Record Straight," "Correspondence," *QST*, Mar 1981, p 56.

[76] L. G. McCoy (W1ICP), "The 50-Ohmer Transmatch," *QST*, Jul 1961, p 30.

[77] D. DeMaw (W1FB), "Ultimate Transmatch Improved," Technical Correspondence, *QST*, Jul 1980, p 39.

[78] J. Rusgrove and G. Woodward, Eds., "A Transmatch for Balanced or Unbalanced Lines," *The Radio Amateur's Handbook*, 58th ed. (Newington, CT: The American Radio Relay League, 1981), p 19-10.

[79] See Fig 26B of *Ref 78*, p 19-12.

[80] J. Nagle (K4KJ), "RF Impedance Bridge Measurement Errors and Corrections," *Ham Radio*, May 1979.

[81] G. L. Hall (K1TD), Ed., *The ARRL Antenna Book*, 14th ed. (Newington, CT: The American Radio Relay League, 1982), p 5-5.

[82] J. Reisert (W1JR), "Simple and Efficient Broadband Balun," *Ham Radio*, Sep 1978, p 12.

[83] W. Orr (W6SAI), "Multiple Dipole for Portable Use," *Ham Radio*, May 1970, p 14.

[84] M. F. DeMaw (W1FB), *Ferromagnetic-Core Design and Application Handbook* (Englewood Cliffs, NJ: Prentice-Hall Inc, 1981), Chap 4.

[85] This reference number not used.

[86] O. M. Woodward, Jr, "Balance Measurements on Balun Transformers," *Electronics*, Sep 1953, p 188.

[87] *Transmission Line Catalog and Handbook*, Pub no. TL-6 (Wallingford, CT: Times Wire and Cable Co, 1972).

[88] R. W. Bricker and H. H. Rickert, "S-Band Resonant Quadrifilar Antenna for Satellite Communications," *RCA Engineer*, Vol 20, no. 5, Feb-Mar 1975. (Reprint from original published in *1974 International IEEE AP-S Symposium Digest*, Atlanta.)

[89] R. W. Bricker, "A Shaped-Beam Antenna for Satellite Data Communication," *1976 International IEEE AP-S Symposium Digest* (Amherst).

[90] G. H. Brown and O. M. Woodward, Jr, "Circularly Polarized Omni-Directional Antenna," *RCA Review*, Vol 8, Jun 1947, pp 259-260.

[91] C. C. Kilgus, "Multi-element Fractional Turn Helices," *IEEE Trans*, Vol AP-16, Jul 1968, pp 499-500.

[92] C. C. Kilgus, "Resonant Quadrifilar Helix," *IEEE Trans*, Vol AP-17, May 1969, pp 349-351.

[93] C. C. Kilgus, "Resonant Quadrifilar Helix Design," *Microwave Journal*, Dec 1970, pp 49-54.

[94] M. W. Maxwell (W2DU), *Some Aspects of the Quadrifilar Helix Antenna* (private distribution).

[95] M. W. Maxwell (W2DU), "Cover Story: 2304 MHz Quadrifilar Antenna for AMSAT-OSCAR 7," *AMSAT Newsletter*, Vol 7, no. 1, Mar 1975. See Fig 22-12.

[96] M. W. Maxwell (W2DU), "Some Aspects of the Quadrifilar Helix Antenna," *Trends in HF and VHF/UHF Antenna Design*, IEEE Electro 77 Professional Program, New York, Apr 19-21, 1977.

[97] *H. Woods (W9IK), "Power in Reflected Waves," *Ham Radio*, Oct 1971, p 49, and Dec 1972, p 76.

[98] *G. T. DeLaMatyr (W5GO), "Reflections on 'Reflected Power,'" Technical Correspondence, *QST*, Nov 1972, p 46.

[99] *P. M. Chamberlain (K5KEO), "SWR and Tank Coil Heating," *CQ*, Aug 1966, p 76.

[100] *R. T. Hart (W5QJR), "The Case for the Half-Wave Feed Line," *73*, Mar 1969, p 58.

[101] *K. (Judge) Glanzer (K7GCO), "More Words on Antennas," *CQ*, Jul 1957, p 48.

[102] *L. R. Houghton (K8ZVF), "Converts SWR Into Watts," *CQ*, Jun 1970, p 36. (See *Ref 55* for Anderson's correct rebuttal.)

103 *W. M. Scherer (W2AEF), "*CQ* Reviews the Knight Kit P-2 SWR Power Meter," *CQ*, Mar 1963, p 31.

104 S. Gibilisco (W1GV), "What Does Your SWR Cost You?", *QST*, Jan 1979, p 19.

105 S. Gibilisco (W1GV), "The Imperfect Antenna and How it Works," *QST*, Jul 1979, p 24.

106 W. B. Bruene (W5OLY, ex-WØTTK), "Distortion in Single-Sideband Linear Amplifiers," *QST*, Nov 1954, p 24.

107 W. B. Bruene (W5OLY), "Introducing the Series-Parallel Network," *QST*, Jun 1986, pp 21-23.

108 E. W. Pappenfus, W. B. Bruene (W5OLY), and E. O. Schoenike, *Single-Sideband Principles and Circuits* (New York: McGraw-Hill Book Co, 1964).

109 D. K. Belcher (WA4JVE), "RF Matching Techniques, Design and Example," *QST*, Oct 1972, p. 24.

110 R. W. P. King, H. R. Mimno and A. H. Wing, *Transmission Lines, Antennas and Waveguides* (New York: Dover Publications, Inc, 1965).

111 G. H. Brown, "The Phase and Magnitude of Earth Currents Near Radio Transmitting Antennas," *Proc IRE*, Feb 1935.

112 L. Varney (G5RV), "The G5RV Multiband Antenna . . . Up-to-Date," *The ARRL Antenna Compendium, Volume 1*, G. L. Hall (K1TD), Ed. (Newington, CT: The American Radio Relay League, 1985), p 86.

113 W. B. Bruene (W5OLY), "How to Design RF Coupling Circuits," *Electronics*, May 1952, p 134.

114 S. Ballantine, "On the Radiation Resistance of a Simple Vertical Antenna at Wave Lengths Below the Fundamental," *Proc IRE*, Dec 1924, p 823.

115 S. Ballantine, "On the Optimum Transmitting Wave Length for a Vertical Antenna Over Perfect Earth," *Proc IRE*, Dec 1924, p 833. (Basis for 5⁄8-λ ~~ÇäçeJ8é-Wavelength Antenna Mistique,~~" ~~*The ARRL Antenna Compendium, Volume 1*, G. L. Hall (K1TD), Ed. (Newington, CT: The American Radio Relay League, 1985), p 101.~~ or Ø.64-λ antenna.)

[116] *T. W. Swafford, Jr (W5HGU), "Improved Coax Feed for Low-Frequency Mobile Antennas, "*QST*, Dec 1951, p 40.

[117] *W. I. Orr (W6SAI), Ed., *The Mobile Handbook*, 1st ed. (Port Washington, NY: Cowan Publishing Corp, 1953) p 100.

[118] R. Lewallen (W7EL), "Baluns: What They Do And How They Do It," *The ARRL Antenna Compendium, Volume 1*, G. L. Hall (K1TD), Ed. (Newington, CT: The American Radio Relay League, 1985), p 157.

[119] D. K. Reynolds, "The 5/8-Wavelength Antenna Mystique," *The ARRL Antenna Compendium, Volume 1*, G. L. Hall (K1TD), Ed. (Newington, CT: The American Radio Relay League, 1985), p 101.

[120] "Microwave Mismatch Analysis," *Hewlett-Packard Application Note 56*, Oct 1967.

[121] R. W. Beatty, "Intrinsic Attenuation," *IEEE Trans on Microwave Theory and Techniques*, Vol MTT-11, no. 3, May 1963, p 179.

[122] F. J. Witt (AI1H), "Broadband Dipoles—Some New Insights," *QST*, Oct 1986, pp 27-37. (Note that a printing error exists on p 31, column 3. The value 2.8 in line 16 should be 2.28.)

[123] F. Witt (AI1H), "The Coaxial Resonator Match and the Broadband Dipole," *QST*, Apr 1989, pp 22-27.

[124] F. Witt (AI1H), "The Coaxial Resonator Match," *The ARRL Antenna Compendium, Volume 2*, G. L. Hall (K1TD), Ed. (Newington, CT: The American Radio Relay League, 1989), pp 110-118.

[125] J. M. Haerle (WB5IIR), *The Easy Way, HF Antenna Systems* (Denton, TX: Overtones, Inc, 1984).

[126] F. E. Terman, *Radio Engineers' Handbook*, 1st ed. (New York and London: McGraw-Hill Book Co, 1943).

[127] A. Roehm (W2OBJ), "Some Additional Aspects of the Balun Problem," *The ARRL Antenna Compendium, Volume 2*, G. L. Hall (K1TD), Ed. (Newington, CT: The American Radio Relay League, 1989), pp 172-174.

[128] *C. R. Ward (WA5LVG), "Transmatches," Technical Correspondence, *QST*, Oct 1984, pp 41-42.

[129] *C. C. Whysall (W8TV), "The 'Double-Bazooka' Antenna," *QST*, Jul 1968, pp 38-39.

[130] R. D. Snyder, "The Snyder Antenna," *RF Design*, Sep/Oct 1984, pp 49-51. Also R. D. Snyder, "Broadband Antennae Employing Coaxial Transmission Line Sections," United States Patent no. 4,479,130, issued Oct 23, 1984.

[131] M. W. Maxwell (W2DU), "Cover Story: 2304 MHz Quadrifilar Antenna for AMSAT-OSCAR 7," *AMSAT Newsletter*, Vol 7, No. 1, Mar 1975.

[132] J. S. Belrose (VE2CV), "Tuning and Constructing Balanced Transmission Lines," Technical Correspondence, *QST*, May 1981, p 43.

[133] R. L. Measures (AG6K), "A *Balanced* Balanced Antenna Tuner," *QST*, Feb 1990, pp 28-32.

[134] J. S. Belrose (VE2CV) and P. Bouliane (VE3KLO), "The Off-Center-Fed Dipole Revisited," scheduled for appearance in *QST*, Aug 1990, or shortly thereafter.

[135] M. W. Maxwell (W2DU), "Helical Coaxial Resonator RF Filter," United States Patent no. 3,437,959, issued Apr 8, 1969. (Invention is an isolation filter which prevents receiver desensitization from adjacent-channel transmitters flying on TIROS weather satellites and on RELAY television satellites.)

REFLECTIONS

PROOF OF PURCHASE

Enlarged version of Figs 3-2 and 9-2.

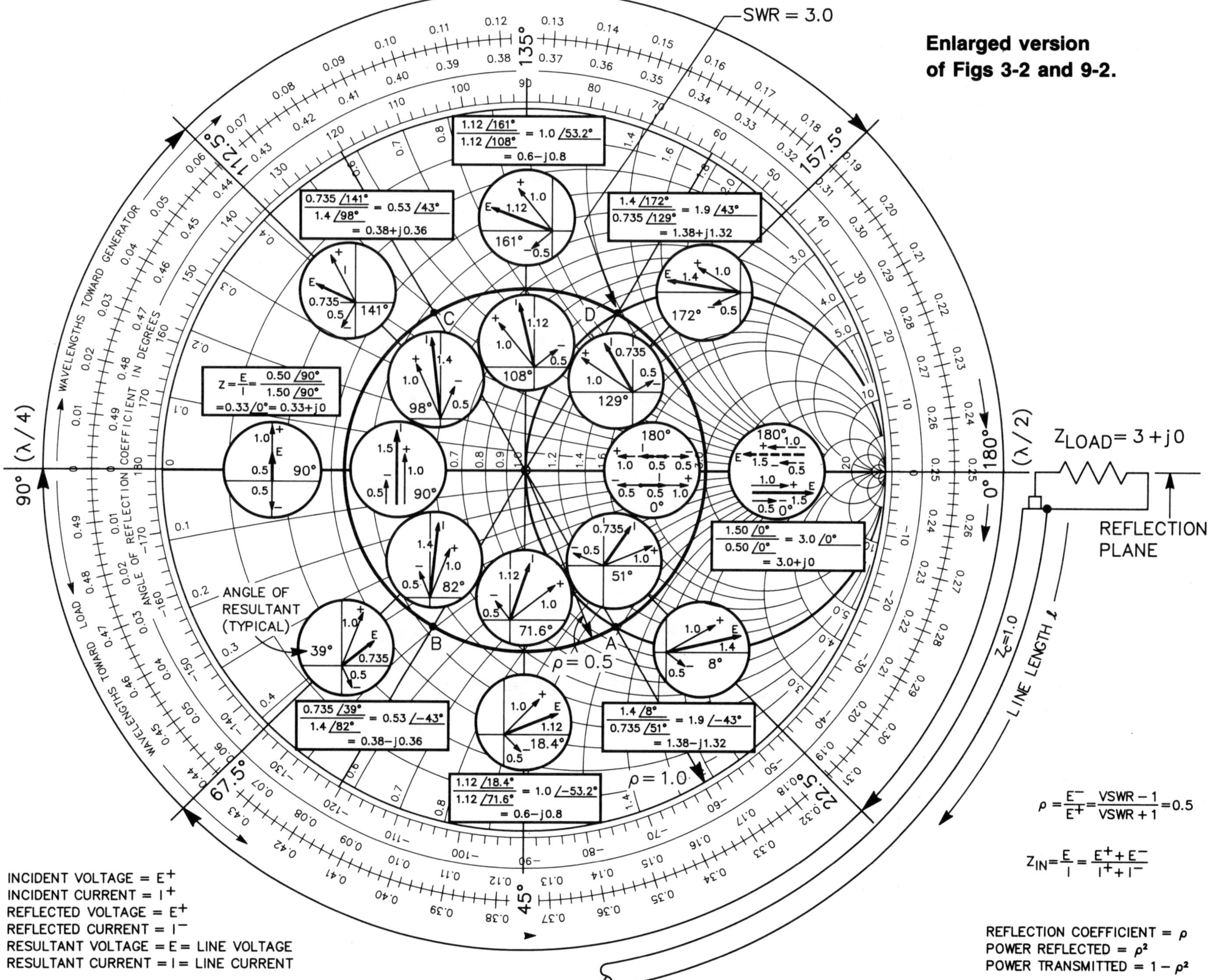

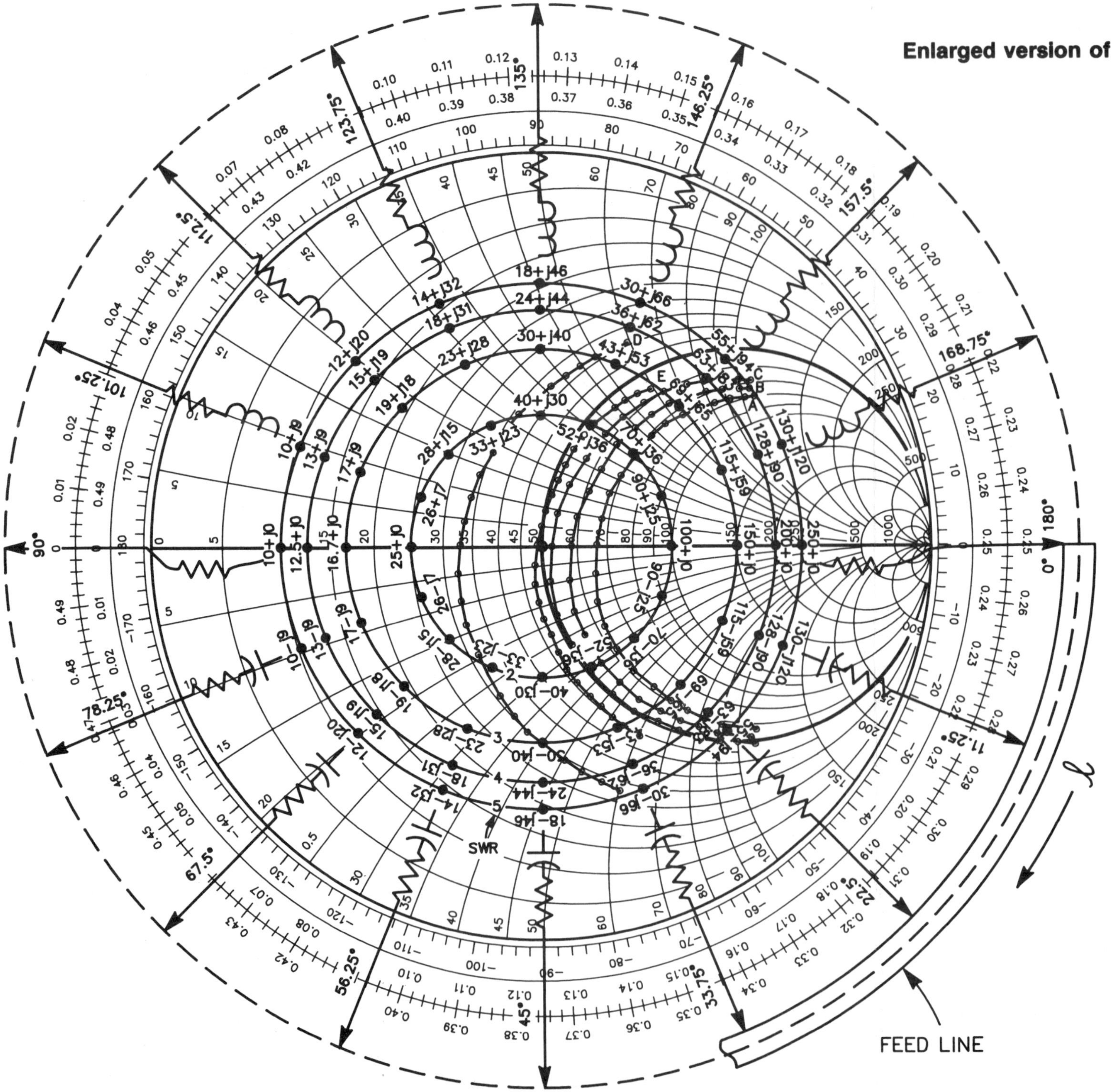

Enlarged version of Fig 10-1.

From________________________________

____

____

Please affix postage. Post Office will not deliver without sufficient postage.

Editor, Reflections
American Radio Relay League
225 Main Street
Newington, CT USA 06111

please fold and tape

FEED BACK

Please use this form to give us your comments on this book. Tell us what you liked best about the book, and what improvements you would like to see us make in future editions.

Name

______________________________ Call sign ____________

Daytime Phone (   ) ____________________ Age ________

Address ____

City, State, Zip ______________________________________

How long have you been licensed? ________________________

Edition 1 2 3 4 5 6 7 8 9 10 11 12

Printing 1 2 3 4 5 6 7 8 9 10 11 12